# EARTHMOVING EQUIPMENT PRODUCTION RATES AND COSTS

**Written by**
**Daniel Benn Atcheson**
*Formerly with the*
*Department of Engineering Technology*
*Texas Tech University*

CONSTRUCTION ESTIMATING INSTITUTE

5011 Ocean Boulevard
Sarasota, FL 34242
941-349-5401
800-423-7058

**Library of Congress Cataloging in Publication Data**

Atcheson, Daniel Benn
Earthmoving Equipment Production Rates and Costs
First Edition
Library of Congress Catalogue Card Number: 93-83233
ISBN 0-9613202-5-7



Printed in the United States of America
Typeset by BizArt, Sarasota, FL
Printed by Newspaper Printing Co., Tampa, FL
Cover design by Christine Page-Gilbert

*To my wife, Cheryl,*
*who helped type this manuscript,*
*took care of things and gave me moral support*
*as I wrote this book.*

*And to my Lord and Savior,*
*Jesus Christ,*
*who gave me eyes to see,*
*hands to type and a mind to think.*

**ACKNOWLEDGEMENTS**

American Society for Testing and Materials, 1916 Race St., Philadelphia, PA 19103

Atcheson, James Edward, A.I.A., 1906-1986

Atcheson, Tami Danielle, Dallas, TX

Athey Products Corp., P.O. Box 669, Raleigh, NC 27602

Balderson Inc., P.O. Box 6, Wamego, KS 66547

Barber-Greene, Division of Caterpillar Paving Products, 3000 Barber-Greene Road, DeKalb, IL 60115

Bucyrus-Erie, P.O. Box 500, South Milwaukee, WI 53172

Caterpillar Inc., 100 N.E. Adams, Peoria, IL 61629

Cedarapids, Inc., 916-16th St., N.E., Cedar Rapids, IA 52402

Ciciarelli, John A., Practical Physical Geology, 1986, College of Earth and Mineral Sciences, The Pennsylvania State University, Monaca, PA

CMI Corporation, P.O. Box 1985, Oklahoma City, OK 73101

Dynapac, P.O. Box 615, Schertz, TX 78154

Erie Strayer Co., P.O. Box 1031, Erie, PA 16512

Fiat-Allis North America, Inc., 245 E. North Ave., Carol Stream, IL 60188

Goodyear Tire and Rubber Co., 1144 East Market St., Akron, OH 44316

Marathon LeTourneau Co., P.O. Box 2307, Longview, TX 75606

Miskin Scraper Works, Inc., P.O. Box 218, Ucon, ID 83454

PCSA (Power Crane and Shovel Association), a Bureau of Construction Industry Manufacturers' Association, Marine Plaza, Suite 1700, 111 E. Wisconsin Ave., Milwaukee, WI 53202

Peerless Chain Co., 1416 E. Sanborn, Winona, MN 55987

Robuck, Roger, Ardaman & Associates, 2500 Bee Ridge Road, Sarasota, FL 34239

Rome Plow Co., P.O. Box 48, Cedartown, GA 30125

Terex Division of Terex Corp., P.O. Box 3107, Tulsa, OK 74101

Townsend, D.L., Queens University, Kingston, Ontario, Canada

U.S. Department of the Army, Washington, D.C.

Wacker Corp., P.O. Box 9007, Menomonee Falls, WI 53052

York Rakes, Division of York Modern Corp., P.O. Box 488, Unadilla, NY 13849

## DISCLAIMER

# TABLE OF CONTENTS



## APPENDICES

## PREFACE

I have written *Earthmoving Equipment Production Rates and Costs* as a companion to my first book, *Estimating Earthwork Quantities.* Both subjects are equally and vitally important to the earthmoving contractor since it is impossible to accurately and competitively bid a project without knowing both the cubic yards of earth to be moved, and the required cost per cubic yard to move the earth.

This book will help the earthmoving contractor determine what machinery is best to use in a given set of circumstances, how long the machinery will be required to work and how much it will cost to move each cubic yard of earth while using a given type of machine.

This book addresses more that the equipment production and cost concerns of the contractor. In fact, one of the problems I had to contend with as this book began to develop was that of its title. I was faced with the problem of giving the book a simple, yet descriptive title that appropriately represented the contents of the book (This dilemma is still unresolved.). A more appropriate, but cumbersome title would have probably been: *Management and Selection of Earthmoving (and Paving) Equipment, Including Production Rates, Hourly Costs and Unit Production Costs.*

I have included the most current information available to me at the time of this book's printing, and I have endeavored to apply the information as accurately as possible while working example problems. I have also attempted to make this book as practical and easy to understand as possible; however, the subject of earthmoving production and costs is not a simple one. With any given machine, we are dealing with a variety of interacting variables which influence production rates, and therefore, unit costs. Some of these variables include soil type, underfoot conditions, grades, excavation depth, haul distance, altitude, operator proficiency, management skill, interaction with other equipment and numerous and unique job site conditions. Therefore, at best, we can only *estimate* production rates and costs, and although the information and example problems represent a fair statement of the results to be achieved under the given circumstances, there is no guarantee that the predicted performance will be achieved on a given job.

I am extremely grateful to Caterpillar Inc. for supplying the bulk of the production data and photographs presented in this book. Throughout this book, I have generally limited the data tables and performance charts to one model for each type of machine, since the purpose of this book is to teach you *how* to use the performance data, and not to provide data on every machine model, which can be obtained from the *Caterpillar Performance Handbook*. If your equipment was not accompanied with performance data, ask the dealer to supply the information, since this is essential for accurate performance (and thus, bidding) calculations.

The performance of all mobile equipment is dependent upon variables discussed in Chapter 2; therefore, my suggestion to you is to start at the beginning of the book and grow with its development.

The contractor who reads *Estimating Earthwork Quantities* and *Earthmoving Equipment Production Rates and Costs* will be on the leading edge of competitiveness, efficiency and knowledge in the earthmoving profession.

Your comments and suggestions will be welcomed.

DBA, 1993

# PEARLS OF WISDOM

## ON ESTIMATING

Luke 14:28-30

(28)" For which of you, desiring to build a tower, does not first sit down and count the cost, whether he has enough to complete it?

(29) Otherwise, when he has laid a foundation, and is not able to finish, all who see it begin to mock him,

(30) saying, 'This man began to build, and was not able to finish.'"

## ON BUILDING

He that is fond of building will soon ruin himself without the help of enemies. (Plutarch)

## IN GENERAL

There's no time lost sharpening. (Anonymous wood carver)

Don't be afraid to ask dumb questions. They're easier to handle than dumb mistakes. (Anonymous)

If you think education is costly, try ignorance. (Anonymous)

A good thing about telling the truth is that you don't have to remember what you said. (Anonymous)

Money will buy a fine dog, but only kindness makes him wag his tail.
(Anonymous)

There's always something to be thankful for. If you can't pay your bills, you can be thankful you're not one of your creditors. (Anonymous)

He who wants to lose a troublesome visitor should lend him money.
(Anonymous)

# *Chapter 1*

# EXCAVATING EQUIPMENT

The excavating equipment that we will discuss in this chapter includes crane shovels, draglines, clamshells, backhoes, mass excavators, front shovels and trenching machines.

## CRANE-SHOVELS

A crane-shovel (revolving shovel) consists of a crane body (mounting and revolving superstructure), boom and front-end attachment controlled by cables (Figures 1-1 and 1-2). A crane-shovel can be converted to a different type of excavator by altering the front-end attachment and boom (Figure 1-2).

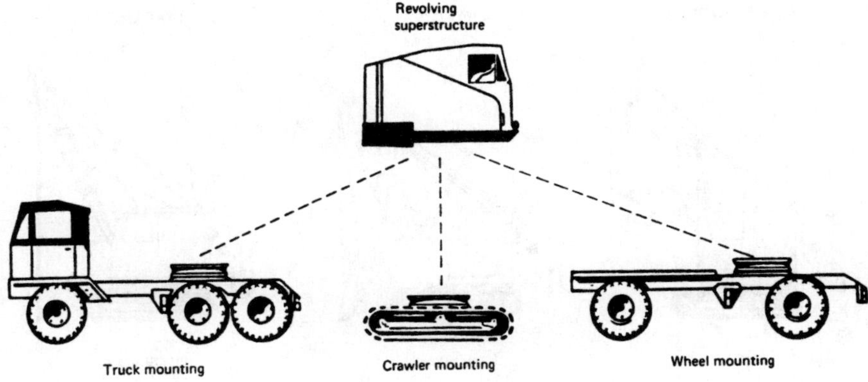


**Figure 1-1. Mountings for the crane-shovel.**
(*Courtesy of the U.S. Department of the Army*)

Crane-shovels are equipped with rubber tires, or crawler tracks (Photo 1-1). Wheel-mounted units can travel fast and are best suited for projects where extensive travel is required to and from, and around the job site, or where track-mounted units are prohibited, such as over a paved road surface. However, rubber tires require a firm underfooting and they are poorly suited for working over sharp-edged rock, especially if the working surface is wet. You can help protect tires by covering them with protective mesh chains (Photo 1-2). If the excavator must be frequently repositioned, a track-mounted vehicle will usually be more efficient, since outriggers on a wheel-mounted unit must be raised and lowered each time the machine is moved. However, a wheel-mounted excavator with outriggers is more stable than a crawler-mounted vehicle, since the crawler unit is not equipped with outriggers. Also, on uneven terrain, independently controlled outriggers can be used to level the machine. Sometimes, a dozer blade is mounted on the rear of a wheel-mounted machine and is used in lieu of rear outriggers.

The machine can then also be used for light dozing and backfilling operations. Another advantage of a wheel-mounted machine is that you have the option of mounting tires that are best suited for the particular project.

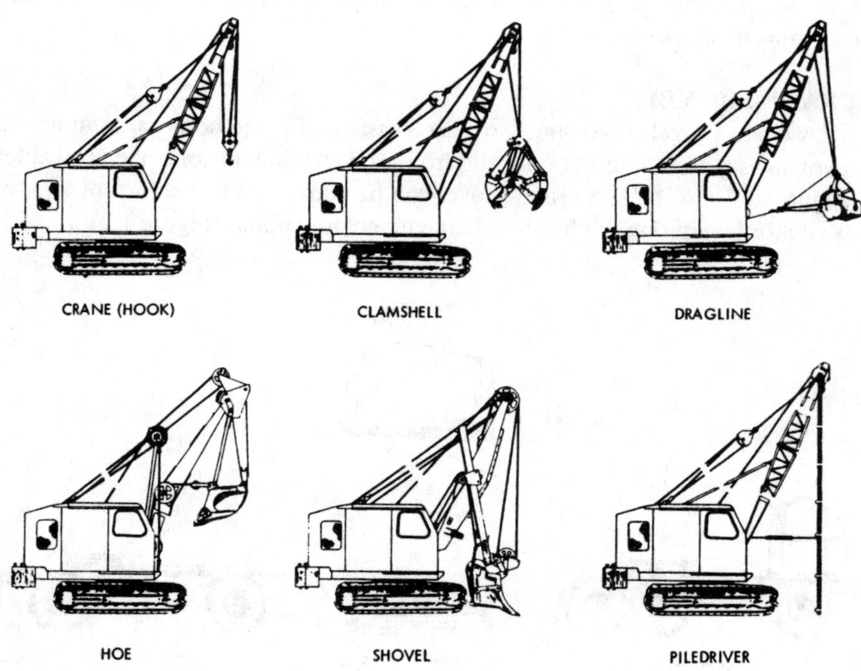


**Figure 1-2. Crane-shovel attachments.** (*Courtesy of the U.S. Department of the Army*)

Crawler-mounted units are best suited for projects where good traction and low ground surface pressures are required, and where the working surface is abrasive enough to cause excessive tire wear. However, a crawler-mounted unit must be transported on a trailer if the travel distance is substantial, or if crawler tracks are prohibited over the travel surface.

Crane-shovels are manufactured as self-propelled, or towed machines (Figure 1-1). Self-propelled units use a single engine and are propelled and operated from a single cab. Truck-mounted units use one engine to propel the unit and another to operate the shovel; thus, two separate cabs are required.

**Photo 1-1. Track-mounted shovel.** (*Courtesy of Bucyrus-Erie*)

**Photo 1-2. Tire protective chains.** (*Courtesy of Peerless Chain Co.*)

## EXCAVATOR PRODUCTION

The production rate of an excavator is determined by its *cycle time* which consists of the following:

1) Load bucket
2) Swing loaded bucket
3) Dump bucket
4) Swing empty bucket

For any given set of job conditions, the cycle time is primarily dependent upon the size of the excavator. Small machines can usually cycle faster than large ones; however, they produce less because of the small bucket size.

Production for any given type and size of excavator depends on the type of soil excavated, the depth of the excavation, the angle of swing, job and management conditions, the size of the hauling units and operator skill.

When excavating hard, tough soils, it takes longer to fill the bucket. Wet, sticky soils take longer to dump and they adhere to the bucket and teeth. This reduces the bucket's capacity and its ability to penetrate the soil. Production can be increased by keeping bucket teeth clean and sharp, and by replacing worn teeth. Never excavate after a tooth has fallen out because this could damage the shank to the extent that it will not be able to hold a new tooth properly. Attempting to overload a bucket will slow production and increase spillage, which must eventually be cleaned up.

Production rates can also be improved by using the correct bucket size and machine power required for the existing conditions. In soft soils, use the largest bucket available that does not exceed the safe lifting capacity of the excavator.

The effects of excavating less than, or greater than optimum depth will be discussed for each specific type of excavator. As depth increases, soil normally is more compact and thus, harder to excavate.

Increasing the swing angle will increase the swing time. Obstacles will increase swing time, as will dumping into trucks located above the excavator. Do not start to swing until the bucket clears the excavation.

Excavated material can be loaded into trucks or dumped into spoil piles. When hauling units are not used, as the excavation progresses, the spoil pile will grow. To avoid the pile, the angle of swing and dump height must often be changed, increasing cycle time. Never swing the bucket over the cab of a truck.

Excavating around existing underground utilities will slow production since care must be exercised not to break a buried line. Excavating around workmen can slow production, especially when they are inside a trench shield within a ditch. Excavating near highlines will also slow production.

The ability of the operator will affect every phase of the total cycle.

## BUCKET CAPACITIES

The *size* of an excavator bucket is usually expressed in cubic yards of struck volume, which is the volume enclosed inside the side plates and rear and front bucket enclosures; that is, the volume of soil beneath the strike-off plane (Figure 1-3).

The *capacity* of an excavator bucket is expressed in cubic yards of heaped capacity, which is the struck volume plus the volume above the strike-off plane, having an angle of repose of 1:1 (Figure 1-3). The angle of repose is the ratio (relationship) of run to rise of the soil pile, and is expressed as run:rise.

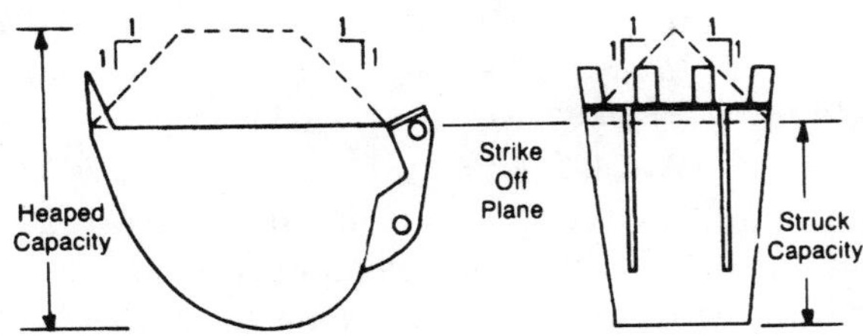


**Figure 1-3. Excavator bucket rating.** (*Courtesy of Caterpillar Inc.*)

Power shovel and dragline bucket capacities are rated by the PCSA (Power Crane and Shovel Association) according to their struck volumes. As the excavator is excavating in-place material, the soil will swell somewhat, but only in an upward direction since the soil is confined by the sides and bottom of the bucket. This upward swelling manifests itself in the heap at the top of the bucket. Heaped capacity often ranges between 1/8 and 1/4 the struck capacity. However, for all practical purposes, the struck capacity of the bucket is equivalent to the heaped capacity expressed in bank (in-place) cubic yards (BCY). Therefore, a 2-C.Y. bucket has a heaped capacity of 2 bank cubic yards of material.

To be more specific, when determining the production rate of an excavator, the bucket *payload* is expressed as a percentage of the heaped capacity. The percentage of heaped capacity is referred to as the *bucket fill factor*, or bucket efficiency factor. This factor varies, depending on the type of soil, or the condition (size) of rock being excavated. Thus, the bucket payload can be determined by:

Bucket Payload = Heaped Volume x Bucket Fill Factor **(Equation 1.1)**

Bucket fill factors for various types of excavators are given in Tables 1-3, 1-16 and 1-20.

The volume of soil in the bucket can be expressed as loose, or bank cubic yards, depending on the physical condition of the soil being excavated. For example, if an excavator is excavating in-place material, its production will be expressed in bank cubic yards (BCY) per hour, but if the excavator is excavating loose, stockpiled material, the production will be expressed in loose cubic yards (LCY) per hour.

The production rates and tables for excavators are normally expressed in bank cubic yards per hour, so we can use factors from Table 1-1 to convert loose-cubic-yard production to bank-cubic-yard production.

Bank cubic yards can be converted to loose cubic yards by:

$$LCY/BCY = \frac{Weight/BCY}{Weight/LCY} \quad \textbf{(Equation 1.2)}$$

Loose cubic yards can be converted to bank cubic yards by:

$$BCY/LCY = \frac{Weight/LCY}{Weight/BCY} \quad \textbf{(Equation 1.3)}$$

Equation 1.3 is defined as the *load factor* of the soil.
Soil swell and shrinkage is discussed in greater detail in Chapter 7.

**Example 1-1:** Use data from Table 1-1 to convert bank cubic yards of loam to loose cubic yards of loam, and vice versa.
**Solution:** A loose cubic yard of loam weighs 2100 pounds and a bank (in-place) cubic yard weighs 2600 pounds; therefore:

$$\text{a) } LCY/BCY = \frac{2600 \text{ lb./BCY}}{2100 \text{ lb./LCY}} = 1.24 \text{ LCY/BCY} \quad \text{(Eq. 1.2; Table 1-1)}$$

Thus, we say that loam swells 24%.

$$\text{b) } BCY/LCY = \frac{2100 \text{ lb./LCY}}{2600 \text{ lb./BCY}} = 0.81 \text{ BCY/LCY} \quad \text{(Eq. 1.3; Table 1-1)}$$

Thus, we say that loam has a load factor of 0.81.

**Example 1-2:** Assuming a swell of 30 % and a bucket fill factor of 90%, determine the BCY payload of a 2-C.Y. power shovel excavating loose soil.
**Solution:** Expressed in bank cubic yards, the bucket payload is:

$$\text{Bucket Payload (BCY)} = \frac{2 \text{ LCY}}{1.3 \text{ LCY/BCY}} \times 0.90 = 1.38 \text{ BCY} \quad \text{(Eq. 1.1)}$$

## Table 1-1: Weights of Materials (*Courtesy of Caterpillar Inc.*)

| WEIGHT* OF MATERIALS | LOOSE kg/m³ | LOOSE lb/yd³ | BANK kg/m³ | BANK lb/yd³ | LOAD FACTORS |
|---|---|---|---|---|---|
| Basalt | 1960 | **3300** | 2970 | **5000** | .67 |
| Bauxite, Kaolin | 1420 | **2400** | 1900 | **3200** | .75 |
| Caliche | 1250 | **2100** | 2260 | **3800** | .55 |
| Carnotite, uranium ore | 1630 | **2750** | 2200 | **3700** | .74 |
| Cinders | 560 | **950** | 860 | **1450** | .66 |
| Clay — Natural bed | 1660 | **2800** | 2020 | **3400** | .82 |
| Dry | 1480 | **2500** | 1840 | **3100** | .81 |
| Wet | 1660 | **2800** | 2080 | **3500** | .80 |
| Clay & gravel — Dry | 1420 | **2400** | 1660 | **2800** | .85 |
| Wet | 1540 | **2600** | 1840 | **3100** | .85 |
| Coal — Anthracite, Raw | 1190 | **2000** | 1600 | **2700** | .74 |
| Washed | 1100 | **1850** | | | .74 |
| Ash, Bituminous Coal | 530-650 | **900-1100** | 590-890 | **1000-1500** | .93 |
| Bituminous, Raw | 950 | **1600** | 1280 | **2150** | .74 |
| Washed | 830 | **1400** | | | .74 |
| Decomposed rock — | | | | | |
| 75% Rock, 25% Earth | 1960 | **3300** | 2790 | **4700** | .70 |
| 50% Rock, 50% Earth | 1720 | **2900** | 2280 | **3850** | .75 |
| 25% Rock, 75% Earth | 1570 | **2650** | 1960 | **3300** | .80 |
| Earth — Dry packed | 1510 | **2550** | 1900 | **3200** | .80 |
| Wet excavated | 1600 | **2700** | 2020 | **3400** | .79 |
| Loam | 1250 | **2100** | 1540 | **2600** | .81 |
| Granite — Broken | 1660 | **2800** | 2730 | **4600** | .61 |
| Gravel — Pitrun | 1930 | **3250** | 2170 | **3650** | .89 |
| Dry | 1510 | **2550** | 1690 | **2850** | .89 |
| Dry 6-50 mm (1/4"-2") | 1690 | **2850** | 1900 | **3200** | .89 |
| Wet 6-50 mm (1/4"-2") | 2020 | **3400** | 2260 | **3800** | .89 |
| Gypsum — Broken | 1810 | **3050** | 3170 | **5350** | .57 |
| Crushed | 1600 | **2700** | 2790 | **4700** | .57 |
| Hematite, iron ore, high grade | 1810-2450 | **4000-5400** | 2130-2900 | **4700-6400** | .85 |
| Limestone — Broken | 1540 | **2600** | 2610 | **4400** | .59 |
| Crushed | 1540 | **2600** | — | — | — |
| Magnetite, iron ore | 2790 | **4700** | 3260 | **5500** | .85 |
| Pyrite, iron ore | 2580 | **4350** | 3030 | **5100** | .85 |
| Sand — Dry, loose | 1420 | **2400** | 1600 | **2700** | .89 |
| Damp | 1690 | **2850** | 1900 | **3200** | .89 |
| Wet | 1840 | **3100** | 2080 | **3500** | .89 |
| Sand & clay — Loose | 1600 | **2700** | 2020 | **3400** | .79 |
| Compacted | 2400 | **4050** | | | |
| Sand & gravel — Dry | 1720 | **2900** | 1930 | **3250** | .89 |
| Wet | 2020 | **3400** | 2230 | **3750** | .91 |
| Sandstone | 1510 | **2550** | 2520 | **4250** | .60 |
| Shale | 1250 | **2100** | 1660 | **2800** | .75 |
| Slag — broken | 1750 | **2950** | 2940 | **4950** | .60 |
| Snow — Dry | 130 | **220** | | | |
| Wet | 520 | **860** | | | |
| Stone — crushed | 1600 | **2700** | 2670 | **4500** | .60 |
| Taconite | 1630-1900 | **3600-4200** | 2360-2700 | **5200-6100** | .58 |
| Top Soil | 950 | **1600** | 1370 | **2300** | .70 |
| Taprock — broken | 1750 | **2950** | 2610 | **4400** | .67 |
| Wood Chips** | — | — | — | — | — |

*Varies with moisture content, grain size, degree of compaction, etc. Tests must be made to determine exact material characteristics.

**Weights of commercially important wood species can be found in the last pages of the Logging & Forest Products section. To obtain wood weights use the following equations: $lb/yd^3 = (lb/ft^3) \times .4 \times 27$

$kg/m^3 = (kg/m^3) \times .4$

**Example 1-3:** Work the previous example assuming that the excavator is excavating in-place soil.

**Solution:** The bucket payload is:

$$\text{Bucket Payload (BCY)} = 2\text{ BCY} \times 0.90 = 1.8\text{ BCY} \qquad \text{(Eq. 1.1)}$$

## CABLE-OPERATED POWER SHOVELS

The cable-operated power shovel was the original member of the crane-shovel family and was invented by William S. Otis in 1836. The power shovel is a crane-shovel equipped with a shovel attachment (Figure 1-2) (Photo 1-3). With the exception of solid rock, a crane-operated power shovel can excavate all types of earth without prior loosening.

A crane-operated power shovel (revolving shovel) consists of a mounting, cab, boom, shipper shaft, dipper stick and a dipper (Figure 1-4). To excavate, the dipper is lowered to the floor of the excavation. By applying tension to the hoist line, the dipper is pulled up through the soil at the face of the excavation. If the depth of the face is optimum for the given shovel, the dipper will be filled as it reaches the top of the face. Although a shovel can dig above or below ground level, shovel excavation is usually most efficient when the natural ground level is at about the same elevation of the shipper shaft. If the depth of the face is less than optimum, it will be difficult, if not impossible, to completely fill the dipper in a single cycle. If the depth of the face is deeper than optimum, the depth of dipper penetration into the face will have to be reduced, or the face of the excavation will have to be removed, starting at a point above the floor. The soil immediately above the floor will then have to be excavated separately.

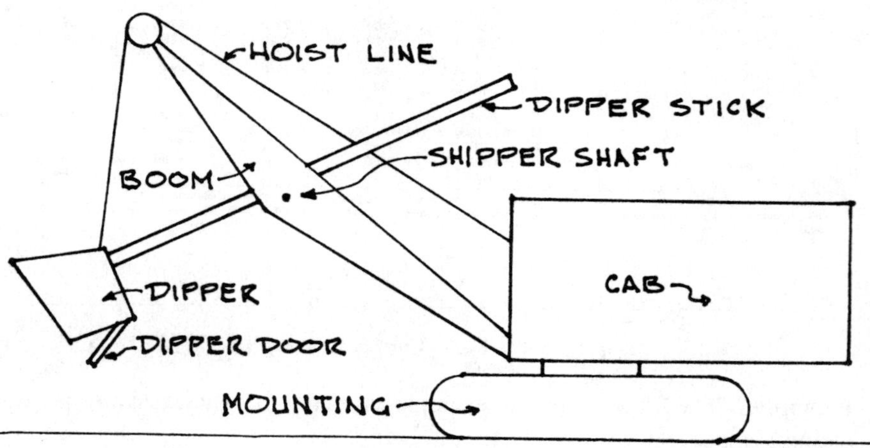


**Figure 1-4. Cable-operated power shovel components.**

Since the shovel can excavate at ground level, it can excavate its own roadway as it advances through the soil. When the excavation is started at ground level, the shovel must dig a ramp down into the soil until it creates a digging face of the proper depth. This procedure is referred to as *ramping down*.

Do not swing the dipper sideways on the ground to move loose material. Oftentimes, it is best to use a small bulldozer to fill holes and keep the pit floor clean. A clean, smooth and well-drained floor will increase power shovel production. Do not spot the shovel too close to the face of the excavation because a retracted dipper can strike and damage the crawlers or boom.

Excavated material can be loaded into trucks or dumped into spoil piles. When hauling units are used, logistics for shovel excavation usually include the frontal approach, or the parallel approach (Figure 1-5). The *frontal approach* allows the shovel to excavate from its optimum digging position, since it can exert the greatest digging force from this location. Trucks are backed into place on each side of the shovel so that the swing angle is approximately 90 degrees (Photo 1-3). In the *parallel approach,* the shovel advances down the face of the excavation as work progresses. This method requires less time for trucks to move into position. However, only one truck can be loaded at one time, and the shovel cannot use optimum digging force; therefore, the parallel approach is normally used only when space is limited. Never swing the bucket of the cab of a truck.

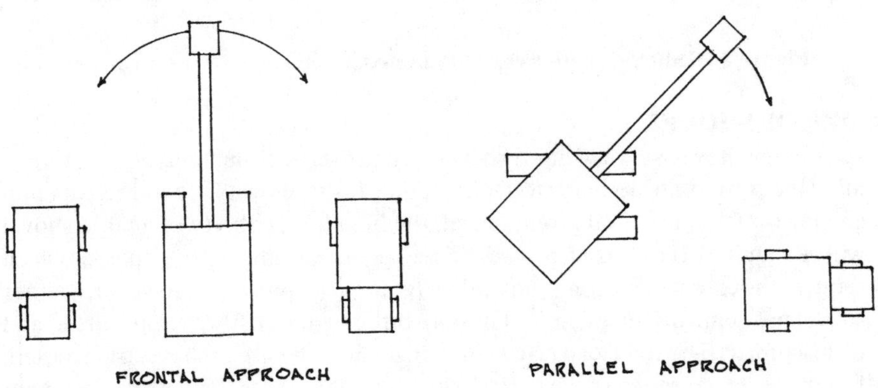


**Figure 1-5. Shovel approach methods.**

The hauling unit should be sized to hold a whole number of bucket loads; otherwise, one of the dumps will be only partially loaded. Figure 1-6 can be used to help interface the excavator with hauling units.

If a power shovel is not available, or the existing shovel breaks down, a backhoe can be used to chip away at the face of the excavation and the loose soil picked up and moved with a front-end loader.

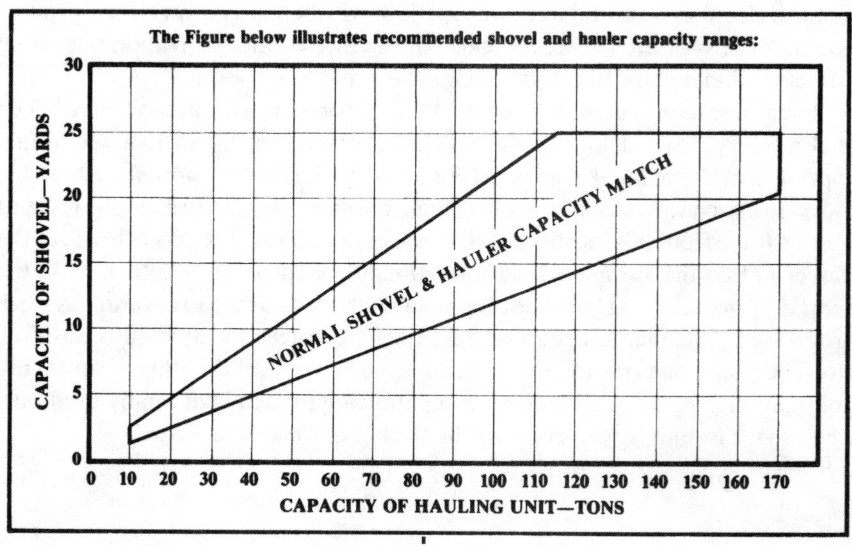


**Figure 1-6. Shovel-hauler capacity match.** (*Courtesy of Terex Corporation*)

## SIZE OF SHOVEL

A large shovel will yield high production and exerts tremendous digging forces, allowing it to excavate hard, tough material. If the shovel is required to dump soil into trucks at natural ground level, the high lifting ability of a large shovel will be required. If rock is excavated, a large shovel can handle larger pieces, which reduces the extent of blasting and drilling required (Photo 1-3). However, a larger shovel has some disadvantages. It is more difficult and costly to transport to and from a project, and road load restrictions might limit the size of shovel transported. Extensive downtime can be expected while waiting for non-standard replacement parts. A larger fleet of hauling units will be required, and the trucks will need to be of adequate size to present a large enough target for the shovel to efficiently dump its load. Large equipment can be hampered by overhead obstructions. Also, the soil bearing power might not be capable of supporting the weight of extremely heavy equipment.

Sometimes, project completion time dictates the size of shovel required. Also, the size of the project will often determine the size of shovel used. To understand this, we must realize that the cost for doing work is based on *unit costs* (or, cost per unit of work). In earthwork, unit costs are commonly expressed in cost per cubic yard. Although it costs more to operate a large machine than a small one, the unit cost for operating the large machine is normally less than that for a small one. This is due primarily to the fact that in proportion to the cost, the larger machine can produce a larger quantity of material per unit of time. Also, the operator's wages are normally the same, no matter what size machine he is operating.

**Photo 1-3. Shovel loading rock into truck using the frontal approach method.**
*(Courtesy of Athey Products Corporation)*

Determining unit costs are discussed in great detail in Chapter ll; however, we can apply the basic concept discussed above to demonstrate how to determine the *break-even point* (point of indifference) for various sized excavators on a given project.

To make a break-even analysis, determine the unit cost of each machine and the cost to mobilize and demobilize each one, including the cost of support equipment. If a machine is expected to be idle for a substantial amount of time, add the equivalent rental rate or hourly owning costs to the mobilization-demobilization costs. This information can then be graphed as shown in Figure 1-7, and the break-even point determined with regard to units of work required (cubic yards excavated).

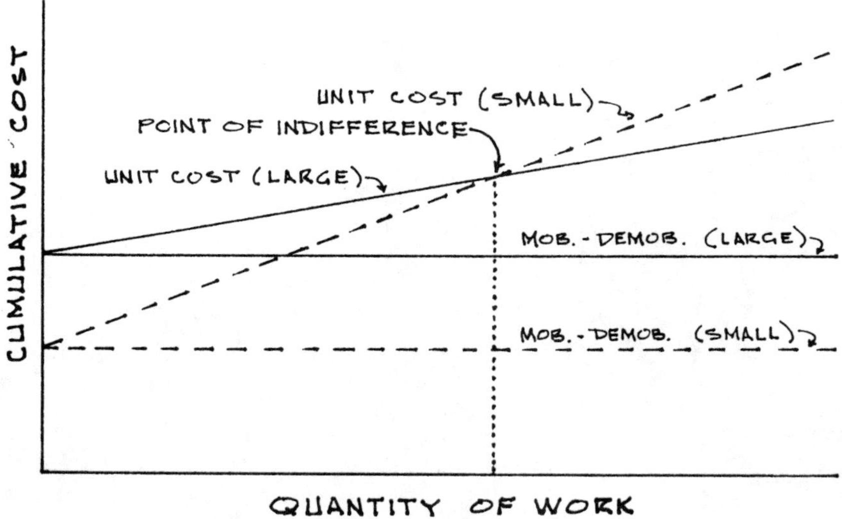


**Figure 1-7. Break-even point for large and small excavators.**

Machine shipping dimensions, operating dimensions, digging ranges and clearances are normally published with manufacturer's specifications for any given excavator (Figure 1-8 and Table 1-2). Of the shovel dimensions given, the most important with regard to job planning and shovel selection include the maximum cutting height, maximum digging radius, maximum dumping radius and the maximum dumping height. The latter two dimensions are very important when the shovel is in a pit loading trucks located at natural ground level.

**Table 1-2:**
**Dimensions and Clearances for Power Shovels at 45-Degree Boom Angle**

| Dipper Size (CY) | Max. Cutting Height (ft.) (A) | Max. Digging Radius (ft.) (B) | Max. Dumping Height (ft.) (C) | Max. Dumping Radius (ft.) (D) |
|---|---|---|---|---|
| 1/2 | 19 | 21 | 14 | 19 |
| 1 | 23 | 31 | 15 | 23 |
| 1-1/2 | 24 | 32 | 18 | 28 |
| 2 | 26 | 33 | 19 | 30 |

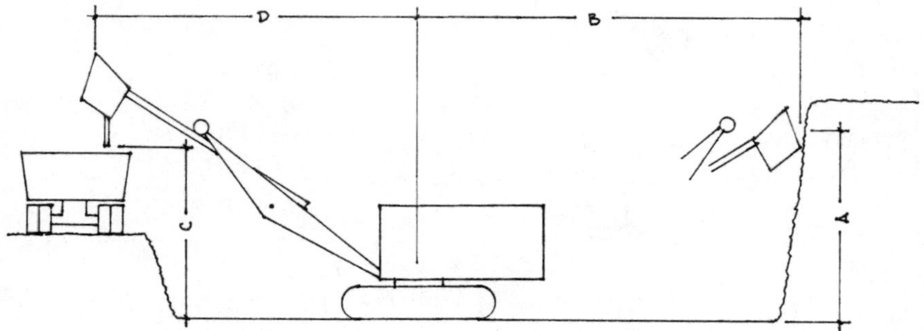


**Figure 1-8. Clearance diagram for a power shovel.**

## POWER SHOVEL PRODUCTION

Power shovel production while working at 100% efficiency can be determined by:

Ideal Shovel Production
(C.Y./Hr.) = Cycles/Hr. x Bucket Payload/Cycle **(Equation 1.4)**

where:

$$\text{Cycles/Hour} = \frac{\text{3600 Sec./Hr.}}{\text{Cycle Time (Sec.)}} \quad \textbf{(Equation 1.5)}$$

or,

$$\text{Cycles/Hour} = \frac{\text{60 Min./Hr.}}{\text{Cycle Time (Min.)}} \quad \textbf{(Equation 1.6)}$$

and,

Bucket Payload (BCY) = Heaped Volume (BCY) x Bucket Fill Factor
(Eq. 1.1)

where: Bucket fill factor (power shovels) is taken from Table 1-3

**Table 1-3: Bucket Fill Factors for Power Shovels**

| Material | Bucket Fill Factor (% of Heaped Bucket Capacity) |
|---|---|
| Sand and gravel | 0.90-1.00 |
| Common earth | 0.80-0.90 |
| Hard clay | 0.65-0.75 |
| Wet clay | 0.50-0.60 |
| Well-blasted rock | 0.60-0.75 |
| Poorly-blasted rock | 0.40-0.50 |

**Example 1-4:** Assuming a cycle time of 22 seconds with no work delays, determine the production rate of a 2-C.Y. cable-operated power shovel excavating common earth.
**Solution:** The bucket payload is:

$$\text{Bucket Payload (BCY)} = 2 \text{ BCY} \times 0.90 = 1.8 \text{ BCY} \qquad \text{(Eq. 1.1; Table 1-3)}$$

The number of cycles per hour is:

$$\text{Cycles/Hour} = \frac{3600 \text{ Sec./Hr.}}{22 \text{ Sec.}} = 164 \text{ cycles/hour} \qquad \text{(Eq. 1.5)}$$

With no delays, the shovel production is:

$$\text{Production (BCY/Hr.)} = 164 \text{ Cycles/Hr.} \times 1.8 \text{ BCY/Cycle} = 295 \text{ BCY/hour} \qquad \text{(Eq. 1.4)}$$

**JOB EFFICIENCY**

No job is ever 100% efficient. Delays, equipment failure, small dump targets, working around highlines, underground utilities and crews, operator skill and other adverse site conditions must be factored into the production equation (Equation 1.4) given above. The ideal production rate can be converted to a more realistic production rate by using an *efficiency factor* as follows:

$$\text{Production} = \text{Cycles/Hour} \times \text{Bucket Payload/Cycle} \times \text{Efficiency Factor} \qquad \textbf{(Equation 1.7)}$$

Job efficiency can be estimated by determining the number of productive minutes per clock hour, or by using efficiency factors given in Table 1-4. However, *never* use both methods simultaneously.

**Table 1-4: Efficiency Factors for Construction Equipment**

| Job Conditions | Management Conditions Excellent | Good | Fair | Poor |
|---|---|---|---|---|
| Excellent | 0.84 | 0.81 | 0.76 | 0.70 |
| Good | 0.78 | 0.75 | 0.71 | 0.65 |
| Fair | 0.72 | 0.69 | 0.65 | 0.60 |
| Poor | 0.63 | 0.61 | 0.57 | 0.52 |

**Example 1-5:** Determine the probable production rate of the power shovel given in the previous example assuming that only 50 minutes out of each clock hour will be productive.
**Solution:** The probable production rate will be:

Production (BCY/Hr.) = 295 BCY/Hr. x 50/60 (Eq. 1.7; Example 1-4)
= 246 BCY/hour

**Example 1-6:** Assuming poor job conditions and good management conditions, use an efficiency factor from Table 1-4 to determine the probable production rate of the power shovel given in Example 1-4.
**Solution:** The probable production will be:

Production (BCY/Hr.) = 295 BCY/Hr. x 0.61 (Table 1-4; Example 1-4)
= 180 BCY/hour

Job efficiency can also be determined by separating the project into three major categories, including labor, management and project attributes. Each major category is further broken down into four sub-attributes as shown in Table 1-5. Each sub-attribute is given equal weight (25%) within each major category. The constant 0.25 is multiplied by an efficiency rating taken from Table 1-6 to obtain a weighted efficiency factor for each sub-attribute. The total weighted efficiency factor for each major category is determined by adding the weighted efficiency factors within that category. The overall job efficiency is determined by obtaining the average of the total efficiency factors of all three major categories.

**Table 1-5: Project Attribute Groups**

| Category | Sub-Attribute | Weight |
|---|---|---|
| Labor | Morale | 0.25 |
| | Work ethic | 0.25 |
| | Health | 0.25 |
| | Training | 0.25 |
| Management | Leadership | 0.25 |
| | Incentive | 0.25 |
| | Experience | 0.25 |
| | Company support | 0.25 |
| Project | Weather | 0.25 |
| | Environment | 0.25 |
| | Interference | 0.25 |
| | Delays | 0.25 |

**Table 1-6: Efficiency Ratings**

| Description of Attributes | Rank | % Efficiency |
|---|---|---|
| Possible only in controlled environment | Highest | 100 |
| Best possible rating on most jobs. | Excellent | 90 |
| Rating in favorable conditions. | Good | 80 |
| Acceptable, but unimpressive rating. | Fair | 70 |
| Rating in deprived areas. | Poor | 60 |
| Rate failing to meet minimum standards. | Bad | 50 |
| Worst possible rating under usual job conditions. | Worse | 40 |
| Rating for typical developing countries. | Lowest | 30 |

**Example 1-7:** Determine the overall job efficiency for a project under the following conditions:

Labor Attributes:

| | | | |
|---|---|---|---|
| Morale | = good | = 0.80 | (Table 1-6) |
| Work ethic | = fair | = 0.70 | |
| Health | = excellent | = 0.90 | |
| Training | = fair | = 0.70 | |

Management Attributes:

| | | |
|---|---|---|
| Leadership | = fair | = 0.70 |
| Incentive | = fair | = 0.70 |
| Experience | = fair | = 0.70 |
| Company support | = poor | = 0.60 |

Project Attributes:

| | | |
|---|---|---|
| Weather | = poor | = 0.60 |
| Environment | = bad | = 0.50 |
| Interference | = bad | = 0.50 |
| Delays | = fair | = 0.70 |

**Solution:** The weighted and total efficiency factors for each major category is determined as follows:

Weighted Labor Attributes:

| | | | |
|---|---|---|---|
| Morale | = 0.25 x 0.80 | = 0.200 | (Table 1-5) |
| Work Ethic | = 0.25 x 0.70 | = 0.175 | |
| Health | = 0.25 x 0.90 | = 0.225 | |
| Training | = 0.25 x 0.70 | = 0.175 | |
| Total Weighted Labor Efficiency | | 0.775 | |

Weighted Management Attributes:

| | | |
|---|---|---|
| Leadership | = 0.25 x 0.70 | = 0.175 |
| Incentive | = 0.25 x 0.70 | = 0.175 |
| Experience | = 0.25 x 0.70 | = 0.175 |
| Company Support | = 0.25 x 0.60 | = 0.150 |
| Total Weighted Management Efficiency | | 0.675 |

Weighted Project Attributes:

| | | |
|---|---|---|
| Weather | = 0.25 x 0.60 | = 0.150 |
| Environment | = 0.25 x 0.50 | = 0.125 |
| Interference | = 0.25 x 0.50 | = 0.125 |
| Delays | = 0.25 x 0.70 | = 0.175 |
| Total Weighted Project Efficiency | | 0.575 |

The overall job efficiency is:

$$\text{Overall Job Efficiency} = \frac{0.775 + 0.675 + 0.575}{3}$$

$$= 0.675$$

$$= 67.5\%$$

## POWER SHOVEL PRODUCTION TABLES

In the previous examples, it was assumed that the shovel cycle time was known, perhaps from field observation, or previous job data for the given conditions. However, factors that affect shovel production, such as excavation depth and swing angle can vary to the extent that previous job data cannot be applied, nor can field observations be made prior to bidding the project. Therefore, the PCSA has developed a production data table for cable-operated power shovels (Table 1-7).

**Table 1-7: Ideal Outputs of Cable-Operated Shovels (BCY/60-Minute Hour)**

| Class of Material | Size of Shovel (C.Y.) | | | | | | | | |
|---|---|---|---|---|---|---|---|---|---|
| | 3/8 | 1/2 | 3/4 | 1 | 1-1/4 | 1-1/2 | 1-3/4 | 2 | 2-1/2 |
| Moist loam or sandy clay | 3.8<br>85 | 4.6<br>115 | 5.3<br>165 | 6.0<br>205 | 6.5<br>250 | 7.0<br>285 | 7.4<br>320 | 7.8<br>355 | 8.4<br>405 |
| Sand and gravel | 3.8<br>80 | 4.6<br>110 | 5.3<br>155 | 6.0<br>200 | 6.5<br>230 | 7.0<br>270 | 7.4<br>300 | 7.8<br>330 | 8.4<br>390 |
| Common earth | 4.5<br>70 | 5.7<br>95 | 6.8<br>135 | 7.8<br>175 | 8.5<br>210 | 9.2<br>240 | 9.7<br>270 | 10.2<br>300 | 11.2<br>350 |
| Hard, tough clay | 6.0<br>50 | 7.0<br>75 | 8.0<br>110 | 9.0<br>145 | 9.8<br>180 | 10.7<br>210 | 11.5<br>235 | 12.2<br>265 | 13.3<br>310 |
| Wet, sticky clay | 6.0<br>25 | 7.0<br>40 | 8.0<br>70 | 9.0<br>95 | 9.8<br>120 | 10.7<br>145 | 11.5<br>165 | 12.2<br>185 | 13.3<br>230 |
| Well-blasted rock | 40 | 60 | 95 | 125 | 155 | 180 | 205 | 230 | 275 |
| Poorly-blasted rock | 15 | 25 | 50 | 75 | 95 | 115 | 140 | 160 | 195 |

*(Based on PCSA data)*

Within each soil classification given in Table 1-7, there are two values shown under each shovel size. The upper number is the optimum depth of cut for the given size of shovel, and the lower number is the ideal production anticipated while excavating at optimum depth and swinging the shovel 90 degrees. Note that Table 1-7 gives production in terms of bank cubic yards per hour. In the event that the machine excavates under different circumstances (which is usually the case), the ideal output given in Table 1-7 must be tempered by using a conversion factor (*swing-depth factor*) obtained from Table 1-8. The percent of optimum depth values given in Table 1-8 can be determined by:

$$\text{Percent of Optimum Depth} = \frac{\text{Actual Depth}}{\text{Optimum Depth}}$$ **(Equation 1.8)**

where: Optimum depth is taken from Table 1-7

**Table 1-8: Swing-Depth Factors for Cable-Operated Power Shovels**

| Percent of Optimum Depth | Angle of Swing (Degrees) | | | | | | |
|---|---|---|---|---|---|---|---|
| | 45 | 60 | 75 | 90 | 120 | 150 | 180 |
| 40 | 0.93 | 0.89 | 0.85 | 0.80 | 0.72 | 0.65 | 0.59 |
| 60 | 1.10 | 1.03 | 0.96 | 0.91 | 0.81 | 0.73 | 0.66 |
| 80 | 1.22 | 1.12 | 1.04 | 0.98 | 0.86 | 0.77 | 0.69 |
| 100 | 1.26 | 1.16 | 1.07 | 1.00 | 0.88 | 0.79 | 0.71 |
| 120 | 1.20 | 1.11 | 1.03 | 0.97 | 0.86 | 0.77 | 0.70 |
| 140 | 1.12 | 1.04 | 0.97 | 0.91 | 0.81 | 0.73 | 0.66 |
| 160 | 1.03 | 0.96 | 0.90 | 0.85 | 0.75 | 0.67 | 0.62 |

*(Based on PCSA data)*

The ideal production obtained from Table 1-7 must also be modified by a job efficiency factor. Thus, we can determine the probable production rate of a shovel as follows:

Production = Ideal Production x Swing-Depth Factor x Efficiency Factor
**(Equation 1.9)**

where: Ideal production is taken from Table 1-7
Swing-depth factor is taken from Table 1-8

and, Efficiency factor is taken from Table 1-4, or is estimated as productive time per clock hour.

**Example 1-8:** Assuming that a 1-C.Y. shovel is excavating sand and gravel at a depth of 8.4 feet and swinging 120 degrees, determine the probable production rate while working a 50-minute hour.
**Solution:** The ideal production rate is:

Ideal Production (BCY/Hr.) = 200 BCY/Hr. (Table 1-7)

The percent of optimum depth is:

$$\% \text{ Optimum Depth} = \frac{8.4'}{6.0'} \qquad \text{(Eq. 1.8; Table 1-7)}$$
$$= 1.4$$
$$= 140\%$$

Using a 120° swing, the swing-depth factor is:

Swing-Depth Factor = 0.81 (Table 1-8)

Job efficiency is:

Efficiency Factor = 50/60 (Given)
= 0.83

Therefore, the probable production rate is:

Production (BCY/Hr.) = 200 BCY/Hr. x 0.81 x 0.83 (Eq. 1.9)
= 134 BCY/hour

There are times when the angle of swing and/or the percent of optimum depth cannot be "pigeonholed" in Table 1-8, and you must interpolate to determine the swing-depth factor. The easiest way to learn how to interpolate is to familiarize yourself with Figure 1-9 and relate it to Equation 1.10; then work some examples.

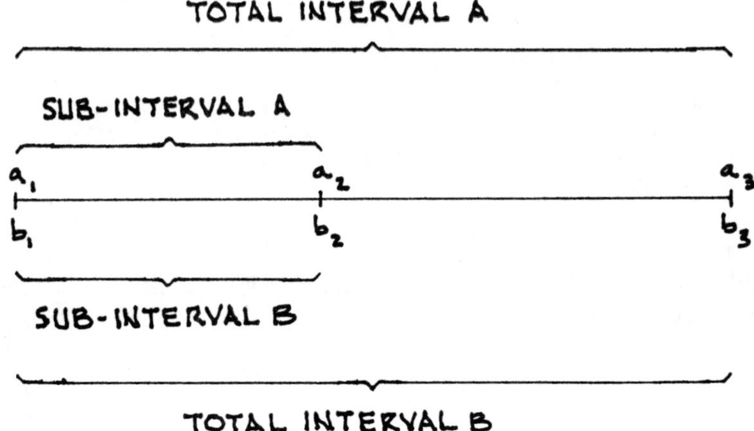


**Figure 1-9. General interpolation diagram.**

The rule of interpolation is:

$$\frac{\text{Sub-Interval A}}{\text{Total Interval A}} = \frac{\text{Sub-Interval B}}{\text{Total Interval B}} \qquad \textbf{(Equation 1.10)}$$

In layman terms, we would say: Sub-interval A is to the total interval A, as sub-interval B is to the total interval B.

**Example 1-9:** Determine the swing-depth factor for a shovel excavating at 40% of optimum depth if the swing angle is 130 degrees.
**Solution:** We can find 40% of optimum depth in Table 1-8, but not 130 degrees; therefore, we must determine the swing-depth factor through interpolation. We will let the A-values in equation 1.10 be the swing-depth factors and the B-values be the angle of swing. Using data from Table 1-8, we can set up an interpolation diagram as shown in Figure 1-10.

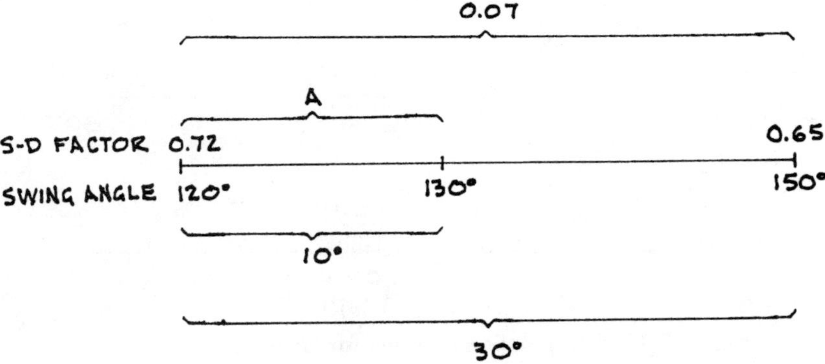


**Figure 1-10. Interpolation diagram for Example 1-9.**

Now, we can use Equation 1.10 to solve for the swing-depth factor:

$$\frac{A}{0.07} = \frac{10}{30} \qquad \text{(Eq. 1.10)}$$

Solving for A, we have:

$$A = \frac{0.07 \times 10}{30}$$

$$= 0.023$$

Since the swing-depth factor decreases as the swing angle increases (Table 1-8), the swing-depth factor for 130 degrees is:

$$\text{Swing-Depth Factor} = 0.720 - 0.023 = 0.697 = 0.70$$

**Example 1-10:** Determine the swing-depth factor for a shovel excavating at 75% of optimum depth if the swing angle is 90 degrees.
**Solution:** We can find a 90-degree swing angle in Table 1-8, but not 75% optimum depth; therefore, we must determine the swing-depth factor through interpolation. We will let the A-values in Equation 1.10 be the swing-depth factors and the B-values be the percent of optimum depth. Using data from Table 1-8, we can set up an interpolation diagram as shown in Figure 1-11.

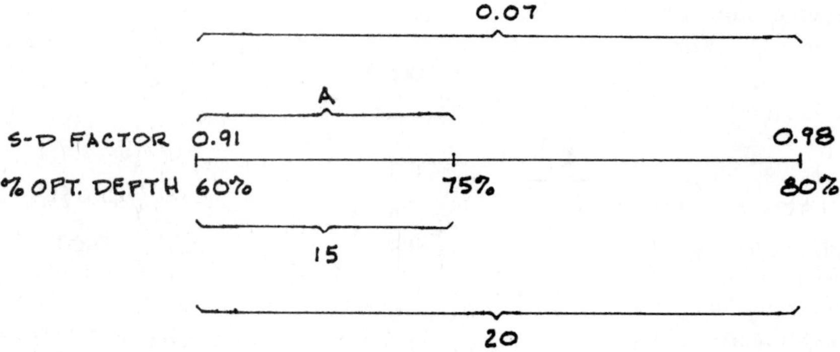


**Figure 1-11. Interpolation diagram for Example 1-10.**

The interpolation equation is:

$$\frac{A}{0.07} = \frac{15\%}{20\%} \qquad \text{(Eq. 1.10)}$$

Solving for A, we have:

$$A = \frac{0.07 \times 15\%}{20\%} = 0.053$$

Since the swing-depth factor increases as the percent of optimum depth increases (Table 1-8), the swing-depth factor for 75% of optimum depth is:

Swing-Depth Factor = 0.910 + 0.053
= 0.963
= 0.96

## ALTITUDE DERATING

Engines work less efficiently at higher altitudes unless they are equipped with a turbocharger. As engine efficiency is reduced, production will also be reduced. The subject of derating engine power while working at high altitudes is discussed in detail in Chapter 2; however, I want you to be aware of this situation at this point in your reading. I have provided the following example to help you understand the concept of altitude derating.

**Example 1-11 :** Assuming that the shovel given in Example 1-8 is working at such a high elevation that the engine is only 85% as efficient as it is at sea level, determine the production rate of the excavator.
**Solution:** The probable production rate at the given altitude would be:

Production (BCY/Hr.) = 134 BCY/Hr. x 0.85
= 114 BCY/hour

## CABLE-OPERATED DRAGLINES

Most crane-shovels up to 2-1/2-C.Y. capacity can be converted to a dragline by replacing the shovel boom with a crane boom, and substituting a dragline bucket (Figure 1-12) (Photo 1-4). The size of the dragline is indicated by the size of the bucket. Although a dragline can be equipped with more than one size of bucket, it is usually outfitted with a bucket the same size as the dipper of the power shovel from which it was converted. As with a shovel bucket, the struck capacity of the bucket is equivalent to the heaped capacity expressed in bank cubic yards.

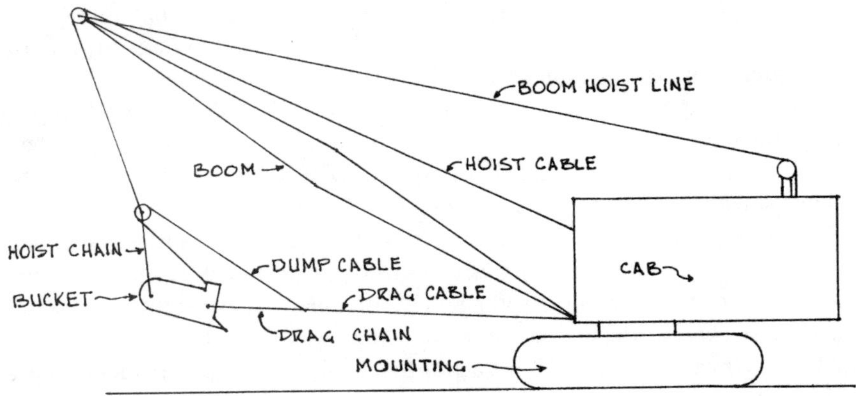


**Figure 1-12. Dragline components.**

**Photo 1-4. Cable-operated dragline.** (*Courtesy of Bucyrus-Erie*)

The dragline bucket is controlled by coordinating the tensions placed on the hoist and drag cables. The bucket is raised vertically by placing tension on the hoist cable while playing out the drag cable. The bucket is lowered vertically by releasing tension on the hoist cable while rolling in the drag cable. The digging depth is controlled by regulating the tension placed on the hoist cable. The bucket is pulled toward the machine by placing tension on the drag cable while slacking off the hoist cable. The bucket is dumped by placing tension on the hoist cable while releasing the drag cable, which also releases tension on the dump cable (Figure 1-15).

The dragline has the advantage of being able to excavate from above the pit at natural ground level. This eliminates the problem of maintaining the pit floor, and hauling units need not enter the pit. In fact, draglines are often used for dredging material from below water. The dragline also has a long digging and dumping reach, and there are instances when the soil can be disposed of in soil piles at some distance from the excavation. This allows the excavation and disposal of material in a single operation, eliminating the need for hauling units.

However, due to a slower cycle time, a dragline has a production rate of only about 75 to 80 % of that of a comparably-sized power shovel. Also, there is less bucket control, and thus, more spillage. The dragline bucket's ability to excavate material is primarily dependent upon the bucket weight, and since the bucket is not firmly attached to the machine, it is difficult to keep it aligned, and the bucket will tend to bounce or move sideways over hard material. Therefore, a dragline should be confined to excavating soft materials, whenever possible, and the soil should be taken off in layers. In hard soil, the use of a smaller bucket will sometimes increase production since there is less digging resistance.

It is more difficult to accurately dump a dragline than a shovel; therefore, large hauling units should be used with dragline excavation (Photo 1-5). The hauling units should be able to hold 5 to 6 times the capacity of the dragline bucket. Also, the hauling unit should be sized to hold a whole number of bucket loads; otherwise, one of the dumps will be only partially loaded. Never swing the bucket over the cab of a truck.

**Photo 1-5. Dragline loading a truck.**
(*Courtesy of Athey Products Corporation*)

Dragline production can be optimized by using the proper size of bucket. Lift and swing simultaneously, but avoid swinging uphill or downhill, and try to use the same digging and dumping radius. This eliminates the need for raising and lowering the boom during each cycle.

Logistics for dragline excavation usually include the in-line approach or parallel approach (Figure 1-13). With the in-line approach, the dragline backs down the centerline of the excavation and dumps material into spoil piles or trucks on one or both sides of the excavation. The sides of the ditch are removed ahead of the center in order to keep the ditch from narrowing. This method is used mostly for ditch and canal construction.

In the parallel approach, the dragline advances down the face of the excavation as work progresses, and dumps material into a spoil bank or trucks at the rear of the machine. This method is used for wide excavations and for sloping embankments.

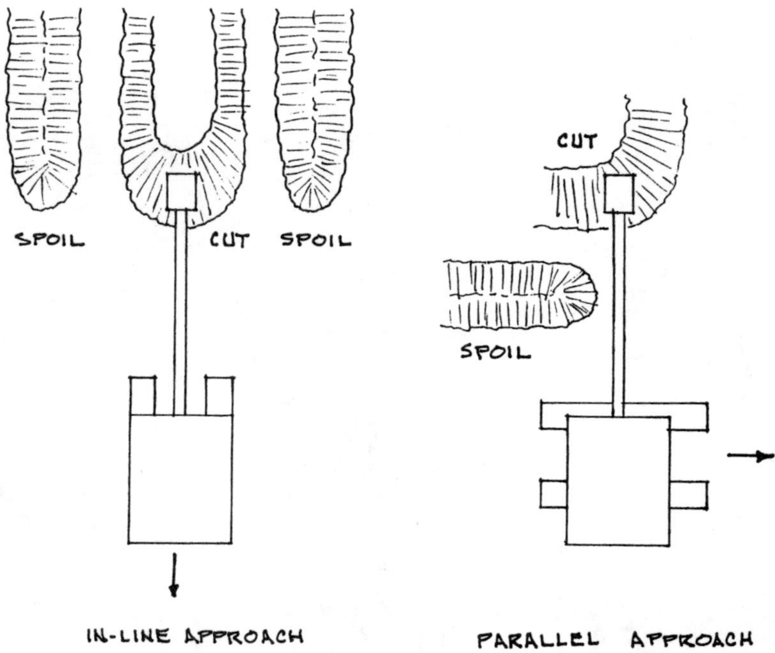


**Figure 1-13. Dragline approach methods.**

Figure 1-14 shows dragline excavating zones. The most efficient digging zone lies within an angle of about 15 degrees in front of and behind a vertical line through the boom point.

Avoid unnecessary bucket casting and hoisting, and avoid swinging the bucket until it clears the ground. Also, do not drag the bucket too close to the machine. This will cause soil to pile up in the front of the excavator and will interfere with the movement of the drag cable. Do not use a bucket that is too large for the machine to handle safely and efficiently. Keep the bucket teeth sharp and clean, and replace worn teeth.

Machine shipping dimensions, operating dimensions, digging ranges and clearances are normally published with manufacturer's specifications for any given excavator (Figure 1-15 and Table 1-9). Of the dimensions given, the most important with regard to job planning and dragline selection include the maximum digging depth, maximum digging reach, maximum dumping radius and maximum dumping height.

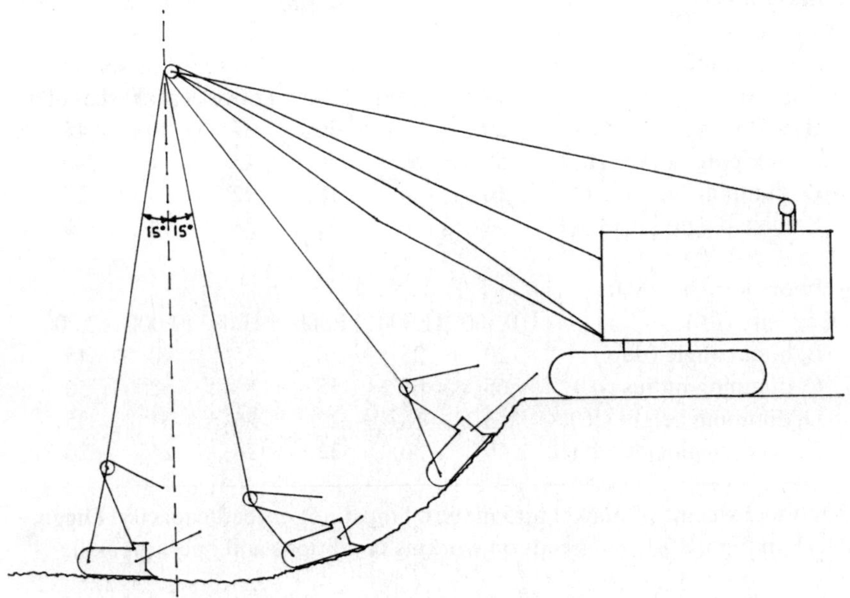


**Figure 1-14. Dragline excavating zones.**

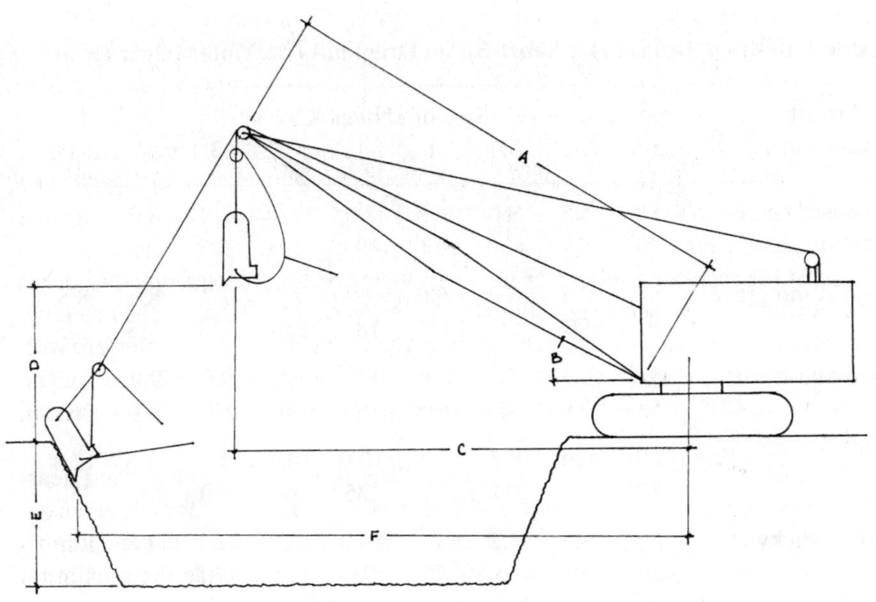


**Figure 1-15. Dragline range diagram.**

**Table 1-9: Working Ranges for a Cable-Operated Dragline With Maximum Counterweights**

| A, boom length = 50 ft. | | | | | | |
|---|---|---|---|---|---|---|
| Capacity (lb.)* | 12,000 | 12,000 | 12,000 | 12,000 | 12,000 | 12,000 |
| B, boom angle (deg.) | 20 | 25 | 30 | 35 | 40 | 45 |
| C, dumping radius (ft.) | 55 | 50 | 50 | 45 | 45 | 40 |
| D, dumping height (ft.) | 10 | 14 | 18 | 22 | 24 | 27 |
| E, max. digging depth (ft.) | 40 | 36 | 32 | 28 | 24 | 20 |
| A, boom length = 60 ft. | | | | | | |
| Capacity (lb.)* | 10,500 | 11,000 | 11,800 | 12,000 | 12,000 | 12,000 |
| B, boom angle (deg.) | 20 | 25 | 30 | 35 | 40 | 45 |
| C, dumping radius (ft.) | 65 | 60 | 55 | 55 | 52 | 50 |
| D, dumping height (ft.) | 13 | 18 | 22 | 26 | 31 | 35 |
| E, max. digging depth (ft.) | 40 | 36 | 32 | 28 | 24 | 20 |

* Combined weight of bucket and material must not exceed capacity. Digging reach (F in Figure 1-15) depends on working conditions and operator skill.

## CABLE-OPERATED DRAGLINE PRODUCTION

Dragline production is determined by the same methods used to determine power shovel production, except that the data tables have different values. Table 1-10 can be used to determine ideal production, which is then tempered with a swing-depth factor taken from Table 1-11, and an efficiency factor.

**Table 1-10: Ideal Outputs for Short-Boom Draglines (BCY/60-Minute Hour)**

| Class of Material | Size of Shovel (C.Y.) | | | | | | | | |
|---|---|---|---|---|---|---|---|---|---|
| | 3/8 | 1/2 | 3/4 | 1 | 1-1/4 | 1-1/2 | 1-3/4 | 2 | 2-1/2 |
| Moist loam or sandy clay | 5.0 | 5.5 | 6.0 | 6.6 | 7.0 | 7.4 | 7.7 | 8.0 | 8.5 |
| | 70 | 95 | 130 | 160 | 195 | 220 | 245 | 265 | 305 |
| Sand and gravel | 5.0 | 5.5 | 6.0 | 6.6 | 7.0 | 7.4 | 7.7 | 8.0 | 8.5 |
| | 65 | 90 | 125 | 155 | 185 | 210 | 235 | 255 | 295 |
| Common earth | 6.0 | 6.7 | 7.4 | 8.0 | 8.5 | 9.0 | 9.5 | 9.9 | 10.5 |
| | 55 | 75 | 105 | 135 | 165 | 190 | 210 | 230 | 265 |
| Hard, tough clay | 7.3 | 8.0 | 8.7 | 9.3 | 10.0 | 10.7 | 11.3 | 11.8 | 12.3 |
| | 35 | 55 | 90 | 110 | 135 | 160 | 180 | 195 | 230 |
| Wet, sticky clay | 7.3 | 8.0 | 8.7 | 9.3 | 10.0 | 10.7 | 11.3 | 11.8 | 12.3 |
| | 20 | 30 | 55 | 75 | 95 | 110 | 130 | 145 | 175 |

*(Based on PCSA data)*

**Table 1-11: Swing-Depth Factors for Draglines**

| Percent of Optimum Depth | Angle of Swing (Degrees) | | | | | | | |
|---|---|---|---|---|---|---|---|---|
| | 30 | 45 | 60 | 75 | 90 | 120 | 150 | 180 |
| 20 | 1.06 | 0.99 | 0.94 | 0.90 | 0.87 | 0.81 | 0.75 | 0.70 |
| 40 | 1.17 | 1.08 | 1.02 | 0.97 | 0.93 | 0.85 | 0.78 | 0.72 |
| 60 | 1.25 | 1.13 | 1.06 | 1.01 | 0.97 | 0.88 | 0.80 | 0.74 |
| 80 | 1.29 | 1.17 | 1.09 | 1.04 | 0.99 | 0.90 | 0.82 | 0.76 |
| 100 | 1.32 | 1.19 | 1.11 | 1.05 | 1.00 | 0.91 | 0.83 | 0.77 |
| 120 | 1.29 | 1.17 | 1.09 | 1.03 | 0.98 | 0.90 | 0.82 | 0.76 |
| 140 | 1.25 | 1.14 | 1.06 | 1.00 | 0.96 | 0.88 | 0.81 | 0.75 |
| 160 | 1.20 | 1.10 | 1.02 | 0.97 | 0.93 | 0.85 | 0.79 | 0.73 |
| 180 | 1.15 | 1.05 | 0.98 | 0.94 | 0.90 | 0.82 | 0.76 | 0.71 |
| 200 | 1.10 | 1.00 | 0.94 | 0.90 | 0.87 | 0.79 | 0.73 | 0.69 |

*(Based on PCSA data)*

**Example 1-12:** Assuming that a 2-C.Y. dragline is excavating common earth at a depth of 7.9 feet and swinging 120 degrees, determine the probable production rate while working a 50-minute hour.
**Solution:** The ideal production rate is:

Ideal Production (BCY/Hr.) = 230 BCY/hr. (Table 1-10)

The percent of optimum depth is:

$$\% \text{ Optimum Depth} = \frac{7.9'}{9.9'} = 0.80 = 80\% \qquad \text{(Eq. 1.8: Table 1-10)}$$

Using a 120° swing, the swing-depth factor is:

Swing-Depth Factor = 0.90 (Table 1-11)

Excluding job efficiency, the production rate is:

Production (BCY/Hr.) = 230 BCY/Hr. x 0.90
= 207 BCY/hour

Job efficiency is:

Efficiency Factor = 50/60 (Given)
= 0.83

Therefore, the probable production rate is:

Production (BCY/Hr.) = 230 BCY/Hr. x 0.90 x 0.83 (Eq. 1.9)
= 172 BCY/hour

**DRAGLINE BUCKETS**

Dragline buckets are available in three basic classes: Type I (light duty), Type II (medium duty) and Type III (heavy duty). Type I buckets are used for excavating easy-to-dig materials such as sandy loam, sandy clay, and stockpiled material. Type II buckets are normally used for excavating clay, soft shale or loose gravel. Type III buckets are recommended for excavating blasted rock, hardpan and for strip mining. Dragline buckets are further classified as solid (Class S), or perforated (Class P), and both are available with, or without teeth (Photo 1-6). The perforated bucket allows excess water to drain from the bucket (Photo 1-5). The toothless bucket is used for stripping topsoil, grading and general cleanup work.

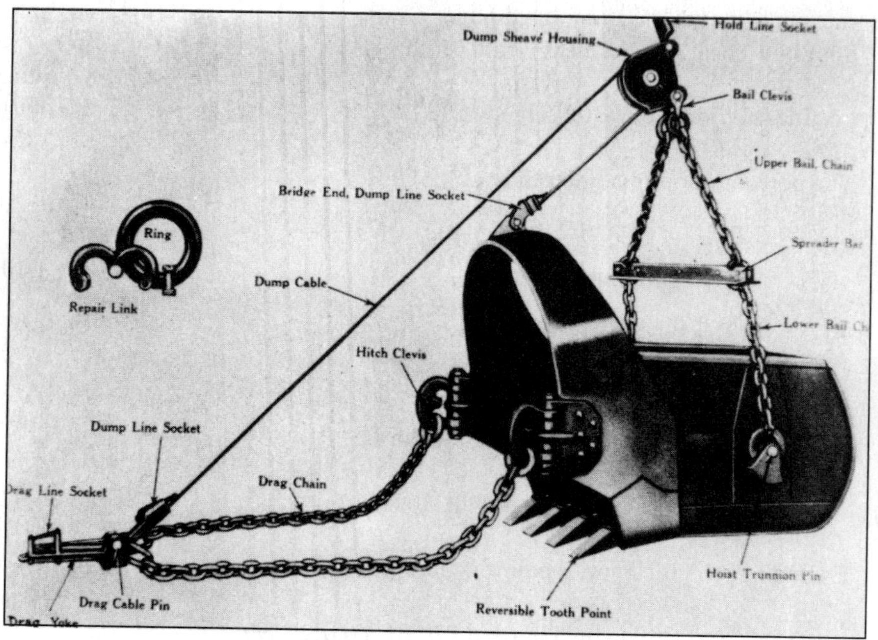


**Photo 1-6. Dragline bucket.**
(*Courtesy of Erie Strayer Company*)

The capacity of a dragline bucket is based on its struck capacity. It is rated nominally in cubic yards, but more accurately in cubic feet (Table 1-12).

**Table 1-12: Capacities and Weights of Dragline Buckets**

| Nominal Size (C.Y.) | Struck Capacity (Cu. Ft.) | Weight of Bucket (lb.) | | |
|---|---|---|---|---|
| | | Light-duty | Medium-duty | Heavy-duty |
| 1/2 | 17 | 1275 | 1460 | 2100 |
| 1 | 32 | 2220 | 2945 | 3700 |
| 1-1/2 | 47 | 3010 | 3750 | 4525 |
| 1-3/4 | 53 | 3375 | 4030 | 4800 |
| 2 | 60 | 3925 | 4825 | 5400 |
| 2-1/2 | 74 | 4310 | 5675 | 6540 |
| 3 | 90 | 5560 | 6660 | 7920 |

## CLAMSHELLS

A crane-shovel can be converted to a clamshell by replacing the shovel boom with a crane boom and substituting a clamshell bucket (Figure 1-2). Clamshells can also be mounted on cranes (Figure 1-16). Figure 1-17 shows details of a clamshell bucket.

The vertical position of the clamshell bucket is controlled by the holding line attached to the top of the bucket. The tag line prevents the bucket from twisting or swinging. The bucket consists of two hinged leaves or scoops and is closed by placing tension on the closing line attached to counterweights at the center of the bucket (Photo 1-7). The bucket is opened by releasing tension on the closing line. When the bucket is open, the weight of the bucket is held by the holding line. Clamshells can also be opened and closed via hydraulic controls (Photo 1-8).

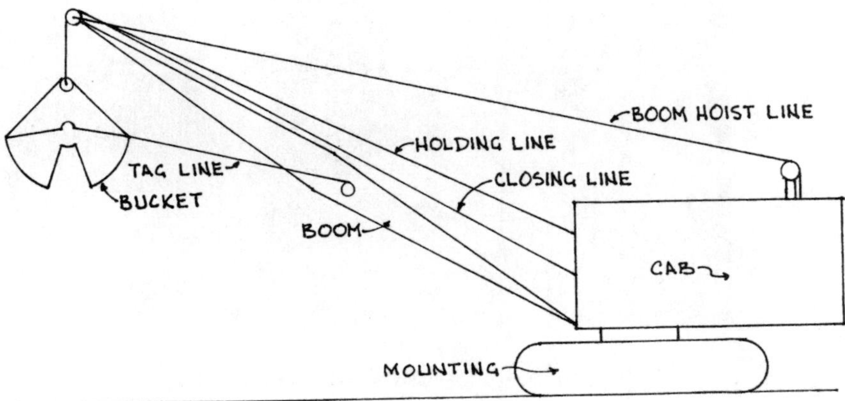


**Figure 1-16. Clamshell components.**

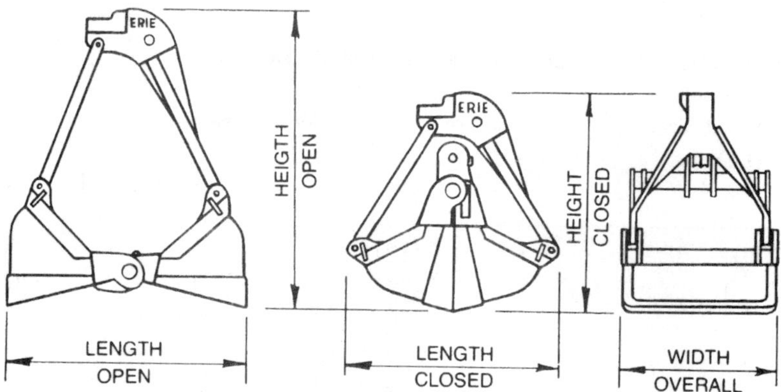


**Figure 1-17. Clamshell details.** (*Courtesy of Erie Strayer Company*)

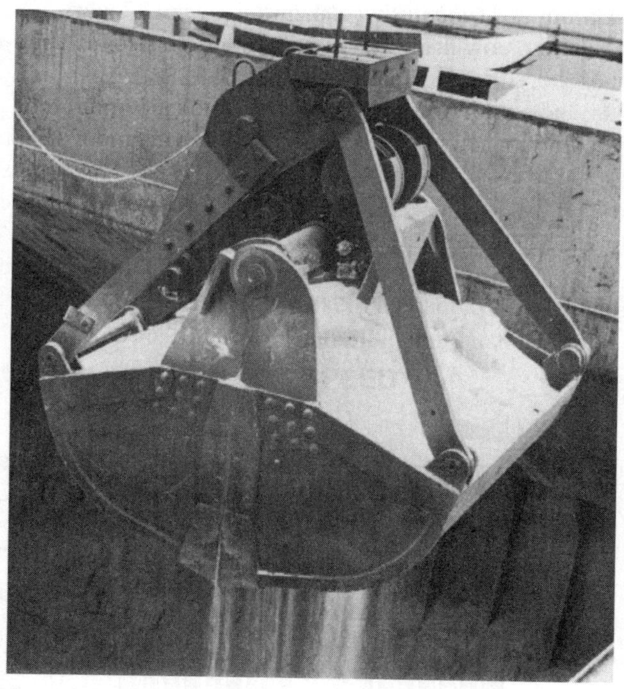

**Photo 1-7. Counter-weighted clamshell bucket.**
(*Courtesy of Erie Strayer Company*)

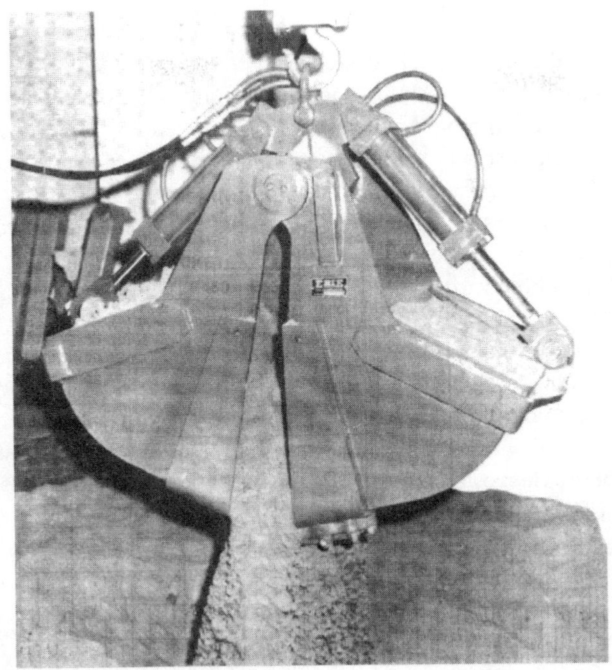

**Photo 1-8. Hydraulically-controlled clamshell bucket.**

*(Courtesy of Erie Strayer Company)*

The clamshell is normally used for excavating vertical shafts and for rehandling loose materials from stockpiles to bins, hoppers or conveyors. A clamshell's digging ability depends primarily on the weight of the bucket.

*Rehandling clamshell buckets* are normally equipped with cutting lips, and are manufactured in four basic types:

1) Extra light rehandlers are designed to move loose, free-flowing material such as ashes, coal, sugar and salt, whose weight does not exceed 75 pounds per cubic foot. These buckets are manufactured with nominal capacities ranging from 3/8 to 10 cubic yards.
2) Light rehandlers are designed to move loose, free-flowing material whose weight does not exceed 100 pounds per cubic foot. Nominal bucket capacities range from 3/8 to 9 cubic yards.
3) Standard rehandlers are designed to move loose materials such as sand and gravel, whose weight does not exceed 130 pounds per cubic foot. Nominal bucket capacities range from 1/4 to 6 cubic yards. The standard rehandler can be optionally equipped with removeable teeth.
4) Wide-scoop rehandling clamshell buckets are manufactured with a large deck area. The *deck area* is the area covered by the bottom of the clamshell bucket when fully open. These buckets are used for rehandling materials from stock piles, barges and gondola cars.

*Excavating clamshell buckets* are normally equipped with teeth, and are manufactured in two basic types:

1) General-purpose buckets are designed to rehandle, dredge and excavate materials in a loose state, whose weight does not exceed 130 pounds per cubic foot. Nominal bucket capacities range from 1/4 to 6 cubic yards.
2) Hard-digging buckets are designed for dredging, excavating, industrial rehandling and rock handling. The weight limitation of the material varies, depending on the strength of the bucket, but it normally ranges from 175 to 250 pounds per cubic foot. Nominal bucket capacities range from 3/8 to 5-1/2 cubic yards.

The size of a clamshell bucket is nominally rated in cubic yards, but more accurately in cubic feet. The cubic-foot capacity is expressed as water-level, plate-line, or heaped volume. The water-level capacity is the volume of liquid the clamshell can hold. The plate-line capacity is the volume of material if struck off at the tops of the clams, and the heaped volume is the capacity when soil is heaped at an angle of repose of 1:1. Clamshell buckets are also rated according to their deck area.

## ORANGE PEEL BUCKETS

A "cousin" of the clamshell is the *orange peel bucket,* which is used for excavating materials that are too hard for the clamshell (Photos 1-9 and 1-10). This type of bucket is also called a *star* or *grapple bucket.* Orange peel buckets are manufactured with full-blade or semi-blade configurations. The blades (tines) of the full-blade model completely close to reduce spillage (photo 1-9), whereas, the semi-blade model is designed with open spaces between the tines to facilitate loading random-shaped materials such as scrap and rock. Some orange peel buckets are manufactured with jaws that work independently, allowing them to pick up irregular-shaped objects (Photo 1-10).

**Photo 1-9. Full-blade orange peel bucket.**
*(Courtesy of Erie Strayer Company)*

Orange peel buckets are manufactured with three tines for light-duty service and four tines for medium-duty service. The three-tine bucket works well for cleaning out material in the corners of gondolas and trucks.

**Photo 1-10. Orange peel bucket loading scrap.**
(*Courtesy of Erie Strayer Company*)

## ROCK GRABS

The *rock grab* is another mechanism that can be installed at the end of a crane boom line to grab and lift rocks and boulders (Photo 1-11). The rock grab is normally manufactured with three grab points equipped with replaceable teeth.

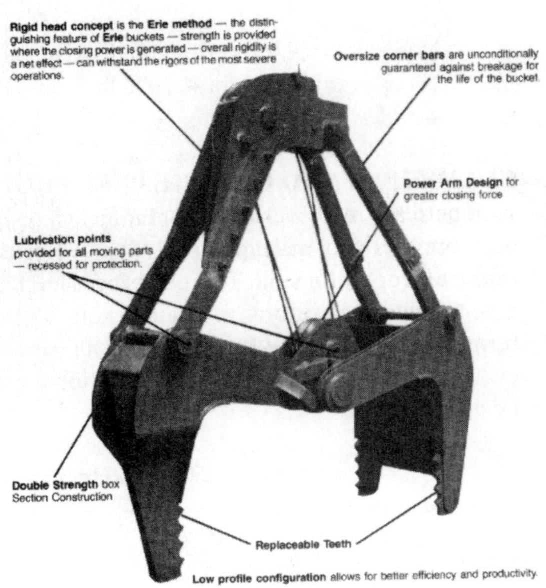


**Photo 1-11. Rock grab.**
(*Courtesy of Erie Strayer Company*)

The lifting capacity of a grab should not be exceeded. The weight of a boulder can be estimated as follows:

1) Measure the dimensions A, B and C as shown in Photo 1-12.
2) Divide the sum of the dimensions by 6.
3) Square the result from Step 2 and multiply by 12.5. This result is the approximate volume of the boulder in cubic feet.
4) Multiply the volume by the density of the rock material. This will give you the weight of the boulder.

**Photo 1-12.**
**Boulder Dimensions.**
*(Courtesy of Erie Strayer Company)*

All of these steps can be condensed into one equation as follows:

$$\text{Weight} = \left(\frac{A + B + C}{6}\right)^2 \text{ x } 12.5 \text{ x Density} \qquad \textbf{(Equation 1.11)}$$

**Example 1-13:** Assuming a density of 170 pounds per cubic foot, determine the weight of a boulder whose dimensions are: A = 8', B = 5' and C = 3'.
**Solution:** The weight of the rock is:

$$\text{Weight} = \left(\frac{8' + 5' + 3'}{6}\right)^2 \text{ x } 12.5 \text{ x } 170 \text{ lb./Cu. Ft.} \qquad \text{(Eq. 1.11)}$$

$$= 88.9 \text{ Cu. Ft. x } 170 \text{ lb./Cu. Ft.}$$
$$= 15{,}113 \text{ lb.} \div 2000 \text{ lb./Ton}$$
$$= 7.6 \text{ tons}$$

## CLAMSHELL AND ORANGE PEEL PRODUCTION

There are many variables that affect clamshell and orange peel production, such as material or soil types, size of load, height of lift, swing angle, size of dump target and operator skill. Therefore, no dependable production tables are available for clamshell or orange peel buckets. The only dependable method for determining production is through previous job data, or by measuring the average cycle time and payload. If this information is available, the production rate can be estimated by:

$$\text{Production (C.Y./Hr)} = \frac{3600 \text{ Sec./Hr. x Bucket Payload x Efficiency Factor}}{\text{Ave. Cycle Time (Sec.)}}$$

**(Equation 1.12)**

**Example 1-14:** Assuming an average cycle time of 30 seconds, determine the production rate of a 1/2-C.Y. clamshell excavating loose material while working a 55-minute hour.
**Solution:** The probable production rate is:

$$\text{Production (LCY/Hr.)} = \frac{3600 \text{ Sec./Hr.} \times 0.5 \text{ LCY} \times 55/60}{30 \text{ Sec.}} \qquad \text{(Eq. 1.12)}$$

$$= 55 \text{ LCY/hour}$$

Clamshell and orange peel production can be optimized by using the proper size of bucket. Lift and swing simultaneously, but avoid swinging uphill or downhill, and try to use the same digging and dumping radius. This eliminates the need for raising and lowering the boom during each cycle. Never swing the bucket over the cab of a truck.

## LIFTING CAPACITY OF A CRANE

Draglines, clamshells and orange peels are cranes equipped with a specialized excavation bucket. Cranes are hydraulically-controlled (Photo 1-13), or cable-operated (Photo 1-14). The safe lifting capacity of a crane is dependent primarily on the *operating radius,* which is the horizontal distance from the center of rotation of the crane to the center of the hoist line. The operating radius is dependent on the boom length and boom angle (Figure 1-18). Other factors that affect lifting capacity include the amount of counterweight, stability of the working surface, the use of outriggers, and the number and strength of the hoist cables.

The *safe* lifting capacity and working range of a crane is furnished in manufacturer's specifications (Table 1-13 and Figure 1-18).

The safe lifting capacity at a specified operating radius is based on a given percentage of *tipping load.* For crawler cranes with no outriggers, the safe lifting capacities are based on 75% of the tipping load, which is defined as a load capable of lifting a crawler roller 2 inches away from the track. For wheel-mounted cranes or cranes with outriggers, the safe lifting capacities are based on 85% of the tipping load. For wheel-mounted cranes, tipping occurs when all tires on one or more wheels leave the supporting surface. The rated loads for crawler- or wheel-mounted cranes are normally based on the direction of least stability for the mounting.

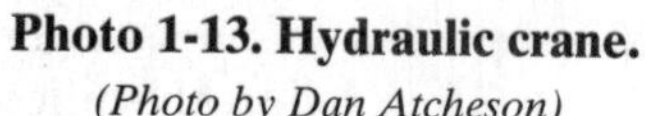

**Photo 1-13. Hydraulic crane.**
*(Photo by Dan Atcheson)*

**Photo 1-14:. Cable operated crane.**
*(Photo by Dan Atcheson)*

**Example 1-15:** Determine the tipping load for the 30-ton crane given in Table 1-13 while using a 40-foot boom at a 20-foot operating radius.
**Solution:** Referring to Table 1-13, the rated (safe) lifting capacity for the crane under the given conditions is 27,000 pounds. The safe lifting capacities given in Table 1-13 are based on 75% of the tipping load; i.e.,

$$\text{Safe Lifting Capacity} = 0.75 \times \text{Tipping Load}$$

Therefore, the tipping load is:

$$\text{Tipping Load} = \frac{\text{Safe Lifting Capacity}}{0.75}$$

$$= \frac{27{,}000 \text{ lb.}}{0.75} \qquad \text{(Table 1-13)}$$

$$= 36{,}000 \text{ pounds}$$

**Table 1-13: Safe Lifting Capacities (lb.) for a 30-Ton Crane (PCSA Class 12-105)**

| Boom Length (Ft.) | Radius (Ft.) | Rated Safe Lifting Capacity (lb.) | | |
|---|---|---|---|---|
| | | Crane | Clamshell* | Dragline* |
| 40 | 12 | 60,000 | | |
| | 15 | 40,000 | | |
| | 20 | 27,000 | | |
| | 25 | 19,900 | 10,300 | |
| | 30 | 15,600 | 10,300 | 8800 |
| | 35 | 12,800 | 10,300 | 8800 |
| | 40 | 10,700 | 9,630 | 8800 |
| 50 | 15 | 40,610 | | |
| | 20 | 26,810 | | |
| | 25 | 19,710 | | |
| | 30 | 15,410 | 10,300 | |
| | 35 | 12,610 | 10,300 | |
| | 40 | 10,510 | 9,460 | 8800 |
| | 45 | 8,930 | 8,035 | 8800 |
| | 50 | 7,730 | 6,955 | 7730 |

*Combined weight of bucket and material must not exceed capacity.

The PCSA and U.S. crane manufacturers have developed a standard method for rating cranes. In this method, a crane is given a nominal safe lifting capacity (in tons) in the direction of least stability (with outriggers set if the crane is so equipped) at a specified radius (usually 12 feet) while equipped with a base boom length. The nominal rating is followed by two numbers separated by a hyphen. The first number indicates the operating radius used for the rated capacity. The second number indicates the rated load in hundreds of pounds (rounded off to the nearest whole number) when the crane is operating at a 40-foot radius using a 50-foot boom. For example, the crane shown in Table 1-13 is classified as 30-ton crane (class 12-105). The crane is so classified because it can safely lift a maximum load of 60,000 pounds (30 tons) at a radius of 12 feet with a basic boom (in this case, 40 feet). While equipped with a 50-foot boom and working at a 40-foot radius, the crane can safely lift 10,510 pounds. The load includes the weight of the item lifted plus the weight of hooks, hook blocks, slings, etc., but excludes the weight of the hoist line.

**Example 1-16:** Assuming the use of a 30-ton crane (PCSA class 12-105) equipped with a 50-foot boom at a radius of 40 feet, determine the largest heavy-duty dragline bucket permitted when excavating soil weighing 90 pounds per cubic foot.
**Solution:** Referring to Table 1-13, the safe lifting capacity of the crane under the given working conditions is 8800 pounds.

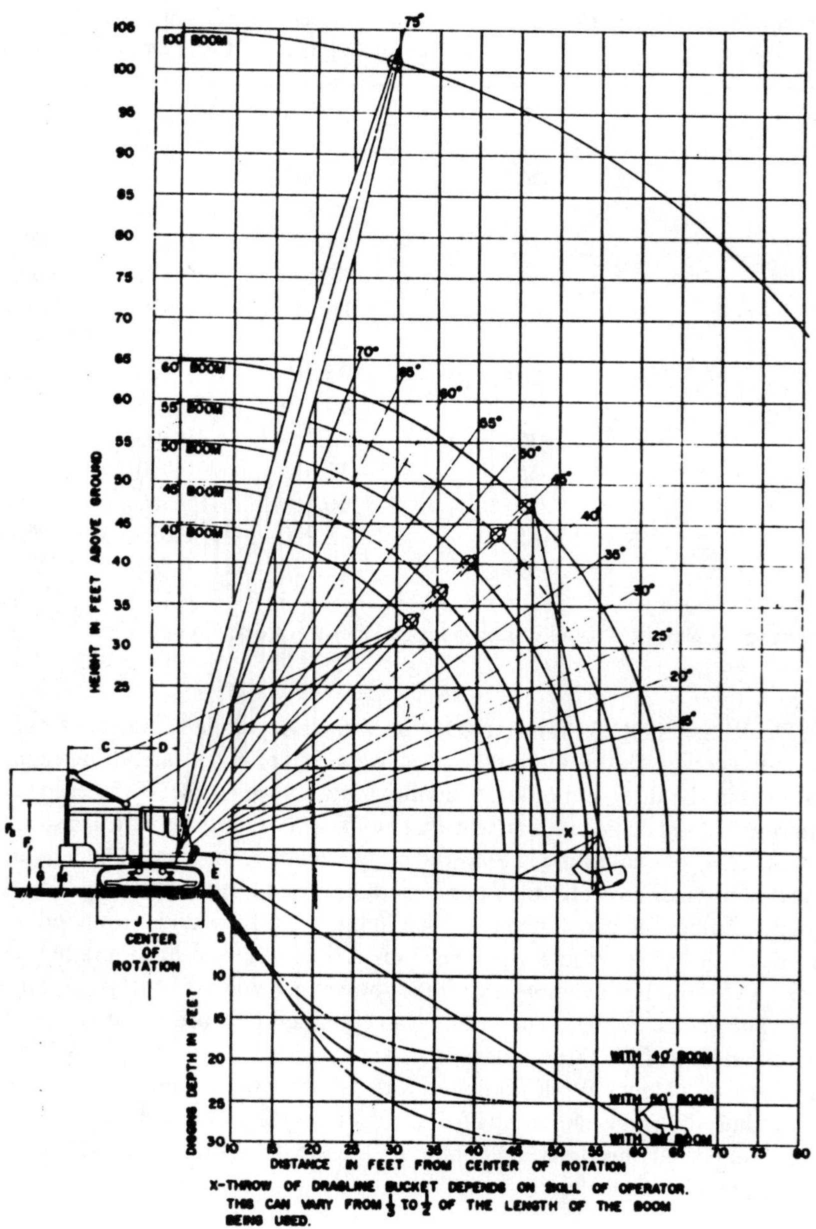


**Figure 1-18. Working ranges for a 30-ton crane (PCSA class 12-105).**

(*Northwest Engineering Company*)

Let's try a 1-3/4-C.Y. bucket. The total load would be:

Load = (53 Cu. Ft. x 90 lb./Cu. Ft.) + 4800 lb. (Table 1-12)
= 9570 pounds

This exceeds the safe lifting capacity. Now, let's try a 1-1/2-C.Y. bucket. The total load would be:

Load = (47 Cu. Ft. x 90 lb./Cu. Ft.) + 4525 lb.
= 8755 pounds

This is within the limits of the crane's safe lifting capacity.

Figure 1-19 can also be used to estimate the safe lifting capacities of various cranes.

**Example 1-17:** Estimate the safe lifting capacity of a 36-ton crane equipped with 1-1/2-C.Y. dragline using a single hoist line and operating at a radius of 50 feet.
**Solution:** Referring to Figure 1-19, the safe lifting capacity under the given conditions is approximately 10,000 pounds.

**Example 1-18:** Determine the minimum boom length and boom angle that will permit a 30-ton (PCSA class 12-105) crane to lift and dump a 1-C.Y. clamshell bucket into a hopper which is 35 feet from the crane's center of rotation and is 25 feet above the surface on which the crane is operating. Assume that the height of the closed clamshell bucket is 7 feet, and allow 5 feet above the bucket for the block and tackle.
**Solution:** The minimum height of the boom point of the crane must be at least 25' + 7' + 5' = 37 feet. Referring to Figure 1-18, the boom will have to be at least 45 feet long while working at an angle of 45 degrees.

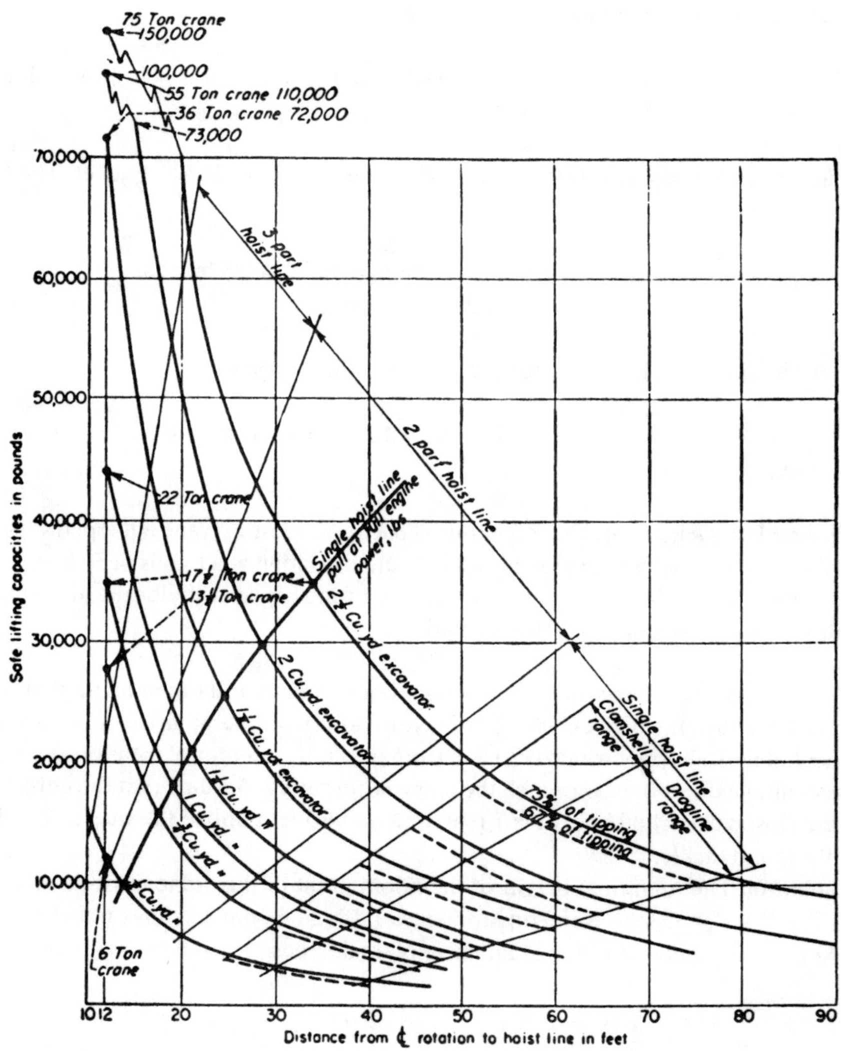


Figure 1-19. Safe lifting capacities of cranes.

## CABLE-OPERATED HOES

A crane-shovel can be converted to a hoe (also referred to as a backhoe, trench hoe, drag shovel, draghoe or pull shovel) by replacing the shovel mechanism with a hoe attachment (Figures 1-2 and 1-20). Unlike hydraulic hoes, most cable-operated hoes do not have a pivoted wrist-action dipper. Instead, they are equipped with a rigid brace.

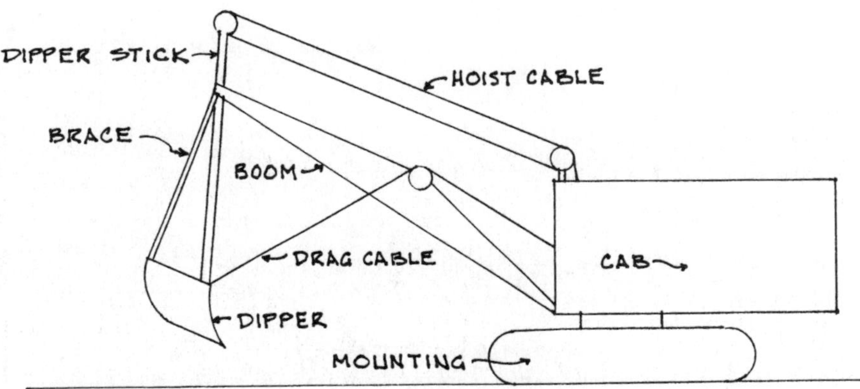


**Figure 1-20. Components of a cable-operated hoe.**

To excavate, the boom is set at the desired angle by controlling the hoist cable. The dipper is extended by placing tension on the hoist cable while playing out the drag cable. The dipper is pulled toward the machine by placing tension on the drag cable while playing out the hoist cable.

The hoe is capable of excavating from above the excavation at natural ground level. A hoe does not have the digging and dumping range of a dragline; however, the rigidity of a hoe allows it to dig more powerfully and accurately. Thus, hoes are well suited for excavating trenches and basements where precision is required.

At moderate depths, a hoe will produce about the same as that of a power shovel of comparable size; however, as the depth increases, hoe production will decrease considerably. Greatest efficiency is obtained when the dipper stick is at right angles to the boom, and when digging near the machine, since this will reduce cycle time.

When trenching, use a bucket width that matches the required width of the trench. Production will be greater when you use a standard bucket, than when you use a small bucket equipped with side cutters to increase the width of the trench.

Various sized dippers can be attached to a cable-operated hoe, and the cycle time does not increase substantially as the size of dipper is increased. Therefore, when excavating a basement, use the largest dipper available, provided it does not exceed the tipping load of the machine.

Machine shipping dimensions, operating dimensions, digging ranges and clearances are normally published with manufacturer's specifications for any given hoe (Figure 1-21 and Table 1-14). Of the dimensions given, the most important with regard to job planning and hoe selection include the maximum digging depth, maximum digging radius and maximum dumping clearance.

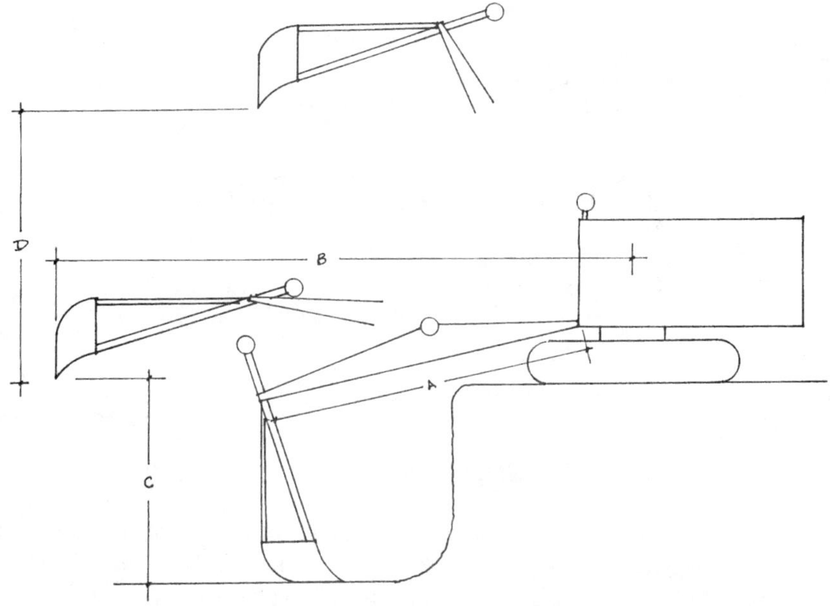


**Figure 1-21. Clearance diagram for a cable-operated hoe.**

**Table 1-14: Dimensions and Clearances for Cable-Operated Hoes**

| Bucket Size (C.Y.) | Boom Length (Ft.) A | Max. Digging Radius (Ft.) B | Max. Digging Depth (Ft.) C | Clearance Under Dipper (Ft.) D |
|---|---|---|---|---|
| 1/2 | 16 | 26 | 15 | 15 |
| 1 | 18 | 30 | 20 | 18 |
| 1-1/2 | 22 | 36 | 25 | 27 |

## HYDRAULIC EXCAVATORS

Hydraulic excavators use oil-filled pistons instead of cables to create machine movement (Figure 1-22). They are more prevalent than cable-operated excavators and are available with a wide assortment of helpful accessories such as one-, or two-piece booms, adjustable booms, long-reach booms, telescopic sticks, quick-release buckets, clamshells, ditch cleaning buckets, grapples, and hammers. They are available with crawler tracks or tires (Photos 1-15 and 1-16). Some tire-mounted models can travel in excess of 20 mph.

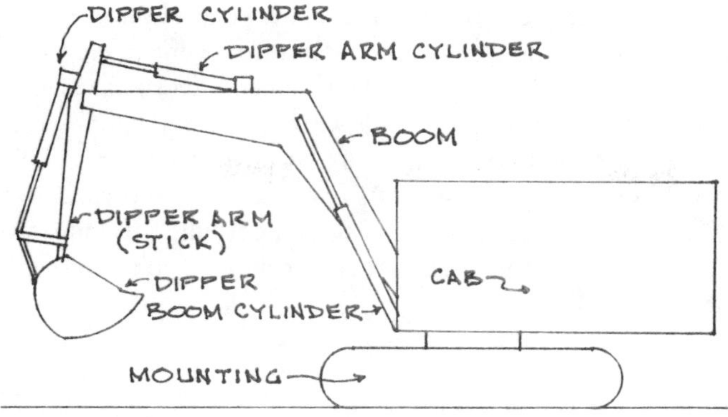


**Figure 1-22. Hydraulic excavator components.**

**Photo 1-15. Track-mounted hydraulic excavator.** (*Courtesy of Caterpillar Inc.*)

**Photo 1-16. Wheel-mounted hydraulic excavator.** (*Courtesy of Caterpillar Inc.*)

Hydraulic excavators are broadly classified as general-purpose excavators, mass excavators and front shovels. As compared to a general-purpose excavator, a mass excavator is equipped with a larger bucket and shorter, but heavier boom and stick (or "crowd"). Front shovels will be discussed later in this chapter.

Machine shipping dimensions (Figure 1-23), digging ranges and clearances (Figure 1-24) are normally published with manufacturer's specifications for any given excavator. Note in Figure 1-24 that some hydraulic excavators manufactured with a one-piece boom are available with short, medium, long and extra long sticks (dipper arms).

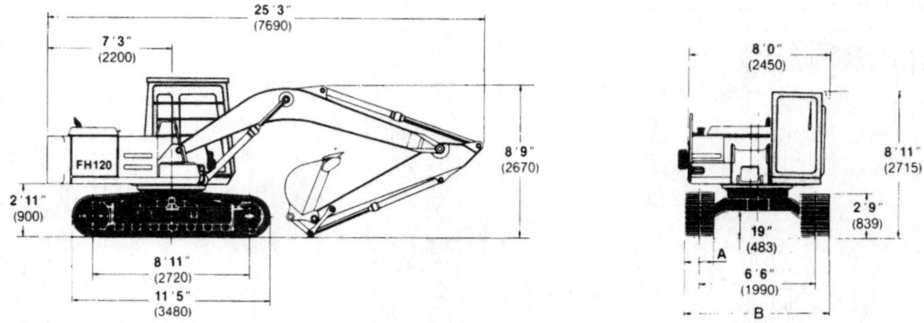


**Figure 1-23. Shipping dimensions of mass excavator.**
(*Courtesy of Fiat-Allis North America, Inc.*)

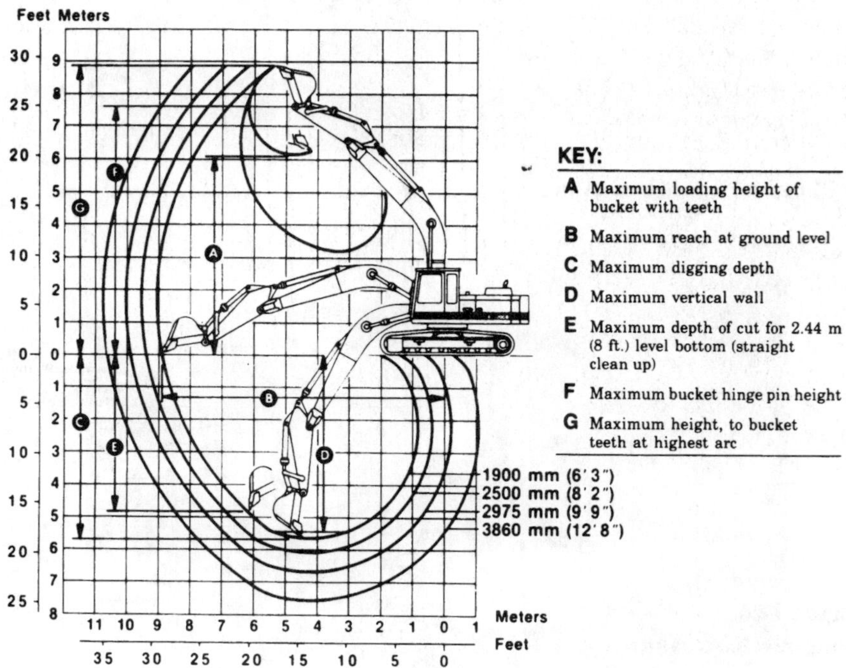


**Figure 1-24. Working ranges of mass excavator.** (*Courtesy of Caterpillar Inc.*)

Some excavators are available with two-piece booms whose boom lengths can be extended or retracted by removing pins, adjusting the foreboom position and replacing the pins (Figure 1-25).

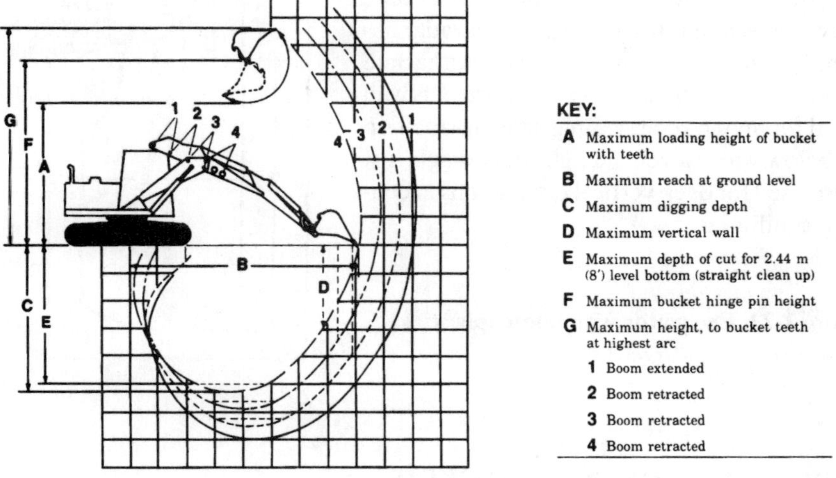


**Figure 1-25. Two-piece-boom excavator.** (*Courtesy of Caterpillar Inc.*)

Long-reach excavators are available for projects requiring reach beyond the range of a normal excavator (Figure 1-26). These excavators are well suited for ditch cleaning, slope finishing, river conservation and other work formerly reserved for draglines.

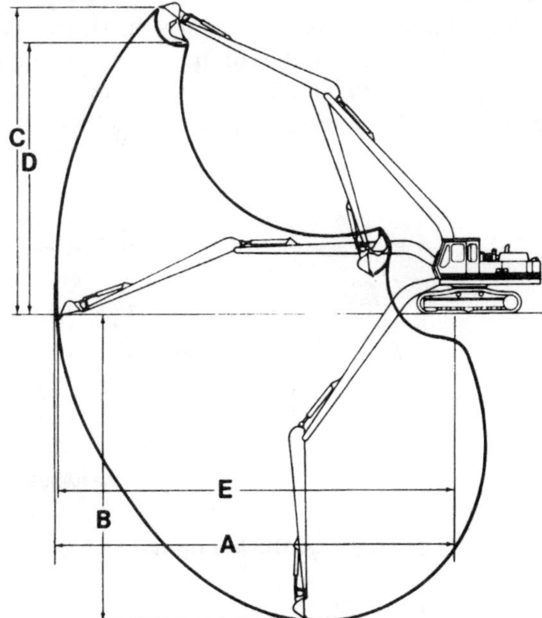


**Figure 1-26. Long-reach excavator.** (*Courtesy of Caterpillar Inc.*)

Some excavators can be equipped with a telescopic stick providing additional digging depth and dumping height (Figure 1-27). The hydraulically extendable and retractable stick consists of an outer tube, middle tube and an inner tube and extends and retracts much like a radio antenna. Normally, a clamshell bucket is attached to the end of the stick. This arrangement is well suited for material rehandling, placing small rip rap below water level and light-duty digging. It is also used for deep vertical excavations in confined conditions.

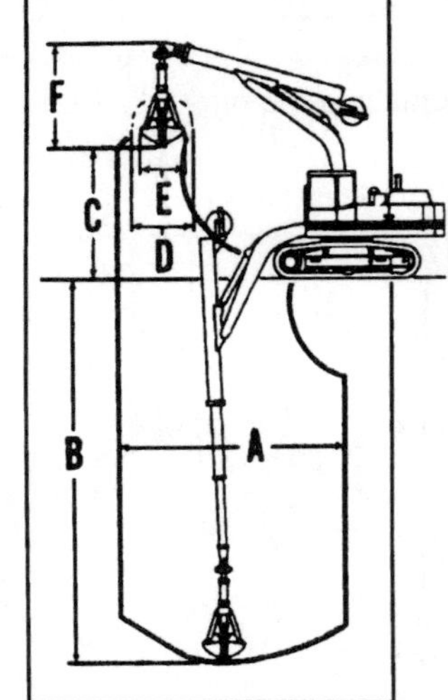


**Figure 1-27. Excavator with telescopic stick.** *(Courtesy of Caterpillar Inc.)*

## HYDRAULIC EXCAVATOR BUCKETS

An excavator bucket should be selected for the type of soil being excavated. The two most important items to be considered is bucket width and bucket tip radius. With a given size excavator, wide buckets are used for excavating easy-to-dig soils, and narrow buckets for harder soils. Also, curling forces are increased with a shorter tip radius; therefore, the shorter the bucket tip radius, the better the bucket will excavate and load (Figure 1-28).

Many buckets are manufactured with two basic profiles. One has a deep pocket profile with a long tip radius for trenching; the other has a shallow profile with a short tip radius for loading and general excavation work (Figure 1-29). From these two basic profiles, several different bucket types are available.

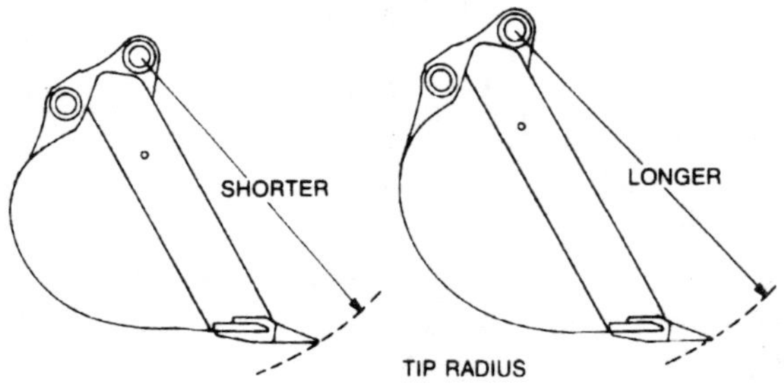


**Figure 1-28. Bucket tip radius.** *(Courtesy of Caterpillar Inc.)*

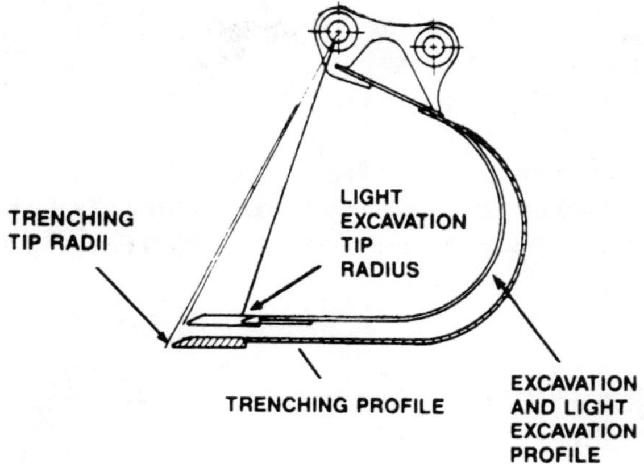


**Figure 1-29. Bucket profiles.** (*Courtesy of Caterpillar Inc.*)

The *trenching* (T) *bucket* has a bite width designed for standard pipe diameters and deep profiles for optimum capacity and performance (Figure 1-30).

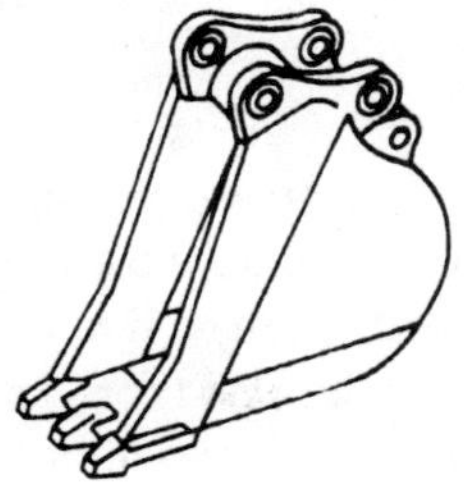

**Figure 1-30. Trenching (T) bucket.** (*Courtesy of Caterpillar Inc.*)

The *extreme service trenching* (ET) *bucket* is designed for excavating tough material such as fragmented rock, frozen ground, caliche, etc. This bucket can be equipped with ripper shanks at the rear of the bucket for use when penetration is impossible with the front cutting edge (Figure 1-31). This bucket is designed with slots for the ripper shanks. Ripper brackets can be welded onto any bucket.

The *excavation* (X) *bucket* is designed with a short tip radius and a wide bite for maximum digging force, easy loading and dumping (Figure 1-32). The *extreme service excavation* (EX) *bucket* is similar to the excavation bucket, but it is manufactured with thicker, tougher and more abrasive materials.

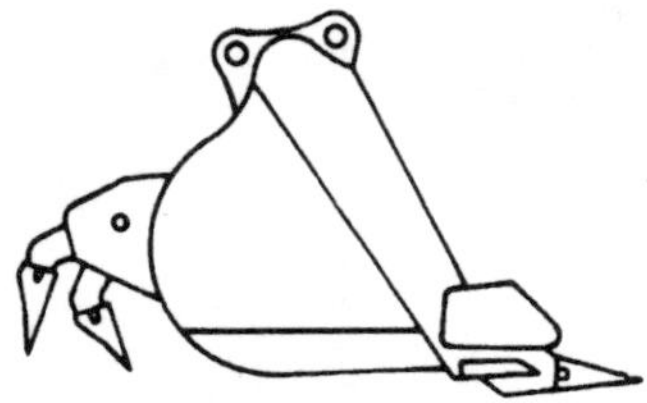

**Figure 1-31. Extreme service trenching (ET) bucket equipped with ripper shanks.**
(*Courtesy of Caterpillar Inc.*)

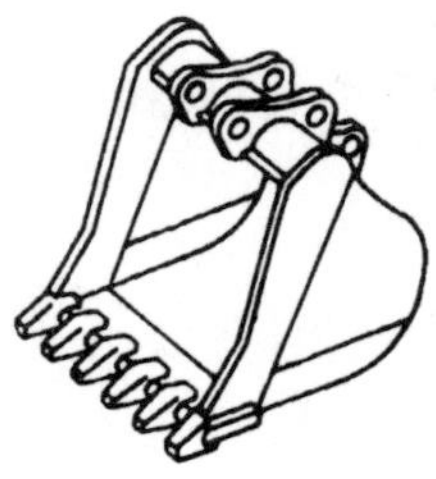

**Figure 1-32. Excavation (X) bucket.**
(*Courtesy of Caterpillar Inc.*)

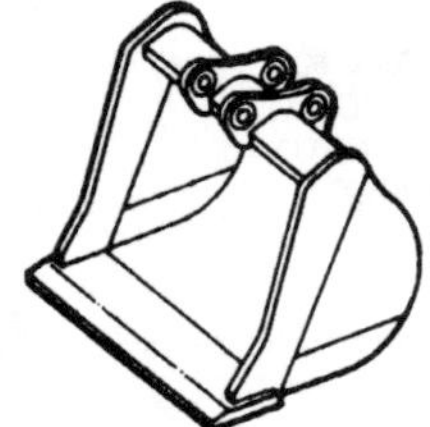

**Figure 1-33. Light material excavation (LX) bucket.**
(*Courtesy of Caterpillar Inc.*)

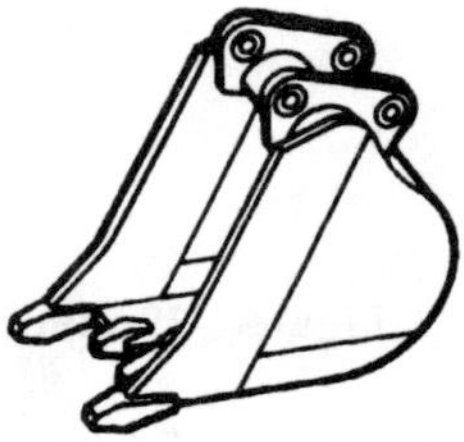

**Figure 1-34. Rock ripping (R) bucket.**
(*Courtesy of Caterpillar Inc.*)

The *light material excavation* (LX) *bucket* is designed for excavating and loading easy-to-dig materials. It can also be used for finishing and cleanup work. It can be equipped with a bottom cutting edge, if necessary (Figure 1-33).

The *rock ripping* (R) *bucket* is designed for excavating rock and other extremely hard materials. The plate thickness is about 75% thicker than that used to build a standard bucket (Figure 1-34).

Any given excavator can often be equipped with a variety of bucket types and sizes as shown in Table 1-15.

When changing buckets, be sure to position the boom and stick with the holes squarely aligned with the bucket so that the bucket pin will slide through with minimum effort.

The capacity of a hydraulic excavator bucket or hydraulic shovel bucket is expressed in loose cubic yards of heaped capacity , which is the struck volume plus the volume above the strike-off plane, having an angle of repose of 1:1 (Figure 1-3). Since various types of soil fill the bucket in varying amounts, the actual bucket payload is expressed as a percentage of heaped capacity.

**Table 1-15:**
**Various Bucket Types and Sizes for the Caterpillar 235D Mass Excavator**

| MODEL | BUCKET TYPE | BUCKET FAMILY | BUCKET BITE WIDTH◄ | | BUCKET TIP RADIUS | | HEAPED CAPACITY | | BUCKET WEIGHT WITH LONG TIPS | |
|---|---|---|---|---|---|---|---|---|---|---|
| | | | mm | in | mm | in | L | $yd^3$ | kg | lb |
| 235D | Trenching | | 777 | 30.5 | 1980 | 78.0 | 1000 | 1.25 | 1110 | 2450 |
| | Trenching | | 927 | 36 | 1980 | 78.0 | 1300 | 1.62 | 1270 | 2800 |
| | Trenching | | 1077 | 42 | 1980 | 78.0 | 1600 | 2.00 | 1415 | 3120 |
| | Rock | | 840 | 33 | 1778 | 70.0 | 800 | 1.00 | 1500 | 3310 |
| | ET | | 861 | 34 | 1985 | 78.1 | 1100 | 1.38 | 1550 | 3400 |
| | ET | | 1087 | 43 | 2060 | 81.1 | 1600 | 2.00 | 1815 | 3993 |
| | Excavation | | 1227 | 48 | 1883 | 74.1 | 1600 | 2.12 | 1430 | 3150 |
| | Excavation | | 1377 | 54 | 1883 | 74.1 | 1900 | 2.50 | 1560 | 3440 |
| | Excavation | | 1600 | 63 | 1883 | 74.1 | 2300 | 3.00 | 1705 | 3758 |
| | Excavation | | 1680 | 66 | 1735 | 68.3 | 2100 | 2.75 | 1690 | 3710 |
| | Excavation | | 1800 | 71 | 1883 | 74.1 | 2680 | 3.50 | 2010 | 4431 |
| | Utility | | 1750 | 69 | 1553 | 61.1 | 2300 | 3.00 | 1520* | 3350* |

(*Courtesy of Caterpillar Inc.*)

The percentage of heaped capacity is referred to as the bucket fill factor or bucket efficiency factor. The factor varies, depending on the type of soil, or the condition (size) of rock being excavated. The bucket fill factor is also dependent upon the bucket configuration. For instance, a deep narrow bucket may fill to only 1/3 the rated capacity. Thus, the bucket payload can be determined by:

Bucket Payload (LCY) = Heaped Volume (LCY) x Bucket Fill Factor (Eq. 1.1)

Bucket fill factors for hydraulic excavators are given in Table 1-16.

**Table 1-16: Bucket Fill Factors for Mass Excavators**

| Material | Fill Factor Range (% of Heaped Bucket Capacity) |
|---|---|
| Moist loam or sandy clay | 100-110 |
| Sand and gravel | 95-110 |
| Hard, tough clay | 80-90 |
| Rock, well-blasted | 60-75 |
| Rock, poorly-blasted | 40-50 |

(*Courtesy of Caterpillar Inc.*)

The safe lifting capacity of the excavator should not be exceeded. Safe lifting capacity should be checked if the excavator is used as a "crane" to lift heavy objects such as pipe suspended from a chain or cable attached to the bucket, or to the bucket pin holes at the end of the stick (Photo 1-17). When "craning," rig the chain or cable as short as possible to prevent the load from swinging out of control. Table 1-17 gives bucket-plus-payload limits for a Caterpillar 235D hydraulic excavator equipped with a one-piece boom.

**Table 1-17: Safe Lifting Capacities of a Caterpillar 235D Mass Excavator**

| Stick | Lifting Capacity (lb.) |
|---|---|
| Short | 15,100 |
| Medium | 14,000 |
| Long | 12,100 |

(*Courtesy of Caterpillar Inc.*)

**Example 1-19:** Assuming a Caterpillar 235D hydraulic excavator is excavating moist loam weighing 2600 pounds per bank cubic yard, determine the largest excavation (X) bucket that can be used if a long stick is required.
**Solution:** While using a long stick, the maximum safe load is:

Maximum Safe Load = 12,100 pounds (Table 1-17)

Assuming the use of a 71-inch type X bucket, the bucket payload is:

Bucket Payload = 3.50 BCY x 1.05 (Eq. 1.1; Tables 1-15 and 1-16)
= 3.68 BCY

The total load is:

Total Load = (3.68 BCY x 2600 lb./BCY) + 4431 lb. (Table 1-15)
= 13,999 pounds

This exceeds the maximum safe load, so let's try a smaller bucket. The payload of a 66-inch type X bucket is:

Bucket Payload = 2.75 BCY x 1.05
= 2.89 BCY

The total load is:

Total Load = (2.89 BCY x 2600 lb./BCY) + 3710 lb. (Table 1-15)
= 11,224 pounds

This bucket is within the safe load limits.

## HYDRAULIC EXCAVATOR PRODUCTION

Hydraulic excavator production can be determined by:

$$\text{Production} = \text{Cycles/Hour x Bucket Payload/Cycle x Efficiency Factor} \qquad \text{(Eq. 1.7)}$$

where:

$$\text{Cycles/Hour} = \frac{\text{3600 Sec./Hr.}}{\text{Cycle Time (Sec.)}} \qquad \text{(Eq. 1.5)}$$

or,

$$\text{Cycles/Hour} = \frac{\text{60 Min./Hr.}}{\text{Cycle Time (Min.)}} \qquad \text{(Eq. 1.6)}$$

and,

$$\text{Bucket Payload} = \text{Heaped Volume x Bucket Fill Factor} \qquad \text{(Eq. 1.1)}$$

where: Bucket fill factor is taken from Table 1-16.

Cycle time for hydraulic excavators can be estimated from Figure 1-35.

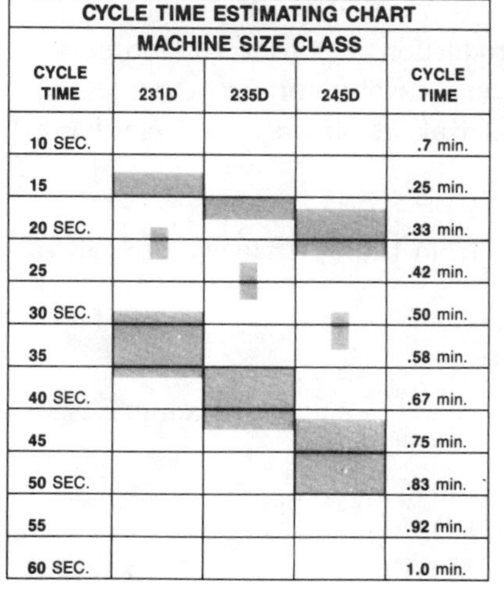

| CYCLE TIME ESTIMATING CHART | | | | |
|---|---|---|---|---|
| | MACHINE SIZE CLASS | | | |
| CYCLE TIME | 231D | 235D | 245D | CYCLE TIME |
| 10 SEC. | | | | .7 min. |
| 15 | | | | .25 min. |
| 20 SEC. | | | | .33 min. |
| 25 | | | | .42 min. |
| 30 SEC. | | | | .50 min. |
| 35 | | | | .58 min. |
| 40 SEC. | | | | .67 min. |
| 45 | | | | .75 min. |
| 50 SEC. | | | | .83 min. |
| 55 | | | | .92 min. |
| 60 SEC. | | | | 1.0 min. |

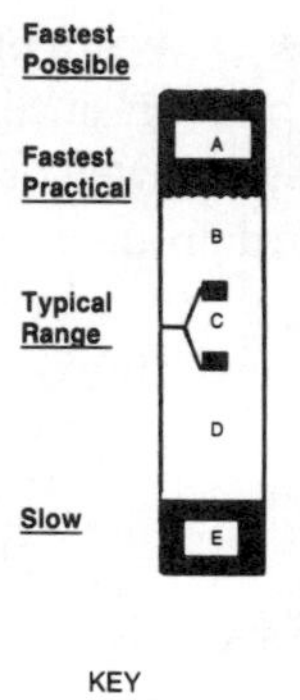


KEY
A — Excellent
B — Above Average
C — Average
D — Below Average
E — Severe

**CYCLE TIME -vs- JOB CONDITION DESCRIPTION**

— Easy digging (unpacked earth, sand gravel, ditch cleaning, etc.). Digging to less than 40% of machine's maximum depth capability. Swing angle less than 30°. Dump onto spoil pile or truck in excavation. No obstructions. Good operator.

— Medium digging (packed earth, tough dry clay, soil with less than 25% rock content). Depth to 50% of machine's maximum capability. Swing angle to 60°. Large dump target. Few obstructions.

— Medium to hard digging (hard packed soil with up to 50% rock content). Depth to 70% of machine's maximum capability. Swing angle to 90°. Loading trucks with truck spotted close to excavator.

— Hard digging (shot rock or tough soil with up to 75% rock content). Depth to 90% of machine's maximum capability. Swing angle to 120°. Shored trench. Small dump target. Working over pipe crew.

— Toughest digging (sandstone, caliche, shale, certain limestones, hard frost). Over 90% of machine's maximum depth capability. Swing over 120°. Loading bucket in man box. Dump into small target requiring maximum excavator reach. People and obstructions in the work area.

**Figure 1-35. Mass excavator cycle time estimating chart.**

(*Courtesy of Caterpillar Inc.*)

**Example 1-20:** Estimate the hourly production rate of a Caterpillar 235D excavator while digging hard tough clay with 71-inch excavation (type X) bucket. Assume average job conditions and working a 50-minute hour.
**Solution:** The cycle time will be approximately:

Cycle Time = 25 seconds (Figure 1-35)

The number of cycles per hour is:

$$\text{Cycles/Hour} = \frac{3600 \text{ Sec./Hr.}}{25 \text{ Sec./Cycle}} \qquad \text{(Eq. 1.5)}$$
$$= 144 \text{ cycles/hour}$$

The average bucket payload is:

Bucket Payload (BCY) = 3.50 BCY x 0.90 (Eq. 1.1, Tables 1-15 and 1-16)
= 3.15 BCY

Therefore, the probable production rate is:

Production (BCY/Hr.) = 144 Cycles/Hr. x 3.15 BCY/Cycle x 50/60 (Eq. 1.7)
= 378 BCY/hour

Some manufacturers express excavator production rates in terms of loose cubic yards per hour (Table 1-18); however, since an excavator normally excavates in-place (bank) material, we must use swell or shrinkage factors to interface bank-yard production with loose-yard production.

**Example 1-21:** Assuming 25% swell, use data from Table 1-18 to work the previous example.
**Solution:** The cycle time is :

Cycle Time = 25 seconds (Example 1-20)

Expressed in loose cubic yards, the bucket payload is:

Bucket Payload (LCY) = 3.15 BCY x 1.25 LCY/BCY (25% Swell)
= 3.94 LCY

Referring to Table 1-18 while using a 24-second cycle time and a 4-LCY payload, we find that the production rate is:

Production (LCY/Hr.) = 600 LCY/hour (Table 1-18)

**Table 1-18: Excavator Production Table (LCY/60-Minute Hour)**

| ESTIMATED CYCLE TIMES | | ESTIMATED BUCKET PAYLOAD** — LOOSE CUBIC YARDS | | | | | | | | | | | | | | | | | | ESTIMATED CYCLE TIMES | |
|---|---|---|---|---|---|---|---|---|---|---|---|---|---|---|---|---|---|---|---|---|---|
| Cycle Time | | | | | | | | | | | | | | | | | | | | | |
| Seconds | Minutes | 0.25 | 0.50 | 0.75 | 1.00 | 1.25 | 1.50 | 1.75 | 2.00 | 2.25 | 2.50 | 2.75 | 3.00 | 3.25 | 3.50 | 3.75 | 4.00 | 4.50 | 5.00 | Cycles Per Min. | Cycles Per Hr. |
| 10.0 | .17 | | | | | | | | | | | | | | | | | | | 6.0 | 360 |
| 11.0 | .18 | | | | | | | | | | | | | | | | | | | 5.5 | 330 |
| 12.0 | .20 | 75 | 150 | 225 | 300 | 375 | | | | | | | | | | | | | | 5.0 | 300 |
| 13.3 | .22 | 67 | 135 | 202 | 270 | 337 | 404 | 472 | 540 | 607 | 675 | 742 | 810 | 877 | 945 | 1012 | 1080 | 1215 | 1350 | 4.5 | 270 |
| 15.0 | .25 | 60 | 120 | 180 | 240 | 300 | 360 | 420 | 480 | 540 | 600 | 660 | 720 | 780 | 840 | 900 | 960 | 1080 | 1200 | 4.0 | 240 |
| 17.1 | .29 | 52 | 105 | 157 | 210 | 262 | 315 | 367 | 420 | 472 | 525 | 577 | 630 | 682 | 735 | 787 | 840 | 945 | 1050 | 3.5 | 210 |
| 20.0 | .33 | 45 | 90 | 135 | 180 | 225 | 270 | 315 | 360 | 405 | 450 | 495 | 540 | 585 | 630 | 675 | 720 | 810 | 900 | 3.0 | 180 |
| 24.0 | .40 | 37 | 75 | 112 | 150 | 187 | 225 | 262 | 300 | 337 | 375 | 412 | 450 | 487 | 525 | 562 | 600 | 675 | 750 | 2.5 | 150 |
| 30.0 | .50 | 30 | 60 | 90 | 120 | 150 | 180 | 210 | 240 | 270 | 300 | 330 | 360 | 390 | 420 | 450 | 480 | 510 | 600 | 2.0 | 120 |
| 35.0 | .58 | 26 | 51 | 77 | 102 | 128 | 154 | 180 | 205 | 231 | 256 | 282 | 308 | 333 | 360 | 385 | 410 | 462 | 513 | 1.7 | 102 |
| 40.0 | .67 | | | | | 112 | 135 | 157 | 180 | 202 | 225 | 247 | 270 | 292 | 315 | 337 | 360 | 405 | 450 | 1.5 | 90 |
| 45.0 | .75 | | | | | | | | | 180 | 200 | 220 | 240 | 260 | 280 | 300 | 320 | 360 | 400 | 1.3 | 78 |
| 50.0 | .83 | | | | | | | | | | | | | | | | | | | 1.2 | 72 |

*(Courtesy of Caterpillar Inc.)*

Unshaded area indicates average production.

Using a 50-minute hour for efficiency, we have an probable production rate of:

$$\text{Production (LCY/Hr.)} = 600 \text{ LCY/Hr.} \times 50/60 = 500 \text{ LCY/hour}$$

Expressed in bank cubic yards per hour, the production rate is:

$$\text{Production (BCY/Hr.)} = \frac{500 \text{ LCY/Hr.}}{1.25 \text{ LCY/BCY}} = 480 \text{ BCY/hour} \qquad \text{(25\% swell)}$$

This is a greater production rate than we obtained in Example 1-20. This is because of the fact that when we used Table 1-18, we "pigeon-holed" a cycle time of 24 seconds and a bucket payload of 4 LCY, as compared to 25 seconds and 3.94 LCY used in Example 1-20.

## REQUIRED PRODUCTION

It is often necessary to determine the production rate required in order to complete a project on schedule. Required production can be determined as follows:

$$\text{Required Production} = \frac{\text{Total Quantity To Be Excavated}}{\text{Total Allowable Project Time}} \qquad \textbf{(Equation 1.13)}$$

**Example 1-22:** Determine the production rate required to excavate 50,000 bank cubic yards in 150 hours.
**Solution:** The required production rate is:

$$\text{Required Production} = \frac{50{,}000 \text{ BCY}}{150 \text{ Hr.}} \qquad \text{(Eq. 1.13)}$$
$$= 334 \text{ BCY/hour}$$

**BUCKET REQUIRED**

Once we have determined the required production rate, we can use the following equation to determine the bucket size needed to achieve the required production rate:

$$\text{Bucket Size} = \frac{\text{Volume Required/Cycle}}{\text{Bucket Fill Factor}} \qquad \textbf{(Equation 1.14)}$$

where:

$$\text{Volume Required/Cycle} = \frac{\text{Volume Required/Hour}}{\text{Cycles/Hour}} \qquad \textbf{(Equation 1.15)}$$

and, Bucket fill factor is taken from Table 1-16.

**Example 1-23:** Assuming a cycle time of 24 seconds while digging hard clay, determine the minimum bucket size required to produce 334 BCY/hour while working a 50-minute hour.
**Solution:** Including job efficiency, we will work at the rate of:

$$\text{Cycles/Hour} = \frac{3600 \text{ Sec./Hr.}}{24 \text{ Sec./Cycle}} \times 50/60 \qquad \text{(Eq. 1.5)}$$
$$= 125 \text{ cycles/hour}$$

Therefore, we must produce:

$$\text{Volume Required/Cycle} = \frac{334 \text{ BCY/Hr.}}{125 \text{ Cycles/Hr.}} \qquad \text{(Eq. 1.15)}$$
$$= 2.67 \text{ BCY/cycle}$$

The bucket size required is:

$$\text{Bucket Size} = \frac{2.67\ \text{BCY/Cycle}}{0.90} \qquad \text{(Eq. 1.14; Table 1-16)}$$
$$= 2.97$$
$$= 3\ \text{cubic yards}$$

## PROJECT DURATION

The total project duration required for any given item of work can be determined as follows:

$$\text{Project Duration (Days)} = \frac{\text{Total Quantity of Work}}{\text{Daily Production}} \qquad \textbf{(Equation 1.16)}$$

or,

$$\text{Project Duration (Hours)} = \frac{\text{Total Quantity of Work}}{\text{Hourly Production}} \qquad \textbf{(Equation 1.17)}$$

**Example 1-24:** Determine the time required to excavate 60,000 bank cubic yards if the production rate is 300 bank cubic yards per hour.
**Solution:** The project duration is:

$$\text{Project Duration (Hr.)} = \frac{60{,}000\ \text{BCY}}{300\ \text{BCY/Hr.}} \qquad \text{(Eq. 1.17)}$$
$$= 200\ \text{hours}$$

## UNIT PRODUCTION COST

The cost per unit of work produced can be determined as follows:

$$\text{Unit Production Cost} = \frac{\text{Hourly Production Cost}}{\text{Hourly Production Rate}} \qquad \textbf{(Equation 1.18)}$$

**Example 1-25:** Determine the unit production cost of an excavator that costs $100.00 per hour and excavates 200 bank cubic yards per hour.
**Solution:** The unit production cost is:

$$\text{Unit Production Cost} = \frac{\$100.00/\text{Hr.}}{200\ \text{BCY/Hr.}} \qquad \text{(Eq. 1.18)}$$

$$= \$0.50/\text{BCY}$$

## EXCAVATOR TRENCHING PRODUCTION

When an excavator is used for trenching operations, it is often convenient to express production rate in terms of lineal feet per hour, or lineal feet per day. Expressed in lineal feet per hour, the production rate can be determined by:

$$\text{Production (L.F./Hr.)} = \frac{\text{C.Y./Hr.}}{\text{Trench Volume (C.Y./L.F.)}} \times \text{Efficiency Factor}$$

**(Equation 1.19)**

This production rate can be converted to lineal feet per day as follows:

$$\text{Production (L.F./Day)} = \text{Production (L.F./Hr.)} \times \text{Hours Worked/Day}$$

**(Equation 1.20)**

**Example 1-26:** Assuming a production rate as given in Example 1-20, determine the production rate in lineal feet per hour if the trench has an average cross-section end area as shown in Figure 1-36. Assume working a 55-minute hour.

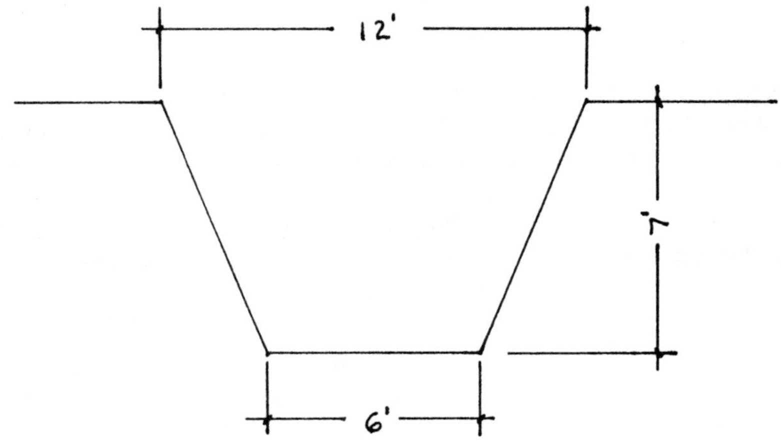


**Figure 1-36. Trench cross section for Example 1-26.**

**Solution:** Excluding job efficiency, the production rate is:

$$\text{Production (BCY/Hr.)} = 144 \text{ Cycles/Hr.} \times 3.15 \text{ BCY/Cycle} \qquad \text{(Example 1-20)}$$
$$= 454 \text{ BCY/hour}$$

The quantity of soil excavated per lineal foot of trench is:

$$\text{Volume/L.F.} = \frac{12' + 6'}{2} \times 7' \qquad \text{(See Equation 10.7 in Chapter 10)}$$
$$= 63 \text{ Cu. Ft./L.F.} \div 27 \text{ Cu. Ft./BCY}$$
$$= 2.33 \text{ BCY/L.F.}$$

Therefore, the production rate is:

$$\text{Production (L.F./Hr.)} = \frac{454 \text{ BCY/Hr.}}{2.33 \text{ BCY/L.F.}} \times 55/60 \qquad \text{(Eq. 1.19)}$$
$$= 179 \text{ L.F./hour}$$

In an 8-hour day, we will excavate:

$$\text{Production (L.F./Day)} = 179 \text{ L.F./Hr.} \times 8 \text{ Hr./Day} \qquad \text{(Eq. 1.20)}$$
$$= 1432 \text{ L.F./day}$$

When loading trenched soil into trucks, there are two truck approaches recommended. If the bucket is small, the truck can line up parallel to the trench so that it can move forward with the excavator while being filled. If the excavation and/or bucket is large enough that the truck can be filled without moving, the truck can be aligned with the swing of the bucket for a more evenly distributed load in the bed of the truck. Never swing the bucket over the cab of a truck.

**TRENCHING WHILE PIPESETTING**

An excavator is often used for laying pipe as well as for excavating (Photo 1-17). In this case, production is dependent upon pipelaying time as well as excavation time. Keep in mind that water pipe can be laid faster than sewer or drain pipe, since the bottom of the trench does not need to be fine-graded and sloped to drain.

**Photo 1-17. Excavator used for laying pipe.**
*(Photo by Dan Atcheson)*

**Example 1-27:** Assuming an excavator that can excavate 400 BCY per hour while digging a trench with an average volume of 4 BCY per lineal foot, determine the daily lineal-foot production rate if the excavator can install a 20-foot length of pipe in 10 minutes. Assume working a 50-minute hour for 10 hours per day.
**Solution:** The lineal-foot excavating production rate is:

$$\text{Excavating Production (L.F./Hr.)} = \frac{\text{400 BCY/Hr.}}{\text{4 BCY/L.F.}} \qquad \text{(Eq. 1.19)}$$

$$= \text{100 lineal feet/hour}$$

The time required to excavate for each section of pipe is:

$$\text{Excavating Time (Min./Pipe)} = \frac{\text{20 L.F./Pipe}}{\text{100 L.F./Hr.}} \times \text{60 Min./Hr.}$$

$$= \text{12 minutes}$$

The total time required per pipe is:

$$\text{Total Time (Min./Pipe)} = \text{12 Min.} + \text{10 Min.} \qquad \text{(Given)}$$

$$= \text{22 minutes}$$

Including job efficiency, the total number of pipe installed per day is:

$$\text{Production (Pipe/Day)} = \frac{60 \text{ Min./Hr.} \times 10 \text{ Hr./Day}}{22 \text{ Min./Pipe}} \times 50/60$$

$$= 23 \text{ pipe/day}$$

The production rate in terms of lineal feet per day is:

$$\text{Production (L.F./Day)} = 23 \text{ Pipe/Day} \times 20 \text{ L.F./Pipe}$$

$$= 460 \text{ lineal feet/day}$$

**SIZE OF TRENCH**

The end area of the trench depends on the size of pipe installed and the safe bank slope required. The bottom width required (A in Figure 1-37) can be determined from Table 1-19.

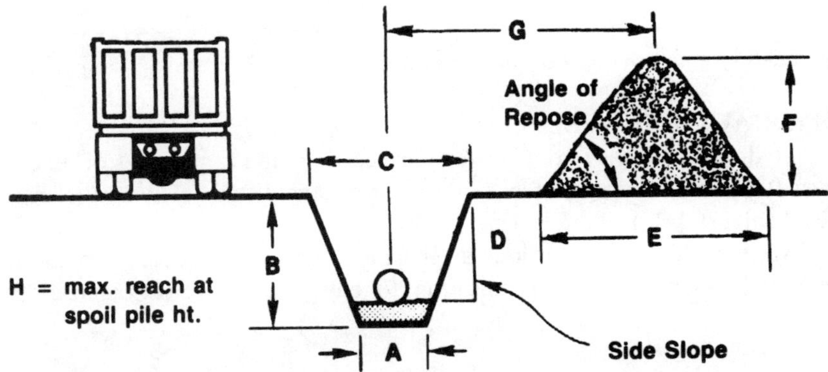


**Figure 1-37. Typical trench cross section.** (*Courtesy of Caterpillar Inc.*)

The methods for determining safe bank slopes, volumes of excavation and spoil piles are presented in my companion book, *Estimating Earthwork Quantities*.

**FELLING TREES**

Hydraulic excavators can also be used to fell trees and to remove tree stumps. Saplings (up to 3 inches in diameter) can often be knocked down by pushing the tree with the bucket. To fell larger trees, use the bucket to cut the roots on the near side of the tree, then push the tree over. The roots on the opposite side of the tree can be cut by using the bucket to pull the felled tree toward the excavator. The felled tree can usually be picked up and moved by curling the bucket around the center of gravity of the tree.

**Table 1-19: Trench Bottom Widths**

| Pipe Diameter | | Trench Width | | Pipe Diameter | | Trench Width | |
|---|---|---|---|---|---|---|---|
| mm | in | m | ft | mm | in | m | ft |
| 102 | **4** | .49 | **1.6** | 1524 | **60** | 2.59 | **8.5** |
| 152 | **6** | .55 | **1.8** | 1676 | **66** | 2.80 | **9.2** |
| 203 | **8** | .61 | **2.0** | 1829 | **72** | 3.05 | **10.0** |
| 254 | **10** | .70 | **2.3** | 1981 | **78** | 3.26 | **10.7** |
| 305 | **12** | .76 | **2.5** | 2134 | **84** | 3.47 | **11.4** |
| 381 | **15** | .91 | **3.0** | 2286 | **90** | 3.69 | **12.1** |
| 457 | **18** | 1.03 | **3.4** | 2438 | **96** | 3.93 | **12.9** |
| 533 | **21** | 1.16 | **3.8** | 2591 | **102** | 4.15 | **13.6** |
| 610 | **24** | 1.25 | **4.1** | 2743 | **108** | 4.36 | **14.3** |
| 686 | **27** | 1.37 | **4.5** | 2896 | **114** | 4.54 | **14.9** |
| 838 | **33** | 1.58 | **5.2** | 3048 | **120** | 4.75 | **15.6** |
| 914 | **36** | 1.70 | **5.6** | 3200 | **126** | 4.99 | **16.4** |
| 1067 | **42** | 1.92 | **6.3** | 3353 | **132** | 5.21 | **17.1** |
| 1219 | **48** | 2.13 | **7.0** | 3505 | **138** | 5.43 | **17.8** |
| 1372 | **54** | 2.38 | **7.8** | 3658 | **194** | 5.64 | **18.5** |

**Note:** Trench widths based on 1.25 Bc + 1.0 where Bc is the outside diameter of the pipe in feet

(*Courtesy of American Concrete Pipe Association*)

## HYDRAULIC SHOVELS

Hydraulic shovels (front shovels) are used for the same types of service as power shovels; however, a front shovel can excavate well below natural ground level (Figure 1-38) (Photo 1-18).

Machine shipping dimensions, digging ranges and clearances are normally published with manufacturer's specifications for any given excavator (Figure 1-39).

**Photo 1-18. Hydraulic shovel with bottom-dump bucket.**

(*Courtesy of Caterpillar Inc.*)

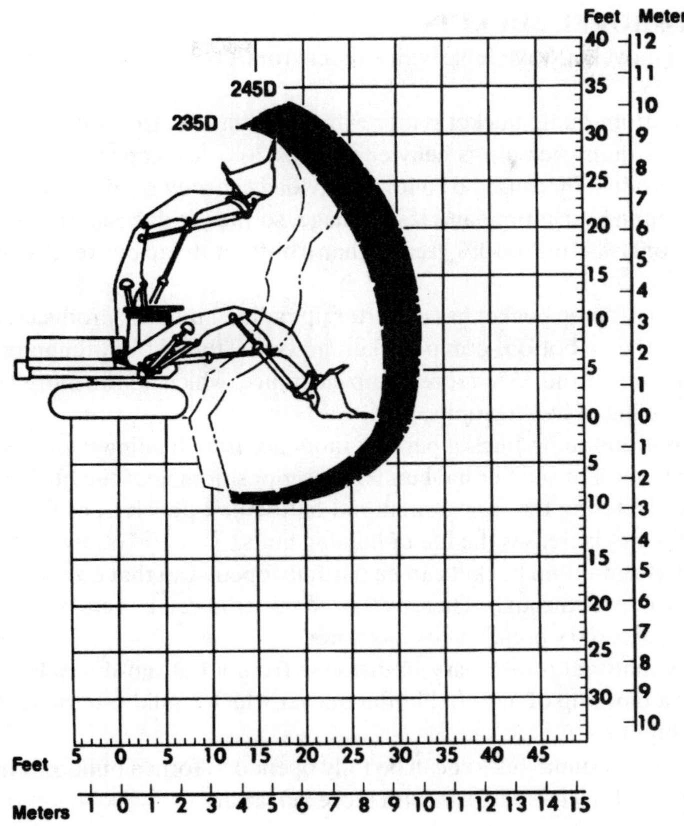


**Figure 1-38. Digging ranges of front shovels.** (*Courtesy of Caterpillar Inc.*)

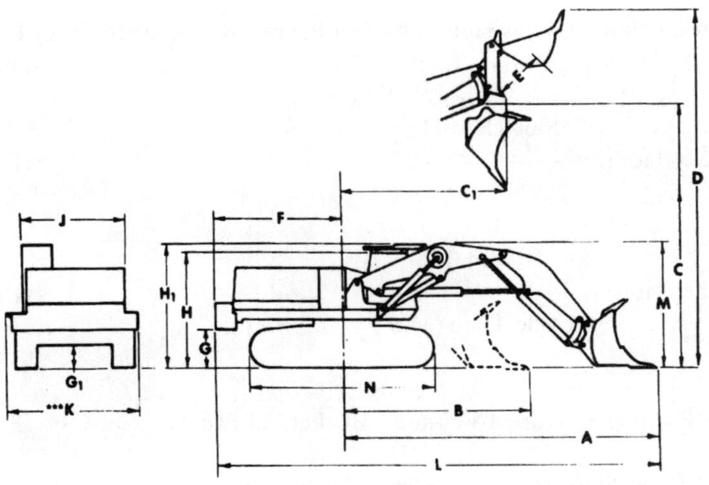


**Figure 1-39. Operating dimensions for front shovels.** (*Courtesy of Caterpillar Inc.*)

## FRONT-SHOVEL BUCKETS

Front shovels are available with either a front-dump, or bottom-dump bucket (Photo 1-18).

The bottom-dump bucket is more versatile than the front-dump bucket, but the bottom-dump version is heavier and has 20% less capacity. However, the reduced production caused by a lower payload capacity is offset somewhat by a faster dumping cycle time and less spillage, so the production of a front-dump bucket is only about 5 to 10% greater than a bottom-dump bucket of comparable size.

The front-dump bucket has a shorter tip radius, and thus, produces more digging force than a bottom-dump bucket; however, the bottom-dump bucket has 17% more reach and 35% more dump clearance, which allows more latitude in the positioning of hauling units.

The bottom-dump bucket permits more accurate loading with less spillage, allowing the use of smaller haul units. Closer positioning of the bucket over the truck and metering the flow of material on the first pass lessens the impact on truck beds and increases the life of hauling units.

The bottom-dump bucket can be partially opened so that only smaller-sized material can pass through. Thus, oversized material can be screened and segregated for secondary breakage by a crusher.

Sticky material is also easy to dislodge from a bottom-dump bucket. This prevents a build-up of material in the bucket which would otherwise lower the production rate.

The bottom-dump bucket can be fully opened to form a bulldozer assembly, allowing the shovel to perform with more versatility.

## FRONT-SHOVEL PRODUCTION

Hydraulic shovel production can be determined by:

$$\text{Production} = \text{Cycles/Hour} \times \text{Bucket Payload/Cycle} \times \text{Efficiency Factor} \qquad \text{(Eq. 1.7)}$$

where:

$$\text{Cycles/Hour} = \frac{\text{3600 Sec./Hr.}}{\text{Cycle Time (Sec.)}} \qquad \text{(Eq. 1.5)}$$

or,

$$\text{Cycles/Hour} = \frac{\text{60 Min./Hr.}}{\text{Cycle Time (Min.)}} \qquad \text{(Eq. 1.6)}$$

and,

$$\text{Bucket Payload} = \text{Heaped Volume} \times \text{Bucket Fill Factor} \qquad \text{(Eq. 1.1)}$$

where: Bucket fill factor is taken from Table 1-20.

## Table 1-20: Front- and Bottom-Dump Shovel Fill Factors

**Excavators — Front Shovels** | Estimating Cycle Time Charts
Bucket Fill Factors

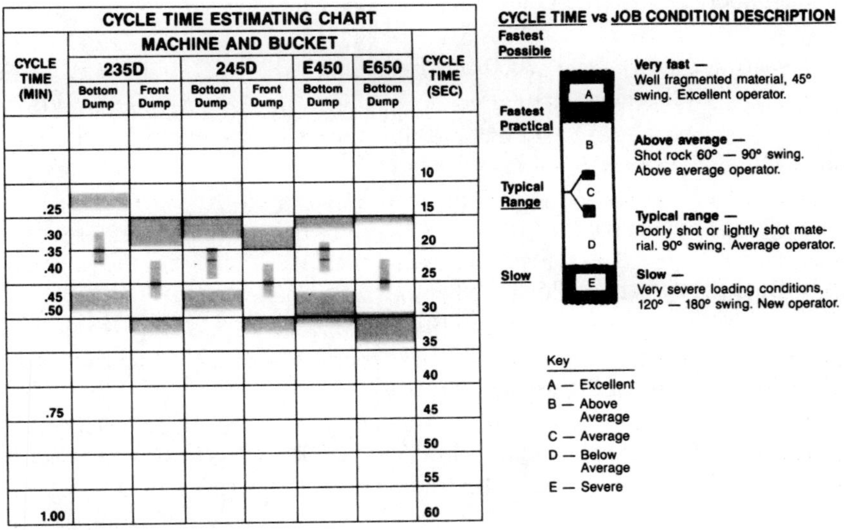

CYCLE TIME ESTIMATING CHART

| CYCLE TIME (MIN) | MACHINE AND BUCKET | | | | | | CYCLE TIME (SEC) |
|---|---|---|---|---|---|---|---|
| | 235D | | 245D | | E450 | E650 | |
| | Bottom Dump | Front Dump | Bottom Dump | Front Dump | Bottom Dump | Bottom Dump | |
| | | | | | | | |
| | | | | | | | 10 |
| .25 | | | | | | | 15 |
| .30 | | | | | | | 20 |
| .35 | | | | | | | |
| .40 | | | | | | | 25 |
| .45 | | | | | | | |
| .50 | | | | | | | 30 |
| | | | | | | | 35 |
| | | | | | | | 40 |
| .75 | | | | | | | 45 |
| | | | | | | | 50 |
| | | | | | | | 55 |
| 1.00 | | | | | | | 60 |

**CYCLE TIME vs JOB CONDITION DESCRIPTION**

Fastest Possible — A — **Very fast —** Well fragmented material, 45° swing. Excellent operator.

Fastest Practical — B — **Above average —** Shot rock 60° — 90° swing. Above average operator.

Typical Range — C, D — **Typical range —** Poorly shot or lightly shot material. 90° swing. Average operator.

Slow — E — **Slow —** Very severe loading conditions, 120° — 180° swing. New operator.

Key
A — Excellent
B — Above Average
C — Average
D — Below Average
E — Severe

**FRONT AND BOTTOM DUMP BUCKET FILL FACTORS**

| Material | Fill Factor* |
|---|---|
| Bank Clay; Earth | 100%-110% |
| Rock-Earth Mixture | 105%-115% |
| Rock — Poorly Blasted | 85%-100% |
| Rock — Well Blasted | 100%-110% |
| Shale, Sandstone — Standing Bank | 85%-100% |

*Percent of heaped bucket capacity.

**Figure 1-40. Front-shovel cycle time estimating chart.**
*(Courtesy of Caterpillar Inc.)*

**Example 1-28:** Estimate the hourly production rate of a 2.38-BCY-capacity Caterpillar 235D bottom-dump shovel while excavating earth. Assume below-average job conditions and working a 50-minute hour.
**Solution:** The cycle time will be approximately:

$$\text{Cycle Time} = 27 \text{ Seconds} \qquad \text{(Figure 1-40)}$$

The number of cycles per hour is:

$$\text{Cycles/Hour} = \frac{3600 \text{ Sec./Hr.}}{27 \text{ Sec./Cycle}} \qquad \text{(Eq. 1.5)}$$
$$= 133 \text{ cycles/hour}$$

The average bucket payload is:

$$\text{Bucket Payload (BCY)} = 2.38 \text{ BCY} \times 1.05 \qquad \text{(Eq. 1.1; Table 1-20)}$$
$$= 2.50 \text{ BCY}$$

Therefore, the probable production rate is:

$$\text{Production (BCY/Hr.)} = 133 \text{ Cycles/Hr.} \times 2.50 \text{ BCY/Cycle} \times 50/60 \qquad \text{(Eq. 1.7)}$$
$$= 277 \text{ BCY/hour}$$

Some manufacturers express shovel production rates in terms of loose cubic yard per hour (Table 1-21); however, since a shovel normally excavates in-place (bank) material, we must use swell or shrinkage factors to interface bank-yard production with loose-yard production.

## Table 1-21: Caterpillar 235D Front Shovel Production

U.S. LCY PER 60 MIN. HOUR*

| ESTIMATED CYCLE TIME | | ESTIMATED BUCKET PAYLOAD** — LOOSE CUBIC YARD | | | | | | | | | ESTIMATED CYCLES | |
|---|---|---|---|---|---|---|---|---|---|---|---|---|
| Cycle Time (Sec) | Cycle Time (Min) | 2 $yd^3$ | 2.5 $yd^3$ | 3 $yd^3$ | 3.5 $yd^3$ | 4 $yd^3$ | 4.5 $yd^3$ | 5 $yd^3$ | 5.5 $yd^3$ | 6 $yd^3$ | Cycles / Minute | Cycles / Hour |
| 15 | .25 | 480 | 600 | 720 | 840 | 965 | 1080 | 1200 | 1320 | 1440 | 4.0 | 240 |
| 18 | .30 | 400 | 500 | 600 | 700 | 800 | 900 | 1000 | 1100 | 1200 | 3.0 | 200 |
| 21 | .35 | 342 | 428 | 513 | 599 | 684 | 770 | 855 | 941 | 1026 | 2.9 | 171 |
| 24 | .40 | 300 | 315 | 450 | 525 | 600 | 675 | 750 | 825 | 900 | 2.5 | 150 |
| 27 | .45 | 266 | 333 | 399 | 466 | 532 | 599 | 665 | 732 | 798 | 2.2 | 133 |
| 30 | .50 | 240 | 300 | 360 | 420 | 480 | 540 | 600 | 660 | 720 | 2.0 | 120 |
| 33 | .55 | 218 | 273 | 327 | 382 | 436 | 491 | 545 | 600 | 654 | 1.8 | 109 |
| 36 | .60 | 200 | 250 | 300 | 350 | 400 | 450 | 500 | 550 | 600 | 1.7 | 100 |

* Actual Hourly Production = (60 Min. Hr. Production) x (Job Efficiency Factor).
** Estimated Bucket Payload = (Heaped Bucket Capacity) x (Bucket Fill Factor).
NOTE: The unshaded areas indicate typical production ranges.

*(Courtesy of Caterpillar Inc.)*

**Example 1-29:** Assuming 25% swell, use data from Table 1-21 to work the previous example.
**Solution:** The cycle time is:

Cycle Time = 27 seconds (Example 1-28)

Expressed in loose cubic yards, the bucket payload is:

Bucket Payload (LCY) = 2.50 BCY x 1.25 LCY/BCY (25% Swell)
= 3.13 LCY

Referring to Table 1-21 while using a 27-second cycle time and a 3-LCY payload, we find that the production rate is:

Production (LCY/Hr.) = 399 LCY/hour (Table 1-21)

Using a 50-minute hour for efficiency, we have a probable production rate of:

Production (LCY/Hr.) = 399 LCY/Hr. x 50/60
= 333 LCY/hour

Expressed in bank cubic yards per hour, the production rate is:

$$\text{Production (BCY/Hr.)} = \frac{\text{333 LCY/Hr.}}{\text{1.25 LCY/BCY}} \quad \text{(25\% swell)}$$
$$= \text{266 BCY/hour}$$

## GRADALL

A Gradall is an excavator mounted on a turntable at the rear end of a truck (Photo 1-19). It is normally used for cutting, cleaning and grading roadside ditches. A Gradall is equipped with a straight, telescoping boom which allows it to safely work under highlines and other overhead obstructions. The boom reach allows it to work over areas too soft or too steep to support heavy equipment. A Gradall can also be used for excavating basements, septic tanks and vertical-faced ditches. The Gradall can be equipped with a blade for pushing soil, a ripper for loosening pavement and hardpan, or it can be used as a crane.

**Photo 1-19. Gradall.** *(Photo by Dan Atcheson)*

## TRENCHING MACHINES

Two basic types of trenching machines are available, including wheel-type and ladder-type (bucket-line) trenchers (Photos 1-20 and 1-21). Trenching machines are used primarily for digging utility trenches in soils stable enough that the banks require no sloping. However, they are not capable of excavating rock. Also, these machines do not perform well where there is ground water combined with unstable soil, sand or mud. Such conditions normally dictate that the trench wall be sloped, requiring the use of some other type of excavator such as a backhoe. Both types of machines can excavate trenches faster than any type of hydraulic excavator.

The actual production of a trenching machine depends on the soil type, depth and width of the trench, the extent and type of shoring required, the extent of vegetation encountered, such as stumps and roots, and the number and types of man-made obstructions, such as buried utilities and sidewalks. Shoring is required in any trench deeper than 5 feet.

Ladder-type trenchers have a smaller digging radius than wheel-type trenchers; therefore the ladder-type trencher can excavate closer to underground obstructions.

In short trenches where many utilities must be jumped, a backhoe will probably perform better than a trenching machine. Also, keep in mind that a backhoe is more versatile than a trencher since it can excavate, install bedding material, set pipe and backfill the trench.

Both types of trenchers require a level travel surface. The terrain can vary along the length of the trench, but it must be level at right angles to the trench to keep the trencher from leaning, which will result in slanted trench walls that tend to cave in.

## WHEEL TRENCHERS

*Wheel trenchers* are manufactured with a maximum digging depth of 8 feet and a maximum digging width of approximately 60 inches (Photo 1-20). Various trenching widths for a given machine can be obtained by using different bucket sizes, or by installing side cutters. The optimum digging depth is 6 feet, and the greatest width readily available is 24 inches. Some wheel trenchers are available with more than 25 digging speeds.

**Photo 1-20. Wheel trencher.** *(Photo by Dan Atcheson)*

Buckets mounted along the circumference of a power-driven wheel excavate as the wheel is lowered and rotated. The soil is lifted by the buckets and discharged onto a conveyor belt which deposits the soil on either side of the trench, leaving very little soil along the top edges of the trench. A long hook installed behind the bucket line is referred to as a *crumbing shoe.* Its purpose is to reduce the amount of loose dirt left behind at the bottom of the trench behind the wheel.

## LADDER TRENCHERS

*Ladder* (bucket-line) *trenchers* are manufactured with a maximum digging depth of more than 30 feet and a maximum digging width of more than 12 feet (Photo 1-21). Various trenching widths for a given machine can be obtained by installing side cutters on the buckets, and production in hard soil can be increased by installing ripper teeth on the backs of the buckets. The digging depth of a given machine can be increased by adding extensions to the boom and installing additional buckets and chain links. Some ladder trenchers are available with more than 30 digging speeds.

**Photo 1-21. Ladder trencher.** *(Photo by Dan Atcheson)*

Buckets are equipped with cutter teeth mounted on two endless chains that travel along the boom and excavate as the boom is lowered into the ground. The soil is lifted by the buckets and discharged onto a conveyor belt which deposits the soil on either side of the trench. A ladder trencher leaves a large amount of loose soil along the top edge of the trench. To minimize this problem, skirts can be installed on each side of the bucket line which drag along the top of the trench to push the loose soil back into the trench.

## BACKHOE LOADERS

The backhoe loader is one of the most versatile excavators in the earthmoving industry (Photo 1-22). It can excavate, carry materials, load or stockpile loose material, grade, backfill, compact, remove trees and perform demolition. It is extremely useful for excavating footing and utility trenches, and for working in relatively tight spaces. However, its performance is limited by its weight and power.

Machine shipping dimensions, digging ranges and clearances are normally published with manufacturer's specifications for any give backhoe loader (Figures 1-41 and 1-42).

**Photo 1-22. Wheel-mounted backhoe loader.** *(Photo by Dan Atcheson)*

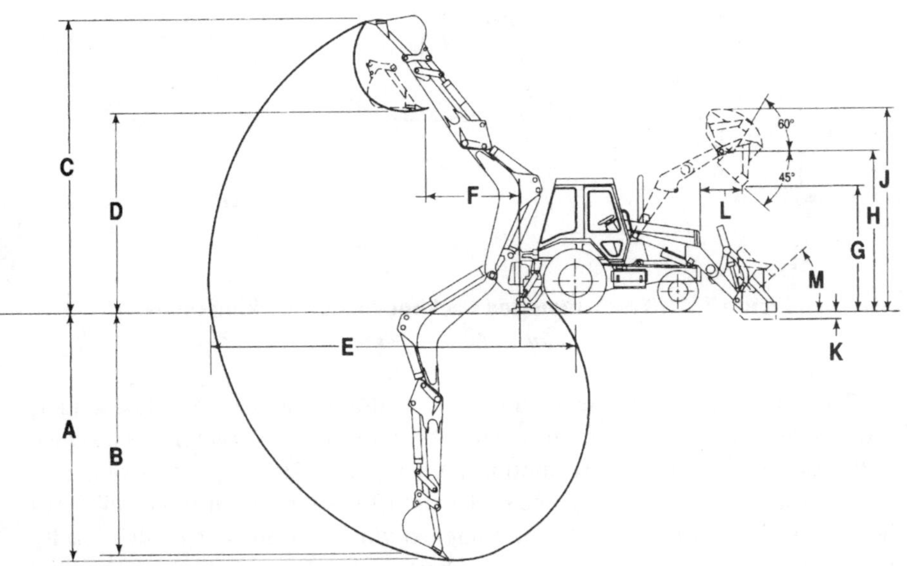


**Figure 1-41. Wheel-mounted backhoe loader working ranges.**
*(Courtesy of Caterpillar Inc.)*

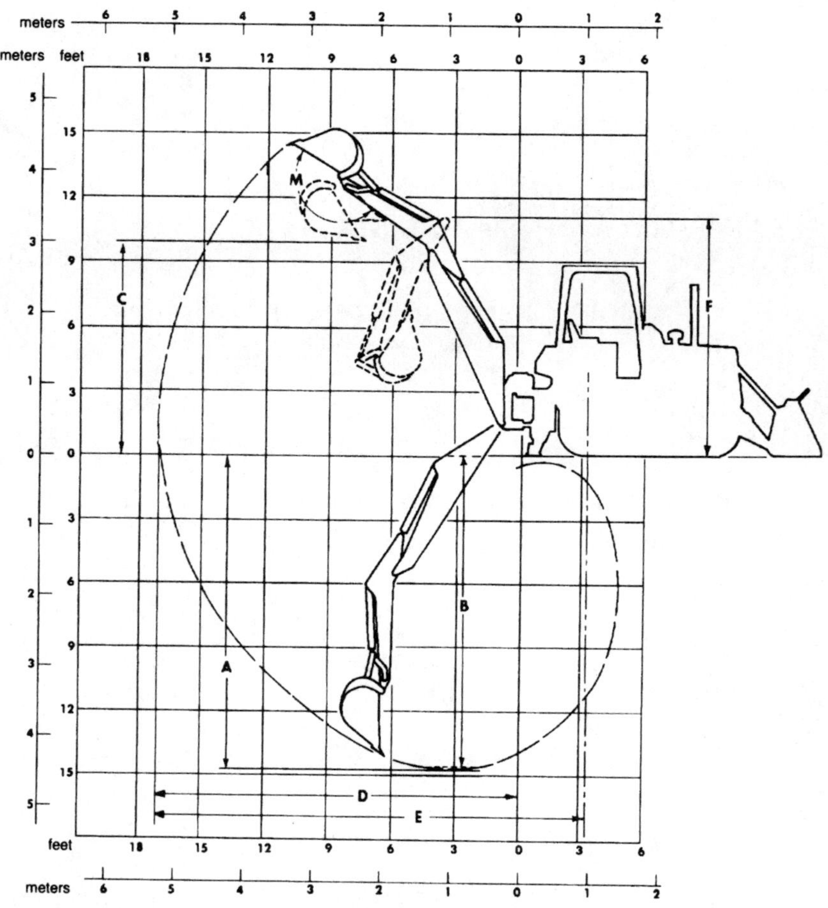


**Figure 1-42. Track-mounted backhoe loader working ranges.**
(*Courtesy of Caterpillar Inc.*)

The lifting capacity of any hydraulic excavator, including a backhoe loader, is dependent upon the weight and center of gravity of the excavator, the distance of the load from the machine centerline (A in Figure 1-43), the position of the lift point (B in Figure 4-43), boom and stick length, bucket position and the hydraulic capacity of the unit. Due to the complexity in determining the lift capacity relative to the machine's lifting position, charts such as Figure 1-44 should be used to determine the *safe* lifting capacity of the excavator.

For any given lift point position, the safe lifting capacity is limited by either the machine's tipping load (discussed previously in this chapter), or hydraulic capacity. The rated loads given in Figure 1-44 do not exceed 75% of the tipping load, or 87% of the machine's hydraulic capacity.

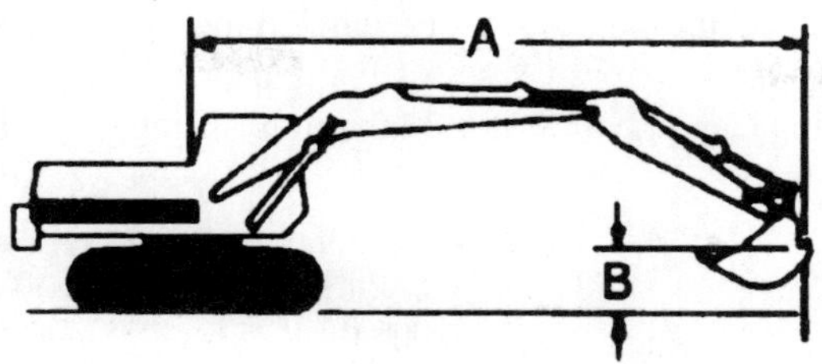


**Figure 1-43. Lift point and lift radius of an excavator.**
(*Courtesy of Caterpillar Inc.*)

KEY
A — Boom lift kg lb
B — Stick lift kg lb

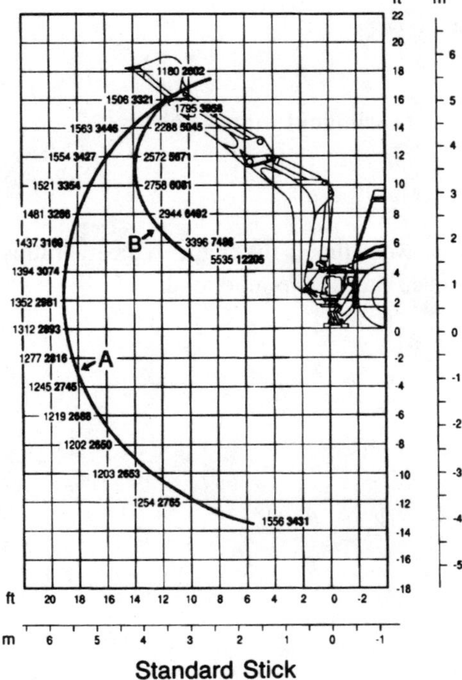


**Figure 1-44. Safe lifting capacities of a Caterpillar 446 backhoe loader equipped with a standard stick.** (*Courtesy of Caterpillar Inc.*)

**Example 1-30:** Assuming the safe lifting capacity of a backhoe loader in a given lift position is 2000 pounds, determine the maximum lifting capacity of the machine.

**Solution:** The safe lifting capacity with regard to the tipping load is:

$$\text{Safe Lifting Capacity} = 0.75 \text{ x Maximum Lifting Capacity}$$

or,

$$\text{Maximum Lifting Capacity} = \frac{2000 \text{ lb.}}{0.75}$$
$$= 2667 \text{ pounds}$$

However, the safe lifting capacity with regard to the hydraulic capacity is:

$$\text{Safe Lifting Capacity} = 0.87 \text{ x Maximum Lifting Capacity}$$

or,

$$\text{Maximum Lifting Capacity} = \frac{2000 \text{ lb.}}{0.87}$$
$$= 2299 \text{ pounds}$$

The "weak link in the chain" is the hydraulic capacity; therefore, we should never exceed the 2299-pound load.

*Chapter 2*

# FACTORS THAT AFFECT THE PERFORMANCE OF MOVING VEHICLES

In Chapter 1, we discussed the performance of equipment with restricted travel within a confined area of a job site. However, when equipment travels even relatively short distances, the performance of the vehicle is affected by rolling resistance, grade resistance and traction.

Before we read further, I want you to know that this is the type of chapter that I would have preferred to place at the end of the book, giving you the option of reading it, or passing it over. However, I am obligated to place it here, because you will never be able to determine the performance of moving vehicles unless you are able to negotiate at least the high points of this chapter. Keep in mind as you study this chapter that you *must* learn how to determine what is referred to as *total resistance* (*effective grade*) before you can use performance curves (Figures 2-6 through 2-10, 4-4, 4-5, 6-7 and 6-8), or travel time charts (Figures 2-11, 3-12, 3-13, 4-2, 4-3, 6-5 and 6-6). So let's "bite the bullet," read and digest this chapter before moving on to the next chapters.

## ROLLING RESISTANCE

*Rolling resistance* (RR) is resistance to movement caused by the road surface over which the vehicle travels. It is the amount of force that must be overcome to roll or pull a wheel over the travel surface. Rolling resistance is independent of the steepness of the grade; however, it is affected by the condition of the travel surface, gross vehicle weight (G.V.W.) and the type of tires or tracks used.

**Figure 2-1. Tire penetration.**
*(Courtesy of Terex Corporation)*

Rolling resistance can vary for wheel-mounted vehicles, depending on the type of tires and the inflation pressure. For instance, on a *hard* road surface, a high-pressure tire with a narrow tread will experience less rolling resistance than will a low-pressure tire with a wide tread. However, the opposite is true on a soft road surface. This is because a high-pressure tire penetrates into the soft road surface and experiences the need to climb out of the road; therefore, the tires are constantly having to travel up grade as well as forward (Figure 2-1). For the same reason, rolling resistance also increases when the road flexes under a load. Internal friction and tire flexing also contribute to rolling resistance (Figure 2-2).

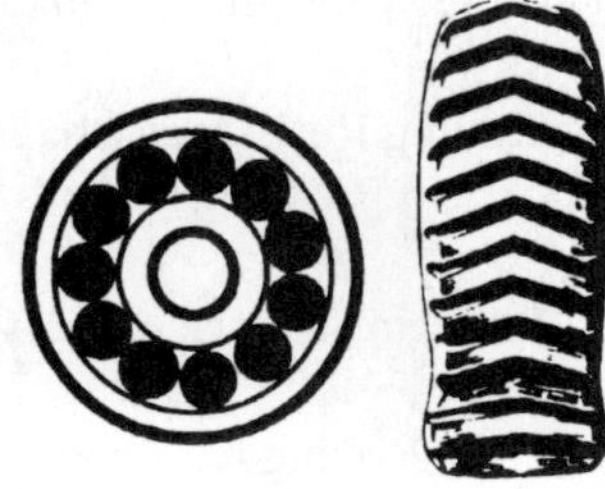

**Figure 2-2. Internal friction and tire flexing.**
*(Courtesy of Terex Corporation)*

Rolling resistance can be reduced by proper equipment selection (track versus wheel-mounted equipment), tire selection (radial tires versus bias or belted tires) and tire inflation pressures. Rolling resistance approaches the minimum value of 40 lb. (bias tires) or 30 lb. (radial tires) (Photo 2-1) only on hard, smooth surfaces with a well-compacted base (Table 2-1).

**Photo 2-1. Off-the-road radial truck tire.**
(*Courtesy of Goodyear Tire and Rubber Company*)

## ROLLING RESISTANCE FOR WHEEL-MOUNTED VEHICLES

Rolling resistance of a wheel-mounted vehicle traveling over a hard, smooth surface is equivalent to about 40 pounds per ton of gross vehicle weight with bias or belted-bias tires, and about 30 pounds per ton with radial tires or dual-tired trucks. The rolling resistance increases by about 30 pounds per ton for each inch of tire penetration into the travel surface. Thus, the rolling resistance for a vehicle equipped with *bias or belted-bias tires* is:

$$\text{Rolling Resistance (lb./Ton)} = 40 \text{ lb./Ton} + (30 \text{ lb./Ton} \times \text{Inches of Penetration}) \qquad \textbf{(Equation 2.1)}$$

Rolling resistance for *dual-tired trucks*, or a vehicle equipped with *radial tires* is:

$$\text{Rolling Resistance (lb./Ton)} = 30 \text{ lb./Ton} + (30 \text{ lb./Ton} \times \text{Inches of Penetration}) \qquad \textbf{(Equation 2.2)}$$

Rolling resistance expressed in pounds of resistance is:

$$\text{Rolling resistance (lb.)} = \text{Rolling Resistance (lb./Ton)} \times \text{G.V.W. (Tons)} \qquad \textbf{(Equation 2.3)}$$

where: G.V.W. = Gross vehicle weight.

**Example 2-1:** Assuming a 20-ton tractor equipped with bias tires is traveling over a soft surface with a tire penetration of 4 inches, determine the rolling resistance in: a) pounds per ton of gross vehicle weight, and b) pounds.
**Solution:** a) The rolling resistance in terms of pounds per ton of gross vehicle weight is:

Rolling Resistance (lb./Ton) = 40 lb./Ton + (30 lb./Ton x 4) (Eq. 2.1)
= 160 pounds/ton

b) The rolling resistance in terms of pounds of resistance is:

Rolling Resistance (lb.) = 160 lb./Ton x 20 Tons (Eq. 2.3)
= 3200 pounds

The rolling resistances for tire-equipped vehicles, expressed in pounds per ton of gross vehicle weight for various travel surfaces are given in Table 2-1.

**Table 2-1: Typical Rolling Resistance Factors***

| Underfooting | RollingResistance (lb. per U.S. Ton) |
|---|---|
| Hard, smooth, stabilized surfaced roadway without penetration under load, watered, maintained | 40 (30)** |
| Firm, smooth, rolling roadway with dirt or light surfacing, flexing slightly under load or undulating, maintained fairly regularly, watered. | 65 (50-55)** |
| Packed snow | 50 |
| Loose snow | 90 |
| Dirt roadway, rutted, flexing under load, little if any maintenance, no water, 1" to 2" tire penetration. | 100 |
| Rutted dirt roadway, soft under travel, no maintenance, no stabilization, 4" to 6" tire penetration. | 150 |
| Loose sand or gravel | 200 |
| Compacted coal | 80 |
| Uncompacted lifts of coal | 120 |
| Uncompacted piles of coal | 200-300 |
| Soft, muddy, rutted roadway, no maintenance. | 200-400 |

*Various tire sizes and inflation pressures will greatly reduce or increase the figures given in this table. The quantities given are sufficiently accurate for estimating purposes when specific information on performance of particular equipment on given soil conditions is not available.
** Radial tires.
(*Courtesy of Caterpillar Inc.*)

**Example 2-2:** Use data from Table 2-1 to determine the rolling resistance in pounds per ton and total pounds for a wheel-mounted vehicle weighing 20 tons traveling over loose sand.

**Solution:** The rolling resistance expressed in terms of pounds per ton of gross vehicle weight is:

Rolling Resistance (lb./Ton) = 200 pounds/ton (Table 2-1)

The rolling resistance in terms of pounds of resistance is:

Rolling Resistance (lb.) = 200 lb./Ton x 20 Tons (Eq. 2.3)
= 4000 pounds

If a vehicle is towing a load, the combined gross weights of the vehicle and the towed load must be used when determining rolling resistance.

**Example 2-3:** A wheel tractor weighing 75,000 pounds is towing a scraper weighing 100,000 pounds. Determine the rolling resistance of this combination if the haul road has a rolling resistance of 100 pounds per ton of gross vehicle weight.

**Solution:** The gross vehicle weight is:

$$\text{G.V.W (Tons)} = \frac{175{,}000 \text{ lb.}}{2000 \text{ lb./Ton}} = 87.5 \text{ tons}$$

The rolling resistance expressed in pounds of resistance is:

Rolling Resistance (lb.) = 100 lb./Ton x 87.5 Tons (Eq. 2.3)
= 8750 pounds

The rolling resistance (in pounds) of a wheel-mounted vehicle traveling over a hard, smooth surface is equivalent to approximately 2% of the gross vehicle weight with bias or belted-bias tires, and about 1.5% with radial tires or dual-tired trucks. The rolling resistance increases by about 1.5% for each inch of tire penetration into the travel surface. Thus, the rolling resistance for a vehicle equipped with *bias or belted-bias tires* is:

Rolling Resistance (lb.) = (0.02 x G.V.W.) + (0.015 x G.V.W. x Inches Penetration) **(Equation 2.4)**

Rolling resistance for a *dual-tired trucks,* or vehicles equipped with *radial tires* is:

Rolling Resistance (lb.) = (0.015 x G.V.W.) + (0.015 x G.V.W. x Inches Penetration) **(Equation 2.5)**

**Example 2-4:** Use Equation 2-4 to determine the rolling resistance of a 20-ton tractor equipped with bias tires travelling over a soft road surface with a tire penetration of 4 inches.
**Solution:** The gross weight of the vehicle is:

$$\begin{aligned} \text{G.V.W.} &= 20 \text{ Tons x } 2000 \text{ lb./Ton} \\ &= 40{,}000 \text{ pounds} \end{aligned}$$

The rolling resistance expressed in pounds of resistance is:

$$\begin{aligned} \text{Rolling Resistance (lb.)} &= (0.02 \text{ x } 40{,}000 \text{ lb.}) + (0.015 \text{ x } 40{,}000 \text{ lb. x } 4) \quad \text{(Eq. 2.4)} \\ &= 3200 \text{ pounds} \end{aligned}$$

Expressed in pounds per ton of gross vehicle weight, the rolling resistance is:

$$\begin{aligned} \text{Rolling Resistance (lb./Ton)} &= \frac{3200 \text{ lb.}}{20 \text{ Tons}} \\ &= 160 \text{ pounds/ton} \end{aligned}$$

## ROLLING RESISTANCE FOR TRACK-MOUNTED VEHICLES

Since the wheels of crawler tractors virtually travel over steel "roads" created by the inside surface of their tracks, the rolling resistance for crawler tractors is often neglected in determining rolling resistance. In reality, the rolling resistance of the traveling surface used when calculating the drawbar pull rating shown on crawler tractor specification sheets and drawbar-pull-versus-ground-speed charts is actually 110 pounds per ton of gross vehicle weight (Figure 2-12).

## GRADE RESISTANCE AND ASSISTANCE

*Grade resistance* is a hindering force that must be overcome in order to move a vehicle up grade. *Grade assistance* is a propelling force that assists a vehicle travelling down grade. In either case, the force is the component of vehicle weight that acts parallel to the inclined travel surface, and affects both wheel- and track-mounted vehicles. Grade resistance or assistance is a force that is independent of rolling resistance, and thus, must be accounted for separately.

Grades are normally measured in terms of slope, or percent grade. *Slope* is defined as the ratio (relationship) of run to rise of a surface, and is expressed as Run:Rise, or:

$$\text{Slope} = \frac{\text{Run}}{\text{Rise}} \qquad \textbf{(Equation 2.6)}$$

**Example 2-5:** Determine the slope of a surface that rises 10 feet vertically through a horizontal distance of 20 feet.
**Solution:** The slope is:

$$\text{Slope} = \frac{20'}{10'} \qquad \text{(Eq. 2.6)}$$

$$= \frac{2}{1}$$

Therefore, the slope is expressed as 2:1, or 2 to 1.

*Percent grade* is defined as:

$$\text{\%-Grade} = \frac{\text{Rise}}{\text{Run}} \qquad \textbf{(Equation 2.7)}$$

**Example 2-6:** Determine the percent grade of the slope given in the previous example.
**Solution:** The percent grade is:

$$\text{\%-Grade} = \frac{10'}{20'} \qquad \text{(Eq. 2.7)}$$

$$= 0.50$$

$$= 50\%$$

You can also refer to Figure 2-3 or Table 2-2 to help you convert slope or percent grade to degrees, and vice versa.

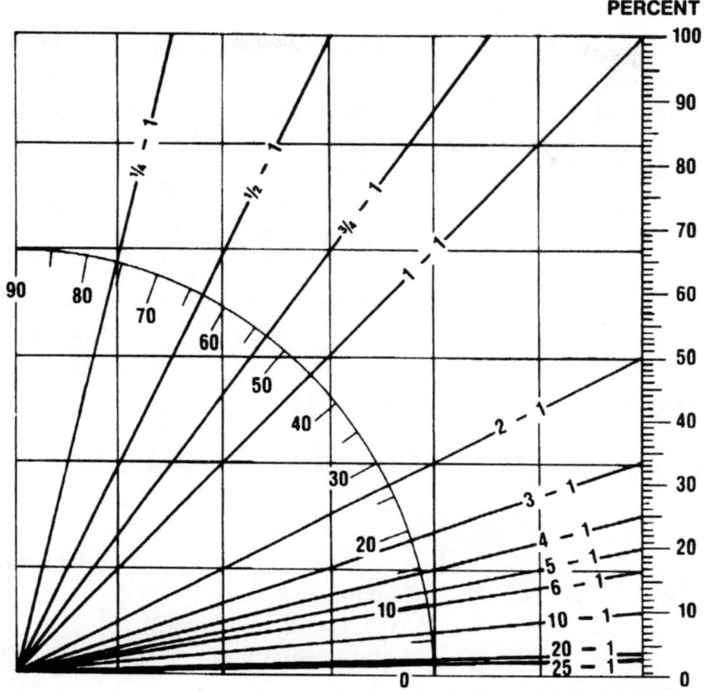


**Figure 2-3. Grade comparison chart (Degrees, %-grade and slope).**
(*Courtesy of Caterpillar Inc.*)

**Table 2-2: Grade in Degrees and Percents**

| Degrees | % Grade | Degrees | % Grade | Degrees | % Grade |
|---|---|---|---|---|---|
| 1 | 1.8 | 16 | 28.7 | 31 | 60.0 |
| 2 | 3.5 | 17 | 30.6 | 32 | 62.5 |
| 3 | 5.2 | 18 | 32.5 | 33 | 64.9 |
| 4 | 7.0 | 19 | 34.4 | 34 | 67.4 |
| 5 | 8.8 | 20 | 36.4 | 35 | 70.0 |
| 6 | 10.5 | 21 | 38.4 | 36 | 72.7 |
| 7 | 12.3 | 22 | 40.4 | 37 | 75.4 |
| 8 | 14.0 | 23 | 42.4 | 38 | 78.1 |
| 9 | 15.8 | 24 | 44.5 | 39 | 81.0 |
| 10 | 17.6 | 25 | 46.6 | 40 | 83.9 |
| 11 | 19.4 | 26 | 48.8 | 41 | 86.9 |
| 12 | 21.3 | 27 | 51.0 | 42 | 90.0 |
| 13 | 23.1 | 28 | 53.2 | 43 | 93.3 |
| 14 | 24.9 | 29 | 55.4 | 44 | 96.6 |
| 15 | 26.8 | 30 | 57.7 | 45 | 100.0 |

Uphill grades ascended by a vehicle are referred to as *adverse grades* and downhill grades descended are called *favorable grades*. For each 1% of adverse slope, 20 pounds of resistance for each gross ton of vehicle weight must be overcome in order to move the vehicle. For each 1% of favorable slope, 20 pounds of assistance for each gross ton of vehicle weight is acquired for aid in moving the vehicle. Thus, grade resistance (assistance) expressed in pounds per ton is defined as:

Grade Resistance (Assistance) (lb./Ton) = 20 lb./Ton x %-Grade

**(Equation 2.8)***

where: %-grade is expressed as a *whole number.*

Grade resistance (assistance) expressed in pounds of resistance is:

Grade Resistance (Assistance) (lb.) = G.V.W.(lb.) x %-Grade

**(Equation 2.9)**

where: %-grade is expressed in *decimal form.*

**Example 2-7:** Determine the grade resistance hindering the movement of a 20-ton tractor ascending a 5% grade.
**Solution:** Expressed in pounds per ton, the grade resistance is:

Grade Resistance (lb./Ton) = 20 lb./Ton x 5% (Eq. 2.8)
= 100 pounds/ton

Expressed in pounds of additional effort required, the grade resistance is:

Grade Resistance (lb.) = 20 Tons x 2000 lb./Ton x 0.05 (Eq 2.9)
= 2000 pounds

**Example 2-8:** Determine the grade assistance contributing to the movement of a 30-ton tractor descending a 4% grade.
**Solution:** Expressed in pounds per ton, the grade assistance is:

Grade Assistance (lb./Ton) = 20 lb./Ton x 4% (Eq 2.8)
= 80 pounds/ton

Expressed in pounds of reduced effort, the grade assistance is:

Grade Assistance (lb.) = 30 Tons x 2000 lb./Ton x 0.04 (Eq. 2.9)
= 2400 pounds

**See Appendix C for proof of Equation 2.8*

**TOTAL RESISTANCE**

*Total resistance* is the combined effect of rolling resistance and grade resistance (Photo 2-2). The total resistance to the movement of a vehicle *ascending* a slope is the *sum* of the rolling resistance and grade resistance; i.e.,

Total Resistance = Rolling Resistance + Grade Resistance **(Equation 2.10)**

The total resistance to the movement of a vehicle *descending* a slope is the *difference* of the rolling resistance and grade assistance; i.e.,

Total Resistance = Rolling Resistance – Grade Assistance **(Equation 2.11)**

Resistances determined from Equations 2.10 or 2.11 can be expressed in terms of pounds per ton of gross vehicle weight, or total pounds of resistance.

**Photo 2-2. Total resistance equals rolling resistance plus grade resistance.**
(*Courtesy of Caterpillar Inc.*)

**Example 2-9:** Determine the total pounds of resistance for a 20-ton wheel-mounted tractor ascending a 5% grade consisting of loose gravel.
**Solution:** The rolling resistance expressed in pounds per ton is:

Rolling Resistance (lb./Ton) = 200 pounds/ton (Table 2-1)

Expressed in pounds, the rolling resistance is:

Rolling Resistance (lb.) = 200 lb./Ton x 20 Tons (Eq. 2.3)
= 4000 pounds

Expressed in pounds, the grade resistance is:

Grade Resistance (lb.) = 20 Tons x 2000 lb./Ton x 0.05 (Eq 2.9)
= 2000 pounds

Expressed in pounds, the total resistance is:

Total Resistance (lb.) = 4000 lb. + 2000 lb. (Eq. 2.10)
= 6000 pounds

**Example 2-10:** Express the total resistance in the previous example in terms of pounds per ton of gross vehicle weight.
**Solution:** The rolling resistance expressed in pounds per ton is:

Rolling Resistance (lb./Ton) = 200 pounds/ton (Example 2-9)

Expressed in pounds per ton, the grade resistance is:

Grade Resistance (lb./Ton) = 20 lb./Ton x 5% (Eq. 2.8)
= 100 pounds/ton

Expressed in pounds per ton, the total resistance is:

Total Resistance (lb./Ton) = 200 lb./Ton + 100 lb./Ton (Eq. 2.10)
= 300 pounds/ton

We could have also determined the total resistance in pounds per ton as follows:

$$\text{Total Resistance (lb./Ton)} = \frac{6000 \text{ lb.}}{20 \text{ Tons}}$$ (Example 2-9)

= 300 pounds/ton

**Example 2-11:** Determine the total pounds of resistance for a 20-ton wheel-mounted tractor descending a 5% grade consisting of loose gravel.
**Solution:** The rolling resistance expressed in pounds per ton is:

Rolling Resistance (lb./Ton) = 200 pounds/ton (Table 2-1)

The rolling resistance expressed in pounds is:

$$\text{Rolling Resistance (lb.)} = 200 \text{ lb./Ton} \times 20 \text{ Tons} = 4000 \text{ pounds} \qquad \text{(Eq. 2.3)}$$

The grade assistance expressed in pounds is:

$$\text{Grade Assistance (lb.)} = 20 \text{ Tons} \times 2000 \text{ lb./Ton} \times 0.05 = 2000 \text{ pounds} \qquad \text{(Eq. 2.9)}$$

Thus, the total resistance is:

$$\text{Total Resistance (lb.)} = 4000 \text{ lb.} - 2000 \text{ lb.} = 2000 \text{ pounds} \qquad \text{(Eq. 2.11)}$$

If a vehicle is towing a load, the combined gross weights of the vehicle and towed load must be used when determining total resistance.

**Example 2-12:** A wheel-mounted tractor weighing 80,000 pounds is towing a scraper weighing 100,000 pounds up a 6% grade. Determine the total pounds of resistance of this combination if the haul road has a rolling resistance of 80 pounds per ton of gross vehicle weight.
**Solution:** The gross vehicle weight is:

$$\text{G.V.W. (Tons)} = \frac{180{,}000 \text{ lb.}}{2000 \text{ lb./Ton}} = 90 \text{ tons}$$

The rolling resistance is:

$$\text{Rolling Resistance (lb.)} = 80 \text{ lb./Ton} \times 90 \text{ Tons} = 7200 \text{ pounds} \qquad \text{(Eq. 2.3)}$$

The grade resistance is:

$$\text{Grade Resistance (lb.)} = 180{,}000 \text{ lb.} \times 0.06 = 10{,}800 \text{ pounds} \qquad \text{(Eq. 2.9)}$$

The total Resistance is:

$$\text{Total Resistance (lb.)} = 7200 \text{ lb.} + 10{,}800 \text{ lb.} = 18{,}000 \text{ pounds} \qquad \text{(Eq. 2.10)}$$

## EFFECTIVE GRADE

The total resistance imposed on a vehicle can be expressed as a percent grade by converting the rolling resistance to percent grade, then adding it to the actual percent grade to obtain total resistance expressed in terms of percent grade (Figures 2-4 and 2-5). Since each 1% of grade is equivalent to a resistance of 20 pounds per ton of gross vehicle weight (see Equation 2.8), rolling resistance can be expressed as percent grade by:

$$\text{Rolling Resistance (\%-Grade)} = \frac{\text{Rolling Resistance (lb./Ton)}}{20}$$

**(Equation 2.12)**

Rolling resistances expressed in terms of %-grade are given in Table 2-3 for various travel-surface conditions.

**Table 2-3: Rolling Resistance (in %-Grade) for Various Travel Surfaces**

| Ground Surface | Rolling Resistance |
|---|---|
| Asphalt | 1.5% |
| Coal, crushed | 5-7% |
| Concrete | 1.5% |
| Dirt-Smooth, hard, dry; well-maintained; free of loose material | 2.0% |
| Dirt-Dry, but not firmly packed; some loose material | 3.0% |
| Dirt-Soft, unplowed; poorly maintained | 4.0% |
| Dirt-Soft, plowed | 8.0% |
| Dirt-Unpacked fills | 8.0% |
| Dirt-Deeply rutted | 16.0% |
| Gravel-Well compacted; dry; free of loose material | 2.0% |
| Gravel-Not firmly compacted; but dry | 3.0% |
| Gravel-Loose | 10.0% |
| Mud-With firm base | 4.0% |
| Mud-With soft, spongy base | 16.0% |
| Sand-Loose | 10.0% |
| Snow-Packed | 2.5% |
| Snow-To 4'' depth; loose | 4.5% |

(*Courtesy of Terex Corporation*)

*Effective grade* (equivalent grade) is the total resistance expressed in terms of percent grade. While *ascending* a grade, the effective grade is:

Effective Grade (%) = Rolling Resistance (%) + Grade Resistance (%)
**(Equation 2.13)**

While *descending* a grade, the effective grade is:

Effective Grade (%) = Rolling Resistance (%) – Grade Assistance (%)
**(Equation 2.14)**

**Example 2-13:** Use data from Table 2-3 to determine the effective grade of a vehicle traveling over loose gravel while: a) ascending a 5% grade, and b) descending a 5% grade.
**Solution:** a) The effective grade while ascending the slope is:

Effective Grade (%) = 10% + 5% (Eq. 2.13; Table 2-3)
= 15%

An illustration of the effective grade is shown in Figure 2-4.

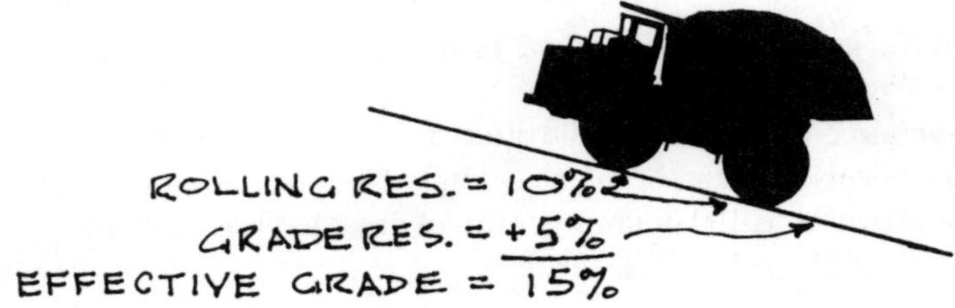


**Figure 2-4. 15% effective grade.**

b) The effective grade while descending the slope is:

Effective Grade (%) = 10% – 5% (Eq. 2.14; Table 2-3)
= 5%

An illustration of the effective grade is shown in Figure 2-5.

**Figure 2-5. 5% effective grade.**

When travel surface conditions differ from those given in Table 2-3, effective grade can be determined as follows:

$$\text{Effective Grade (\%)} = \frac{\text{Rolling Resistance (lb./Ton)}}{20} \pm \text{\%-Grade}$$

**(Equation 2.15)**

where: (+) is used for adverse grades and (–) is used for favorable grades;

or,

$$\text{Effective Grade (\%)} = \frac{\text{Total Resistance (lb./Ton)}}{20}$$ **(Equation 2.16)**

Effective grade (total resistance) is a component of performance curves and travel time charts for wheel-mounted vehicles such as tractor-scrapers (Figures 2-6, 2-8 through 2-11, and 6-5 through 6-8), wheel loaders (Figures 3-12 and 3-13) and trucks (Figures 2-7, and 4-2 through 4-5).

**Example 2-14:** Determine the effective grade for a 30-ton wheel-mounted tractor while ascending, then descending a 4% grade if the rolling resistance of the travel surface is 100 pounds per ton of gross vehicle weight.
**Solution:** The rolling resistance is:

$$\text{Rolling Resistance} = 100 \text{ lb./ton}$$ (Given)

Expressed as %-grade, the rolling resistance is:

$$\text{Rolling Resistance (\%)} = \frac{100 \text{ lb./ton}}{20} = 5\%$$ (Eq. 2.12)

The grade resistance (vehicle ascending) and grade assistance (vehicle descending) is:

$$\text{Grade Resistance (Assistance)} = 4\%$$ (Given)

While ascending, the effective grade is:

$$\text{Effective Grade (Ascending)} = 5\% + 4\% = 9\%$$ (Eq. 2.13)

While descending, the effective grade is:

$$\text{Effective Grade (Descending)} = 5\% - 4\% = 1\% \quad \text{(Eq. 2.14)}$$

If a vehicle is towing a load, the combined gross weights of the vehicle and towed load must be used when determining the effective grade.

**Example 2-15:** A wheel-mounted tractor weighing 70,000 pounds is towing a scraper weighing 100,000 pounds up a 5% grade. Determine the effective grade if the haul load has a rolling resistance of 80 pounds per ton of gross vehicle weight.
**Solution:** Expressed as %-grade, the rolling resistance is:

$$\text{Rolling Resistance (\%)} = \frac{80 \text{ lb./Ton}}{20} = 4\ \% \quad \text{(Eq. 2.12)}$$

Thus, the effective grade is:

$$\text{Effective Grade (\%)} = 4\% + 5\% = 9\ \% \quad \text{(Eq. 2.13)}$$

**Example 2-16:** Use Equation 2.16 to solve the previous example.
**Solution:** The rolling resistance is:

$$\text{Rolling Resistance} = 80 \text{ pounds/ton} \quad \text{(Example 2-15)}$$

The grade resistance is:

$$\text{Grade Resistance} = 20 \text{ lb./Ton} \times 5\% = 100 \text{ pounds/ton} \quad \text{(Eq. 2.8)}$$

Thus, the effective grade is:

$$\text{Effective Grade (\%)} = \frac{80 \text{ lb./Ton} + 100 \text{ lb./Ton}}{20} = 9\ \% \quad \text{(Eq. 2.16)}$$

**Example 2-17:** An off-highway truck weighs 50,000 pounds empty and carries a payload of 100,000 pounds. The haul route has a rolling resistance of 80 pounds per ton of gross vehicle weight and requires the loaded truck to travel down a 3% grade and return empty on the same route. Determine the total resistance and effective grade for each portion of the haul route.

**Solution:** The gross vehicle weight with payload is:

$$\text{G.V.W.} = \frac{150{,}000 \text{ lb.}}{2000 \text{ lb./Ton}} = 75 \text{ tons}$$

The rolling resistance and grade assistance while descending with payload are:

$$\text{Rolling Resistance (lb.)} = 80 \text{ lb./Ton x } 75 \text{ Tons} = 6000 \text{ pounds} \qquad \text{(Eq. 2.3)}$$

$$\text{Grade Assistance (lb.)} = 150{,}000 \text{ lb. x } 0.03 = 4500 \text{ pounds} \qquad \text{(Eq. 2.9)}$$

The total resistances imposed on the descending truck are:

$$\text{Total Resistance (lb.)} = 6000 \text{ lb.} - 4500 \text{ lb.} = 1500 \text{ pounds} \qquad \text{(Eq. 2.11)}$$

$$\text{Total Resistance (lb./Ton)} = \frac{1500 \text{ lb.}}{75 \text{ Tons}} = 20 \text{ pounds/ton}$$

Thus, the effective grade that the truck descends is:

$$\text{Effective Grade (\%)} = \frac{20 \text{ lb./Ton}}{20} = 1\% \qquad \text{(Eq. 2.16)}$$

or,

$$\text{Effective Grade} = \frac{80 \text{ lb./Ton}}{20} - 3\% = 1\% \qquad \text{(Eq. 2.15)}$$

The gross vehicle weight empty is:

$$\text{G.V.W.} = \frac{50{,}000 \text{ lb.}}{2000 \text{ lb./Ton}} = 25 \text{ tons}$$

The rolling and grade resistances while ascending without payload are:

$$\text{Rolling Resistance (lb.)} = 80 \text{ lb./Ton} \times 25 \text{ Tons} = 2000 \text{ pounds} \quad \text{(Eq. 2.3)}$$

$$\text{Grade Resistance (lb.)} = 50{,}000 \text{ lb.} \times 0.03 = 1500 \text{ pounds} \quad \text{(Eq. 2.9)}$$

The total resistances imposed on the ascending truck are:

$$\text{Total Resistance (lb.)} = 2000 \text{ lb.} + 1500 \text{ lb.} = 3500 \text{ lb.} \quad \text{(Eq. 2.10)}$$

$$\text{Total Resistance (lb./Ton)} = \frac{3500 \text{ lb.}}{25 \text{ Tons}} = 140 \text{ pounds/ton}$$

Thus, the effective grade that the truck ascends is:

$$\text{Effective Grade (\%)} = \frac{140 \text{ lb./Ton}}{20} = 7\% \quad \text{(Eq. 2.16)}$$

or,

$$\text{Effective Grade (\%)} = \frac{80 \text{ lb./Ton}}{20} + 3\% = 7\% \quad \text{(Eq. 2.15)}$$

Referring to Equations 2.8 and 2.12, we can see that rolling resistance can be expressed as:

$$\text{Rolling Resistance (lb./Ton)} = \text{Rolling Resistance (\%-Grade)} \times 20$$

**(Equation 2.17)**

where: %-grade is expressed as a *whole number.*

**Example 2-18:** Use data from Table 2-3 to determine the rolling resistance in terms of pounds per ton and total pounds for a wheel-mounted vehicle weighing 15 tons traveling over loose gravel.
**Solution:** The rolling resistance expressed as %-grade is:

Rolling Resistance (%-Grade) = 10% (Table 2-3)

The rolling resistance expressed in pounds per ton of gross vehicle weight is:

Rolling Resistance (lb./Ton) = 10% x 20 (Eq. 2.17)
= 200 pounds/ton

The rolling resistance expressed in pounds of total resistance is:

Rolling Resistance (lb.) = 200 lb./Ton x 15 Tons (Eq. 2.3)
= 3000 pounds

## RETARDER CURVES

As a wheel-mounted vehicle descends a grade, it is possible for the grade assistance to be greater than the rolling resistance, resulting in a negative total resistance and negative effective grade. When this happens, the vehicle will accelerate (sometimes dangerously) as it descends unless the brakes are applied, a lower gear is engaged, the engine is accelerated, or oil is pumped into the hydraulic retarder to absorb horsepower through the transmission cooling system.

Retarder curves are used to determine the maximum safe descending speed and gear range of a vehicle without the use of the service brake, with the retarder fully on (Figure 2-6). The total braking horsepower indicated on some retarder curves is a combination of retarder braking power and engine friction while the throttle is closed. The retarder can also be used to slow a vehicle traveling over level ground. Speeds obtained from retarder curves do not take acceleration or deceleration into consideration. The effects of acceleration and deceleration will be discussed in Chapters 4 and 6. The following example demonstrates the use of a retarder curve.

**Example 2-19:** Assuming a rolling resistance of 100 pounds per ton, determine the maximum safe speed and gear range with maximum retarder effort for a Caterpillar 621E tractor-scraper loaded with a 48,000-pound payload as it descends a 19% grade.
**Solution:** Expressed as percent grade, the rolling resistance is:

$$\text{Rolling Resistance (\%)} = \frac{100 \text{ lb./Ton}}{20} \qquad \text{(Eq. 2.12)}$$

$$= 5\%$$

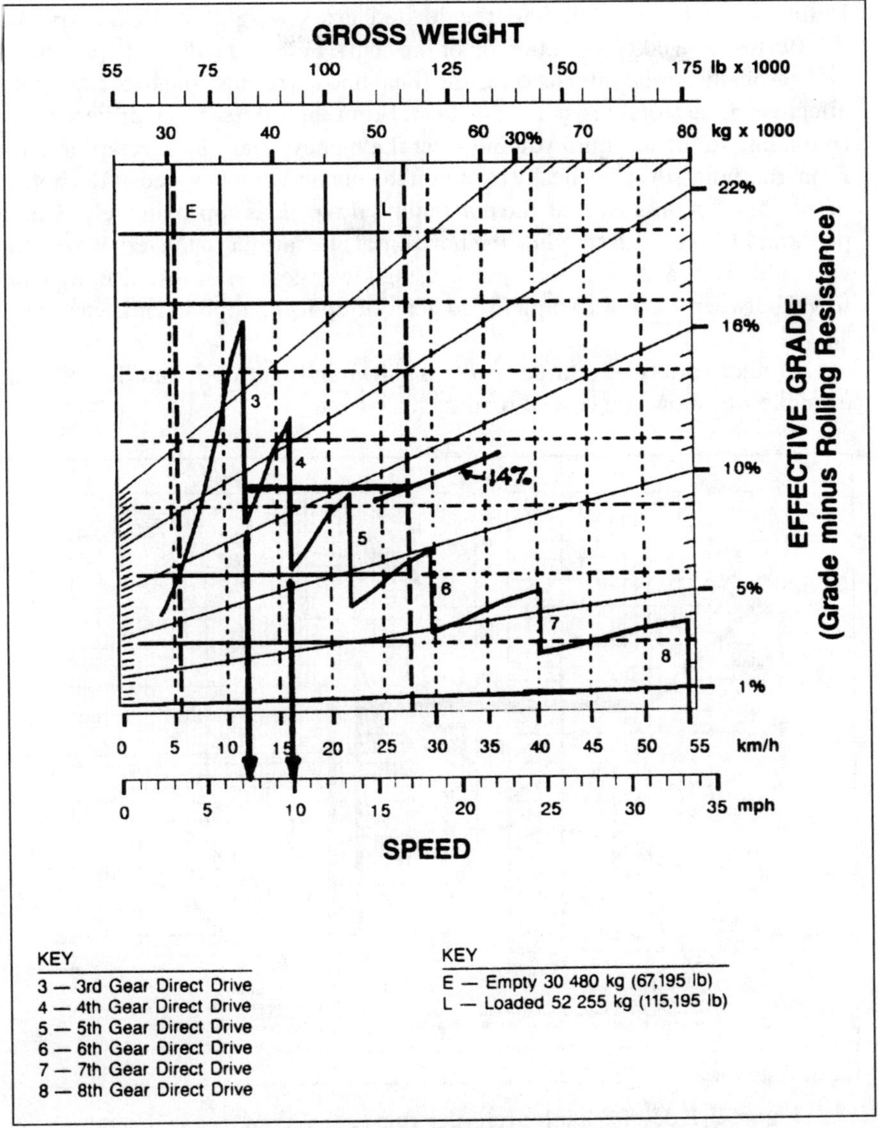


**Figure 2-6. 621E scraper retarder curve.** *(Courtesy of Caterpillar Inc.)*

The grade assistance is:

$$\text{Grade Assistance} = 19\% \qquad \text{(Given)}$$

Thus, the effective grade is:

$$\begin{aligned}\text{Effective Grade} &= 5\% - 19\% \qquad \text{(Eq. 2.14)}\\ &= -14\%\end{aligned}$$

Referring to Figure 2-6, find the loaded gross weight of 115,195 pounds (67,195 lbs. + 48,000 lbs.) at the top of the chart. Follow the dashed line labelled "L" vertically until it intersects the diagonal line corresponding to the favorable effective grade (total resistance) of 14%. From this intersection, project a horizontal line to the left until you intersect the highest gear of the retarder curve. From this point, drop vertically to obtain the maximum safe speed at the bottom of the chart. In this instance, the maximum safe speed is approximately 10 miles per hour in 4th gear. Extending the horizontal line farther to the left, we see that we could also travel at slower speeds using lower gears. For instance, we could travel between 7.5 and 10 mph in 3rd gear, or up to 7.5 mph in 2nd gear.

Another variety of retarder curve is shown in Figure 2-7, and the following example will show you how to read such a curve.

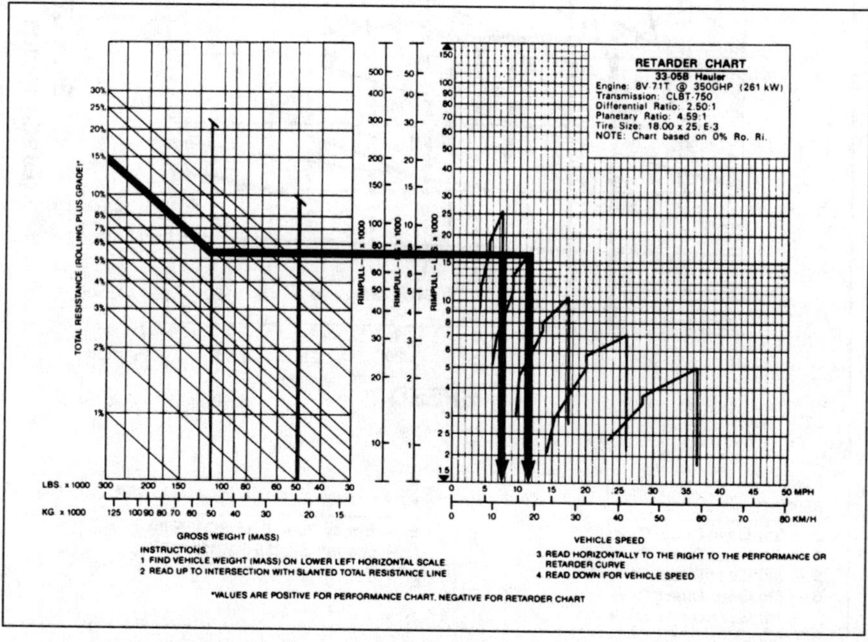


**Figure 2-7. 33-05B hauler retarder curve.**(*Courtesy of Terex Corporation*)

**Example 2-20:** Assuming a 15% effective grade, use Figure 2-7 to determine the maximum safe speed of a loaded Terex 33-05B hauler with a total weight of 114,000 pounds.

**Solution:** Referring to Figure 2-7, find the loaded gross vehicle weight of 114,000 pounds at the bottom of the chart. From this point, follow a vertical line until it intersects the diagonal line corresponding to a total resistance (favorable effective grade) of 15%. From this intersection, project a horizontal line to the right until you intersect the highest gear of the retarder curve. From this point, drop vertically to obtain the maximum safe speed at the bottom of the chart. In this instance, the maximum safe speed is 11.5 mph in 2nd gear, but we could also travel at a speed of up to 7.5 mph in 1st gear.

## RIMPULL AND DRAWBAR PULL

*Rimpull* is the tractive force (measured in pounds at the wheel rim) available between the rubber tires of the driving wheels and the travel surface. Rimpull is an important factor in determining the performance of wheel-mounted vehicles such as tractor-scrapers (Figures 2-8, 2-9 and 2-10) and off-highway trucks (Figures 2-7 and 4-5).

*Drawbar pull* is the pulling force (measured in pounds at a crawler-tractor's hitch) available for towing a load. Drawbar pull is an important factor in determining the performance of track-mounted vehicles such as crawler tractors (Figure 2-12).

Assuming sufficient traction, the *maximum* available rimpull or drawbar pull is dependent upon the power of the engine. The *available* rimpull or drawbar pull produced at any given time by a moving vehicle is dependent upon the gear used and the vehicle speed.

## REQUIRED PULL FOR WHEEL-MOUNTED VEHICLES

The *required* rimpull or drawbar pull of a vehicle is dependent upon the gross vehicle weight, rolling resistance and grade resistance. The effects of rolling and grade resistances on rimpull are incorporated into the performance curves (rimpull-speed-gradeability performance charts) of many types of wheel-mounted vehicles in terms of total (grade plus rolling) resistance expressed as effective grade (Figures 2-8, 2-9 and 2-10). Speeds obtained from rimpull-speed-gradeability performance charts do not take acceleration and deceleration into consideration. The effects of acceleration and deceleration will be discussed in Chapters 4 and 6. The following example demonstrates the use of a performance chart.

**Example 2-21:** Assuming a rolling resistance of 100 pounds per ton of gross vehicle weight, determine the rimpull required for an empty Caterpillar 621E tractor-scraper ascending a 3% grade. Also, determine the maximum speed attainable under the given conditions.

**Solution:** The total resistance expressed as percent grade is:

$$\text{Effective Grade } (\%) = \frac{100 \text{ lb./Ton}}{20} + 3\% \qquad \text{(Eq. 2.15)}$$

$$= 5\% + 3\%$$

$$= 8\%$$

Find the empty gross vehicle weight at the top of Figure 2-8 (67,195 lb.). Follow the vertical dashed line labelled "E" until it intersects the diagonal line corresponding to the total resistance (adverse effective grade) of 8%. Then move horizontally from this intersection to the left of the chart and read the total rimpull

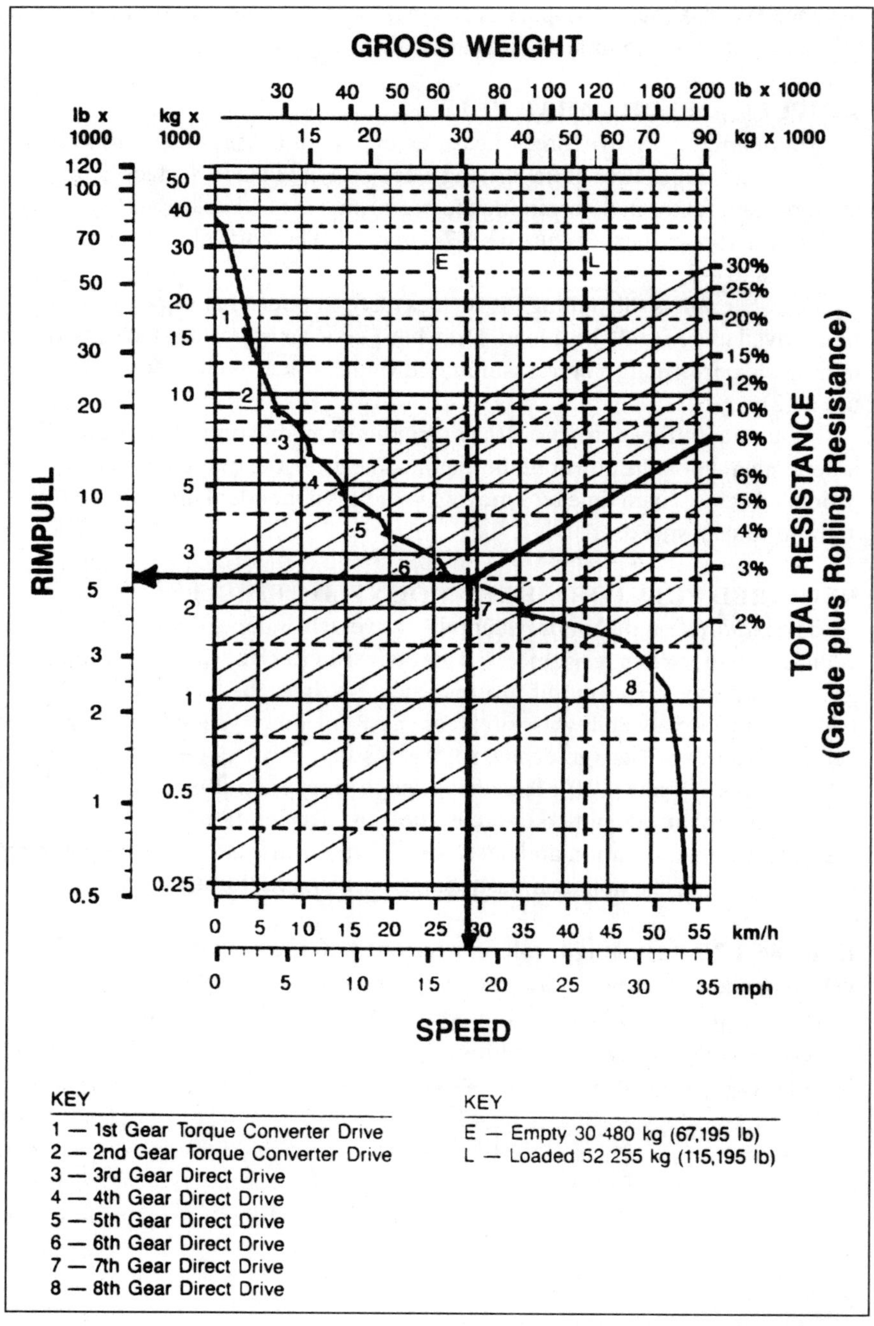


**Figure 2-8. 621E scraper rimpull-speed-gradeability performance chart.**

(*Courtesy of Caterpillar Inc.*)

required. In this case, the required rimpull is approximately 5500 pounds. Where this horizontal line intersects the highest gear of the performance curve, read down vertically to obtain the maximum attainable speed. In this case, the highest attainable speed is approximately 18 miles per hour in 7th gear.

Another variety of rimpull-speed-gradeability performance chart is shown in Figure 2-9, and the following example will show you how to read such a curve.

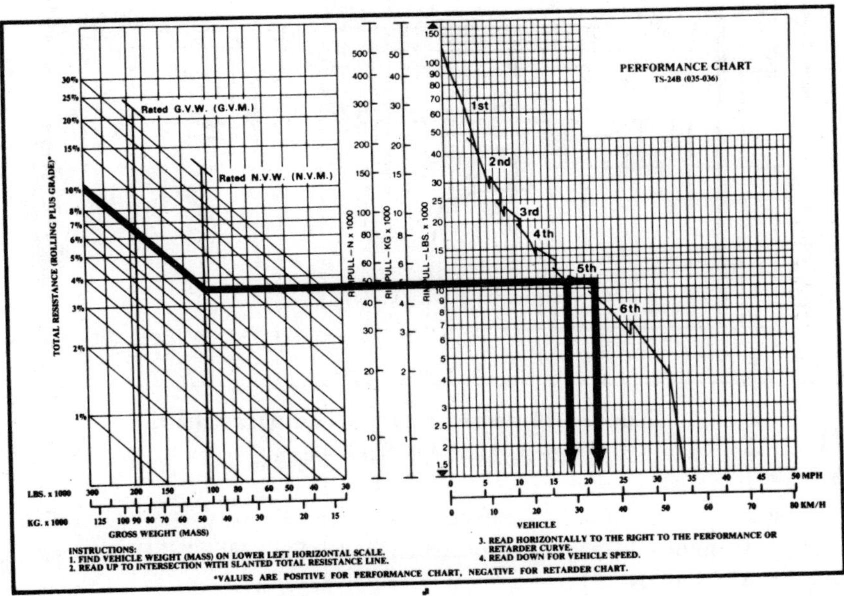


**Figure 2-9. TS-24B scraper performance chart.**

*(Courtesy of Terex Corporation)*

**Example 2-22:** Assuming 10% effective grade, use Figure 2-9 to determine the required rimpull and maximum speed of an empty Terex TS-24B scraper.

**Solution:** Referring to Figure 2-9, find the empty (net) vehicle weight of 108,315 pounds at the bottom of the chart. From this point, follow a vertical line labelled N.V.W. (net vehicle weight) until it intersects the diagonal line corresponding to a total resistance (adverse effective grade) of 10%. From this intersection, project a horizontal line to the right until you intersect the highest gear of the performance curve. From this point, drop vertically to obtain the maximum vehicle speed at the bottom of the chart. In this instance, the maximum speed is 21.5 mph in 5th gear using automatic lock-up drive, and the required rimpull is 10,500 pounds. If slightly more drawbar pull is required, we could travel at 17.5 mph in 5th gear using torque converter drive.

Another type of rimpull-speed-gradeability performance chart is shown in Figure 2-10, and the following example will demonstrate how to use such a curve.

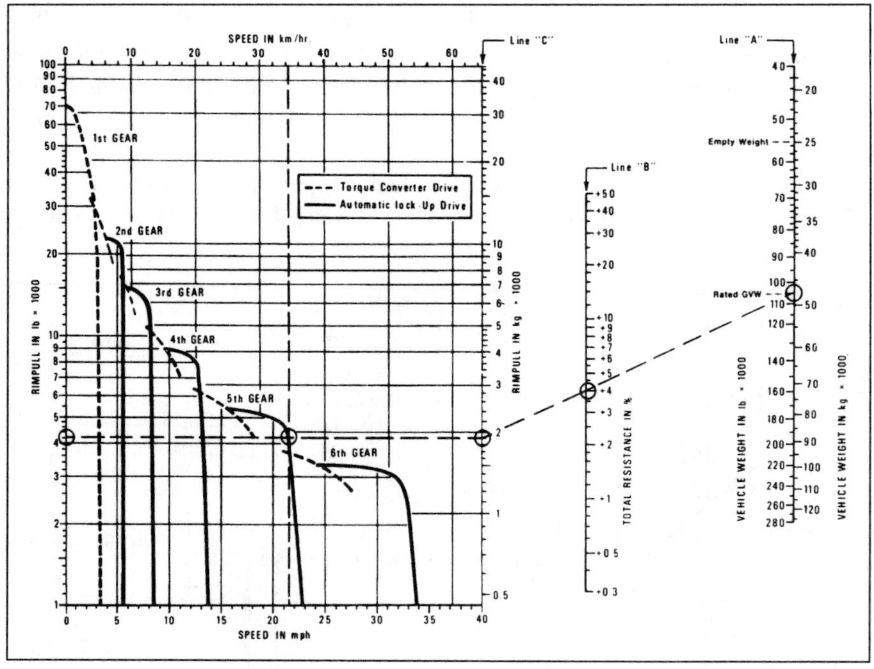


**Figure 2-10. 260-B scraper performance chart.**
*(Courtesy of Fiat-Allis North America, Inc.)*

**Example 2-23:** Assuming 4% effective grade, use Figure 2-10 to determine the required rimpull and maximum speed of a loaded Fiat-Allis 260-B scraper weighing 108,650 pounds.

**Solution:** Referring to Figure 2-10, find the loaded weight of 108,650 pounds on Line A, and the total resistance (adverse effective grade) of 4% on Line B. Extend a line connecting these points to Line C. From this point, project a horizontal line to the left until you intersect the highest gear of the performance curve. From this intersection, drop vertically to obtain the maximum vehicle speed at the bottom of the chart. In this instance, the maximum speed is 21.4 mph in 5th gear, and the required rimpull is 4200 pounds.

## READING TRAVEL TIME CHARTS

Travel time charts are used to determine the required travel time for fast-moving vehicles transporting huge payloads, such as dump trucks, tractor-scrapers and large-capacity wheel loaders. Two travel time charts are shown for any given type of vehicle; one for the vehicle hauling its rated payload (not shown here), and the other for the empty vehicle (Figure 2-11). Travel time charts take acceleration and deceleration of the vehicle into consideration at the loading and dumping areas. To use a travel time chart, enter the chart from the left and find the one-way travel distance. Project a horizontal line until it intersects the appropriate diagonal line representing total resistance (effective grade). From this intersection, drop vertically to obtain the travel time at the bottom of the chart.

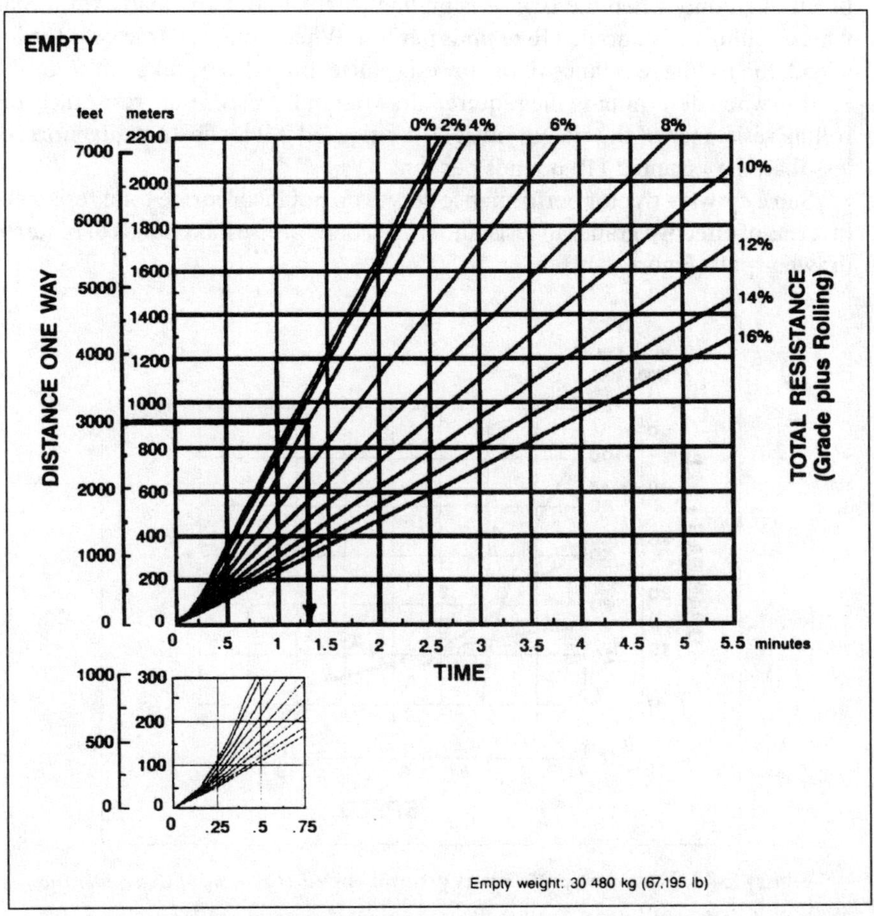


**Figure 2-11. 621E scraper travel time chart (empty).**
*(Courtesy of Caterpillar Inc.)*

**Example 2-24:** Assuming a 3000-foot one-way haul distance and an effective grade of 4%, determine the travel time for an empty Caterpillar 621E tractor-scraper.
**Solution:** Referring to Figure 2-11, find the one-way distance of 3000 feet at the left side of the chart. Project a horizontal line to the right until it intersects the diagonal line corresponding the total resistance (effective grade) of 4%. From this intersection, project a vertical line to the bottom of the chart to obtain the travel time. In this instance, the travel time is 1.3 minutes.

## REQUIRED PULL FOR CRAWLER-MOUNTED VEHICLES

Drawbar-pull vs. ground-speed charts such as Figure 2-12 are based on the assumption that the rolling resistance of the haul road is 110 pounds per ton. The maximum speeds for a Caterpillar D8N crawler tractor in 1st, 2nd and 3rd gears are 2.2, 3.9 and 6.7 miles per hour, respectively (Figure 2-12). These speeds can be attained only when the tractor is pulling itself without any load over a road whose rolling resistance is 110 pounds per ton. When a crawler-tractor is towing a load, the rolling resistance of the towed vehicle must also be taken into consideration when determining the required drawbar pull. Also, grade resistance and rolling resistance of the tractor must be considered if it is substantially more, or less than the assumed 110 pounds per ton.

Since crawler-tractor performance curves do not incorporate total resistance in terms of effective grade, all resistances must be expressed as pounds of required drawbar pull (Figure 2-12).

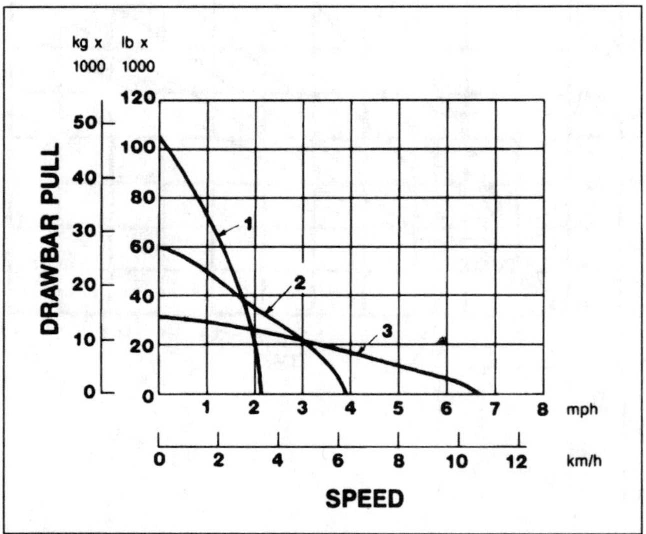


**Figure 2-12. Drawbar pull versus ground speed.** (*Courtesy of Caterpillar Inc.*)

**Example 2-25:** Assuming a rolling resistance of 200 pounds per ton, determine the required drawbar pull of a 69,187-pound Caterpillar D8N crawler tractor as it ascends a 5% grade while towing a 75,000-pound load. Also, determine the maximum speed attainable under the given conditions.

**Solution:** The weight of the crawler tractor in tons is:

$$\text{G.V.W. of Tractor (Tons)} = \frac{69{,}187 \text{ lb.}}{2000 \text{ lb./Ton}}$$

$$= 34.6 \text{ tons}$$

The rolling resistance of the haul road in excess of 110 pounds per ton is:

$$\text{Excess Rolling Resistance (lb./Ton)} = 200 \text{ lb./Ton} - 110 \text{ lb./Ton}$$
$$= 90 \text{ pounds/ton}$$

The excess rolling resistance imposed on the tractor, expressed in pounds is:

$$\text{Excess Rolling Resistance (lb.)} = 90 \text{ lb./Ton x } 34.6 \text{ Tons} \qquad \text{(Eq. 2.3)}$$
$$= 3114 \text{ pounds}$$

The weight of the towed load expressed in tons is:

$$\text{G.V.W. of Load (Tons)} = \frac{75{,}000 \text{ lb.}}{2000 \text{ lb./Ton}}$$
$$= 37.5 \text{ Tons}$$

The rolling resistance imposed on the towed load, expressed in pounds is:

$$\text{Rolling Resistance of Load (lb.)} = 200 \text{ lb./Ton x } 37.5 \text{ Tons} \qquad \text{(Eq. 2.3)}$$
$$= 7500 \text{ pounds}$$

The total gross vehicle weight of this equipment combination is:

$$\text{Total G.V.W.} = 69{,}187 \text{ lb.} + 75{,}000 \text{ lb.}$$
$$= 144{,}187 \text{ pounds}$$

The grade resistance expressed in pounds is:

$$\text{Grade Resistance (lb.)} = 144{,}187 \text{ lb. x } 0.05 \qquad \text{(Eq. 2.9)}$$
$$= 7209 \text{ pounds}$$

The total required drawbar pull is:

$$\text{Required Drawbar Pull} = 3114 \text{ lb.} + 7500 \text{ lb.} + 7209 \text{ lb.}$$
$$= 17{,}823 \text{ pounds}$$

Referring to Figure 2-12, enter the chart from the left using 17,823 pounds for drawbar pull and project a line to the right until you intersect the performance curves. Drop a vertical line from each intersection and read the corresponding speeds. We can haul the load at approximately 2.1 mph in 1st gear, or 3.3 mph in 2nd, but the fastest attainable speed is 4.2 mph in 3rd gear.

In many cases, the crawler tractor's rolling resistance in excess of 110 pounds per ton can be omitted without a substantial change in the performance calculations. For instance, in the last example, the excess resistance amounted to only 3114 pounds. Had we omitted this calculation, the required drawbar pull would have been:

Required Drawbar Pull = 7500 lb. + 7209 lb.
= 14,709 pounds

Referring to Figure 2-12, we can see that this required drawbar pull would allow us to haul the load at a maximum speed of about 4.5 mph in 3rd gear. However, if the haul road had a rolling resistance of 350 pounds per ton, the excess rolling resistance would be substantial.

**Example 2-26:** Work the previous example assuming a rolling resistance of 350 pounds per ton of gross vehicle weight.
**Solution:** The rolling resistance of the haul road in excess of 110 pounds per ton is:

Excess Rolling Resistance (lb./Ton) = 350 lb./Ton – 110 lb./Ton
= 240 pounds/ton

The excess rolling resistance imposed on the tractor, expressed in pounds is:

Excess Rolling Resistance of Tractor (lb.) = 240 lb./Ton x 34.6 Tons (Eq. 2.3)
= 8304 pounds

The rolling resistance imposed on the towed load is:

Rolling Resistance of Load (lb.) = 350 lb./Ton x 37.5 Tons (Eq. 2.3)
= 13,125 pounds

The grade resistance would remain:

Grade Resistance (lb.) = 144,187 lb. x .05 (Eq. 2.9)
= 7209 pounds

The total required drawbar pull is:

Required Drawbar Pull (lb.) = 8304 lb. + 13,125 lb. + 7209 lb.
= 28,638 pounds

Referring to Figure 2-12, we could haul the load at a maximum speed of 2.6 mph in 2nd gear. Had we neglected the 8304 pounds of excess drawbar pull required for the tractor, we would have deceived ourselves by showing a required drawbar pull of only:

Required Drawbar Pull = 13,125 lb. + 7209 lb.
= 20,334 pounds

Referring to Figure 2-12, this required drawbar pull would falsely indicate that we could haul the load at about 3.3 mph is 3rd gear. This seems like only a slight discrepancy; however, the error could become substantial during an extended project requiring hundreds of lengthy haul trips.

On the other hand, if the rolling resistance is substantially less than 110 pounds per ton, the deficient rolling resistance must be deducted when determining the required drawbar pull.

**Example 2-27:** Work the previous example assuming a rolling resistance of 50 pounds per ton.
**Solution:** The rolling resistance of the haul road less than 110 pounds per ton is:

Deficient Rolling Resistance (lb./Ton) = 110 lb./Ton – 50 lb./Ton
= 60 pounds/ton

The deficient rolling resistance expressed in pounds is:

Deficient Rolling Resistance (lb.) = 60 lb./Ton x 34.6 Tons (Eq. 2.3)
= 2076 pounds

The rolling resistance of the towed load is:

Rolling Resistance of Load (lb.) = 50 lb./Ton x 37.5 Tons (Eq. 2.3)
= 1875 pounds

The grade resistance will remain:

Grade Resistance (lb.) = 144,187 lb. x 0.05 (Eq. 2.9)
= 7209 pounds

The total required drawbar pull is:

Required Drawbar Pull = 7209 lb. + 1875 lb. – 2076 lb.
= 7008 pounds

Referring to Figure 2-12, we could haul the load at a maximum speed of 6 mph in 3rd gear.

If the available rimpull or drawbar pull of a vehicle is not known, it can be estimated as follows:

$$\text{Available Pull (lb.)} = \frac{375 \text{ x HP x Efficiency}}{\text{Speed (mph)}} \qquad \textbf{(Equation 2.18)}$$

where: Efficiency of the power train for most trucks and tractors ranges from 80 to 85 % of gross engine horsepower.

**Note:** The number 375 was derived as follows:

$$1 \text{ h.p.} = \frac{550 \text{ Ft.-lb./Sec. x 60 Sec./Min.}}{88 \text{ Ft./Min./mph}}$$
$$= 375 \text{ lb.-mph}$$

**Example 2-28:** Assuming an 80% power-train efficiency, estimate the probable drawbar pull for a 165-hp Caterpillar D6D SA agricultural tractor for the maximum speed in each forward gear if maximum speed is: 1st = 2.5 mph, 2nd = 3.0 mph, 3rd = 3.4 mph, 4th = 4.0 mph, 5th = 4.6 mph and 6th = 5.5 mph.
**Solution:** The probable available pull in 1st gear is:

$$\text{Available Pull (lb.) (1st Gear)} = \frac{375 \text{ x } 165 \text{ x } 0.80}{2.5} \qquad \text{(Eq. 2.18)}$$
$$= 19{,}800 \text{ pounds}$$

The probable available pull in 2nd gear is:

$$\text{Available Pull (lb.) (2nd Gear)} = \frac{375 \text{ x } 165 \text{ x } 0.80}{3.0}$$
$$= 16{,}500 \text{ pounds}$$

The probable available pulls for the remaining gears are as follows:

Available Pull (lb.) (3rd Gear) = 14,559 pounds
Available Pull (lb.) (4th Gear) = 12,375 pounds
Available Pull (lb.) (5th Gear) = 10,761 pounds
Available Pull (lb.) (6th Gear) = 9000 pounds

Keep in mind that Equation 2.18 is used for estimating available pull. Power-train efficiency varies with the RPM of the engine, and the actual drawbar pulls for a D6D SA agricultural tractor available at the rated RPM's are:

| Gear | MPH | Drawbar Pull(lb.) | Power-Train Efficiency |
|---|---|---|---|
| 1 | 2.5 | 19,840 | 80.2 |
| 2 | 3.0 | 16,690 | 80.9 |
| 3 | 3.4 | 14,060 | 77.3 |
| 4 | 4.0 | 11,810 | 76.3 |
| 5 | 4.6 | 10,140 | 75.4 |
| 6 | 5.5 | 8,120 | 72.2 |

(*Courtesy of Caterpillar Inc.*)

The horsepower required to produce a given drawbar pull or rimpull can be estimated as follows:

$$\text{Horsepower Required} = \frac{\text{Required Pull x Speed (mph)}}{\text{375 x Efficiency}} \qquad \textbf{(Equation 2.19)}$$

**Example 2-29:** Assuming 85% power-train efficiency, determine the horsepower required to move a 10,000-pound load at a speed of 3 mph.
**Solution:** The required horsepower is:

$$\text{Horsepower Required} = \frac{10{,}000 \text{ x } 3}{375 \text{ x } 0.85} \qquad \text{(Eq. 2.19)}$$
$$= 94 \text{ horsepower}$$

The top speed in any given forward gear can be estimated as follows:

$$\text{Speed (mph)} = \frac{\text{375 x HP x Efficiency}}{\text{Available Pull}} \qquad \textbf{(Equation 2.20)}$$

**Example 2-30:** Assuming 80% power-train efficiency, determine the top speed in 3rd gear for the tractor given in Example 2-28.
**Solution:** The probable top speed is:

$$\text{Speed (mph)} = \frac{375 \text{ x } 165 \text{ x } 0.80}{14{,}559} \qquad \text{(Eq. 2.20)}$$
$$= 3.4 \text{ miles per hour}$$

## GRADEABILITY

*Gradeability* is the steepest slope (expressed in %-grade) that a vehicle can ascend at a uniform speed. Gradeability can be specified for any given forward gear. Provided that sufficient traction is available, the maximum grade that a crawler tractor can ascend is:

$$\text{Maximum Grade (\%)} = \frac{\text{Pull Available for Grade}}{\text{Pull Required for 1\% Grade}} \qquad \textbf{(Equation 2.21)}$$

where:

Pull Available for Grade = Maximum Available Pull (lb.)
– Rolling Resistance (lb.) **(Equation 2.22)**

and,

Pull Required for 1% Grade = 20 lb./Ton x G.V.W. (Tons) **(Equation 2.23)**

To provide a margin of safety, no more than 85% of the available drawbar pull of a crawler tractor should be used when determining the gradeability of the vehicle.

**Example 2-31:** Determine the gradeability of a Caterpillar D8N crawler tractor towing a loaded scraper under the following conditions:

Tractor weight = 69,187 pounds = 34.6 Tons
Scraper weight (loaded) = 80,000 pounds = 40 Tons
Drawbar pull (1st gear) = 105,000 pounds
Rolling Resistance of travel surface = 200 lb./Ton

**Solution:** The rolling resistance of the haul road in excess of 110 pounds per ton is:

Excess Rolling Resistance (lb./Ton) = 200 lb./Ton – 110 lb./Ton
= 90 pounds/ton

The excess rolling resistance imposed on the tractor, expressed in pounds is:

Excess Rolling Resistance (lb.) = 90 lb./Ton x 34.6 Tons (Eq. 2.3)
= 3114 pounds

The rolling resistance imposed on the scraper, expressed in pounds is:

Rolling Resistance (Scraper; lb.) = 200 lb./Ton x 40 Tons (Eq. 2.3)
= 8000 pounds

The combined rolling resistance of both units is:

$$\text{Total Rolling Resistance} = 3114 \text{ lb.} + 8000 \text{ lb.}$$
$$= 11{,}114 \text{ pounds}$$

The drawbar pull available to overcome grade is:

$$\text{Pull Available for Grade} = 105{,}000 \text{ lb.} - 11{,}114 \text{ lb.} \qquad \text{(Eq. 2.22)}$$
$$= 93{,}886 \text{ pounds}$$

The total weight of the equipment is:

$$\text{Total Weight} = 34.6 \text{ Tons} + 40 \text{ Tons}$$
$$= 74.6 \text{ tons}$$

The pull required to overcome 1% grade is:

$$\text{Pull Required for 1\% Grade} = 20 \text{ lb./Ton} \times 74.6 \text{ Tons} \qquad \text{(Eq. 2.23)}$$
$$= 1492 \text{ pounds}$$

Assuming adequate traction, the maximum grade that this combination can ascend is:

$$\text{Maximum Grade (\%)} = \frac{93{,}886 \text{ lb.}}{1492 \text{ lb.}} \qquad \text{(Eq. 2.21)}$$
$$= 63\%$$

Allowing an 85% safety margin, the maximum grade is:

$$\text{Maximum Safe Grade (\%)} = 0.85 \times 63\%$$
$$= 54\%$$

The Gradeability of a wheel-mounted machine can be determined from the performance curves.

**Example 2-32:** Determine the gradeability of a loaded Caterpillar 621E tractor-scraper traveling in 4th gear under the following conditions:

Total Vehicle Weight (Loaded) = 115,195 pounds = 57.6 tons
Rolling Resistance = 200 pounds/ton

**Solution:** Referring to Figure 2-8, find the maximum rimpull attainable in 4th gear by projecting a line from the top of the 4th-gear performance curve to the left side of the chart. The maximum pull allowed in 4th gear is approximately 13,500 pounds.

The Rolling resistance is:

$$\text{Rolling Resistance (lb.)} = 200 \text{ lb./Ton} \times 57.6 \text{ Tons} = 11{,}520 \text{ pounds} \quad \text{(Eq. 2.3)}$$

The rimpull available to overcome grade is:

$$\text{Pull Available for Grade} = 13{,}500 \text{ lb.} - 11{,}520 \text{ lb.} = 1980 \text{ pounds} \quad \text{(Eq. 2.22)}$$

The pull required to overcome 1% grade is:

$$\text{Pull Required for 1\% Grade} = 20 \text{ lb./Ton} \times 57.6 \text{ Tons} = 1152 \text{ pounds} \quad \text{(Eq. 2.23)}$$

Assuming adequate traction, the maximum grade that the machine can ascend in 4th gear is:

$$\text{Maximum Grade (\%)} = \frac{1980 \text{ lb.}}{1152 \text{ lb.}} = 1.7\% \quad \text{(Eq. 2.21)}$$

We could have also solved the previous example as follows: Referring to Figure 2-8, find the required rimpull of 13,500 pounds at the left side of the chart. Project a horizontal line until it intersects the dashed vertical line labelled "L." Then determine the diagonal effective grade line passing through this point. In this case, the effective grade is approximately 11.7 %. Effective grade is defined as:

$$\text{Effective Grade (\%)} = \text{Rolling Resistance (\%)} + \text{\%-Grade} \quad \text{(Eq. 2.13)}$$

Therefore, the maximum grade that can be ascended is:

$$\text{Maximum Grade (\%)} = \text{Effective Grade (\%)} - \text{Rolling Resistance (\%)}$$

**(Equation 2.24)**

Expressed as percent grade, the rolling resistance is:

$$\text{Rolling Resistance (\%)} = \frac{200 \text{ lb./Ton}}{20} = 10\% \quad \text{(Eq. 2.12)}$$

Thus, the maximum grade that the machine can ascend in 4th gear is:

$$\text{Maximum Grade (\%)} = 11.7\% - 10\% = 1.7\% \qquad \text{(Eq. 2.24)}$$

There is a third method that can be used to determine gradeability. The total resistance can be expressed as:

$$\text{Total Resistance (lb.)} = \text{Rolling Resistance (lb.)} + \text{Grade Resistance (lb.)} \qquad \text{(Eq. 2.10)}$$

Also, grade resistance is:

$$\text{Grade Resistance (lb.)} = \text{G.V.W. (lb.)} \times \text{\%-Grade} \qquad \text{(Eq. 2.9)}$$

where: %-grade can be considered to be maximum grade.

Substituting Equation 2.9 into Equation 2.10, we have:

$$\text{Maximum Pull (lb.)} = \text{Rolling Resistance (lb.)} + (\text{G.V.W. (lb.)} \times \text{\%-Grade}) \qquad \textbf{(Equation 2.25)}$$

Solving for %-grade, we have:

$$\text{\% Grade (Maximum Grade)} = \frac{\text{Maximum Pull (lb.)} - \text{Rolling Res. (lb.)}}{\text{G.V.W. (lb.)}} \qquad \textbf{(Equation 2.26)}$$

**Example 2-33:** Use Equation 2.26 to work the previous example.
**Solution:** The maximum pull of 13,500 pounds in 4th gear can be considered to be the total resistance that the machine must overcome. Also, the loaded gross vehicle weight is 115,195 pounds and the rolling resistance is 11,520 pounds; therefore, the maximum grade that the machine can ascend in 4th gear is:

$$\text{Maximum Grade (\%)} = \frac{13{,}500 \text{ lb.} - 11{,}520 \text{ lb.}}{115{,}195 \text{ lb.}} = 0.017 = 1.7\% \qquad \text{(Eq. 2.26)}$$

## USABLE POWER AND USABLE PULL

Now that we know the mechanics involved in reading performance curves, we must consider three factors that limit the *usable power* and *usable pull* of a vehicle. Engine power is affected by altitude and ambient temperature, and usable pull is affected by engine power and traction.

## ALTITUDE DERATING

Internal combustion engines produce less power as the altitude increases because of a decreased air density at higher elevations. A specific ration of fuel and oxygen must be present in each cylinder for maximum efficiency and engine power. At higher altitudes, the given volume of air drawn into the cylinder contains less oxygen due to a reduction in air density. Since the ratio of fuel and oxygen in the cylinder must remain constant, the quantity of fuel delivered to the cylinder at altitude must be reduced by adjusting the carburetor in order to maintain normal engine life. The end result is a reduction of power.

A two-cycle engine suffers less power loss than a four-cycle engine at higher altitudes because the two-cycle engine's air is supplied under pressure by a blower, whereas the four-cycle engine relies on suction produced by the pistons. Some engines are equipped with a turbocharger which retains the rated power up to altitudes of around 10,000 feet above sea level. Some manufacturers publish altitude derating tables regarding the performance of their machines at various altitudes (Table 2-4).

**Table 2-4: Percent Flywheel Horsepower Available at Specified Altitudes**

| Model | Altitude Above Sea Level (Ft.) | | | | | |
|---|---|---|---|---|---|---|
| | 0-2500 | 2500-5000 | 5000-7500 | 7500-10,000 | 10,000-12,500 | 12,500-15,000 |
| 621E | 100 | 100 | 94 | 87 | 80 | 74 |
| 769C | 100 | 100 | 100 | 97 | 89 | 82 |
| 936F | 100 | 100 | 100 | 98 | 90 | 83 |
| D8N | 100 | 100 | 100 | 100 | 98 | 90 |
| D9N | 100 | 100 | 100 | 96 | 89 | 82 |

(*Courtesy of Caterpillar Inc.*)

In the absence of altitude derating tables, the loss of power for a naturally-aspirated *two-cycle* engine can be estimated at 1% for each 1000 feet of altitude beyond 5000 feet above sea level. Therefore, the percent of power loss due to altitude is:

$$\text{Loss of Power (\%)} = \frac{0.01 \times (\text{Altitude} - 5000')}{1000'}$$

$$= 0.001 \times (\text{Altitude} - 5000') \qquad \textbf{(Equation 2.27)}$$

Thus, the *usable* power expressed as a percent of *rated* power is:

Usable Power (%) = 100% - [0.001 x (Altitude – 5000')] **(Equation 2.28)**

The loss of power for a naturally-aspirated *four-cycle* engine can be estimated at 3% for each 1000 feet of altitude beyond 5000 feet above sea level. Therefore, the percent of power loss due to altitude is:

Loss of Power (%) = 0.003 x (Altitude – 5000') **(Equation 2.29)**

Thus, the usable power is:

Usable Power (%) = 100% - [0.003 x (Altitude – 5000')] **(Equation 2.30)**

**Example 2-34:** A 350-hp four-cycle engine produces 100,000 pounds of drawbar pull at sea level. Determine the probable usable power and usable pull at an elevation of 10,000 feet.
**Solution:** The usable power expressed as a percent of rated horsepower is:

Usable Power (%) = 100% - [0.003 x (10,000' – 5000')] (Eq. 2.30)
= 85%

The probable derated engine power is:

Available Power (hp) = 0.85 x 350 hp
= 298 horsepower

The probable drawbar pull at the given altitude is:

Usable Pull (lb.) = 0.85 x 100,000 lb.
= 85,000 pounds

Use Equations 2.28 and 2.30 only as a last resort for estimating the performance of equipment derated for altitude. If available, a manufacturer's derating tables such as that shown in Table 2-4 are much more reliable.

**Example 2-35:** A 450-horsepower Caterpillar D8N tractor is working under conditions that require a total drawbar pull of 40,000 pounds. Compare the performance of the tractor working at 5000 feet above sea level with that of the tractor working at 13,000 feet.
**Solution:** Referring to Table 2-4, we can see that the tractor performs at 100% flywheel horsepower at 5000 feet above sea level; therefore, the tractor can tow its load at a maximum speed of 1.7 mph in 1st or 2nd gear (Figure 2-12). However, the tractor has only 90% of its rated power at 13,000 feet (Table 2-4); therefore, the tractor will have an equivalent required rimpull of:

$$\text{Equivalent Required Rimpull (13,000')} = \frac{40{,}000 \text{ lb.}}{0.90}$$
$$= 44{,}444 \text{ pounds}$$

Referring to Figure 2-12, the tractor will tow its load at a maximum speed of 1.4 mph in 1st gear.

**Example 2-36:** Compare the sea level performance of a loaded Caterpillar 621E tractor-scraper ascending a 10% effective grade with the performance of the machine working at 9,000 feet above sea level.
**Solution:** Referring to Figure 2-8, the tractor-scraper at sea level has a required rimpull of 12,000 pounds and can travel up a 10% effective grade at a maximum speed of 8.5 mph in 4th gear. However, the machine works at only 87% of its rated power at 9,000 feet (Table 2-4); therefore, the unit will have an equivalent required rimpull of:

$$\text{Equivalent Required Rimpull (9,000')} = \frac{12{,}000 \text{ lb.}}{0.87}$$
$$= 13{,}793 \text{ pounds}$$

Referring to Figure 2-8, the tractor-scraper will travel at a maximum speed of 6.5 mph in 3rd gear.

**Example 2-37:** A bulldozer can move 1000 LCY per hour at sea level. Determine the production at 8000 feet above sea level assuming that only 85% of the engine's rated power is available.
**Solution:** The probable production rate is:

$$\text{Production (8000')} = 0.85 \times 1000 \text{ LCY}$$
$$= 850 \text{ LCY/hour}$$

## COMBINED EFFECT OF PRESSURE AND TEMPERATURE

When an internal-combustion engine is rated for power, tests are conducted under a set of standard conditions. Standard conditions for testing a four-cycle are engine normally considered to be at a temperature of 77° Fahrenheit, and an atmospheric pressure of 29.61 inches of mercury. Some performance tests also include relative humidity as a standard condition. The available horsepower of a four-cycle engine under non-standard conditions can be estimated by using the following equation:

$$H_A = H_S \times \frac{P_A}{P_S} \times \sqrt{\frac{T_S}{T_A}} \qquad \textbf{(Equation 2.31)}$$

where: $H_A$ = Actual horsepower

$H_S$ = Horsepower under standard testing conditions

$P_A$ = Actual atmospheric pressure

$P_S$ = Atmospheric pressure under standard testing conditions
= 29.61 in. Hg

$T_A$ = Actual absolute temperature = 460° F. + Actual Temp., degrees F.

and, $T_S$ = Actual absolute temperature under standard testing conditions
= 460° F. + 77° F.
= 537° F.

Average barometric pressures for various altitudes above sea level are given in Table 2-5. However, the barometric pressure at any given altitude can vary with prevailing climatic conditions.

**Table 2-5: Average Barometric Pressures at Various Altitudes Above Sea Level**

| Altitude Above Sea Level (Ft.) | Barometric Pressure (In., Hg) |
|---|---|
| 0 | 29.92 |
| 1,000 | 28.86 |
| 2,000 | 27.82 |
| 3,000 | 26.80 |
| 4,000 | 25.82 |
| 5,000 | 24.87 |
| 6,000 | 23.95 |
| 7,000 | 23.07 |
| 8,000 | 22.21 |
| 9,000 | 21.36 |
| 10,000 | 20.55 |

**Example 2-38:** A four-cycle diesel engine is rated at 150 horsepower under standard testing conditions. Determine the probable horsepower of the engine at an altitude of 5000 feet where the average daily temperature (for this time of year) is 85°F.
**Solution:** The probable available horsepower is:

$$H_A = 150 \times \frac{24.87}{29.61} \times \sqrt{\frac{537}{460 + 85}} \qquad \text{(Eq. 2.31; Table 2-5)}$$

$$= 150 \times 0.84 \times 0.99$$

$$= 125 \text{ horsepower}$$

**Example 2-39:** Work the previous example assuming an ambient temperature of 35° F.
**Solution:** The probable available horsepower is:

$$H_A = 150 \times 0.84 \times \sqrt{\frac{537}{460 + 35}} \qquad \text{(Eq. 2.31)}$$

$$= 131 \text{ horsepower}$$

Thus, an internal-combustion engine will develop a greater horsepower at a low ambient temperature than at a high ambient temperature.

Most horsepower ratings reflect deductions for power lost due to the air cleaner and standard accessories such as fuel, lubricating oil and jacket water pumps. Power required for auxiliaries such as cooling fans, air compressors, charging alternators, etc. must also be deducted to arrive at the net power available.

The power of an engine is normally determined with a brake (expressed as *brake horsepower*) or a dynamometer (expressed as *flywheel horsepower*).

Horsepower ratings are often expressed as intermittent or continuous horsepower. *Intermittent* horsepower is the horsepower and speed capability (RPM's) of the engine that can be used for about one hour, followed by an hour of operation at, or below the continuous rating. *Continuous* horsepower is the horsepower and speed capability that can be used without interruption or load cycling.

## TRACTION LIMITATIONS

In numerous problems throughout this chapter, I have often interjected the statement: "*assuming adequate traction.*" This is because of the fact that regardless of engine power, the lack of traction will result in the tracks or wheels of a vehicle spinning, producing no forward movement. Thus, drawbar pull and rimpull is limited by traction. The maximum tractive force that a vehicle can produce before slipping occurs is defined as *maximum usable pull*, and this tractive force can be determined by the following equation:

$$\text{Maximum Usable Pull} = \text{Coefficient of Traction} \times \text{Weight on Drivers}$$

**(Equation 2.32)**

Assuming sufficient power and load-bearing capability of the travel surface, the traction (and thus, maximum usable pull) of a vehicle is dependent basically upon the type of tire (or track) used, the type and condition of the travel surface and the weight imposed on the drive wheels. The weight actually resting over the drive wheels is a critical factor because this portion of the total machine weight holds the drive wheels against the ground to be converted into machine movement. Weight resting over non-drive wheels only increases the resistance.

The weight of the drive wheels can be increased by filling them with water or a solution of calcium chloride and water *(hydroflating)*. The calcium chloride solution is heavier and reduces the danger of freezing. Hydroflating increases traction, usable drawbar pull and machine stability while reducing tire wear by decreasing slippage and bounce.

The *coefficient of traction* is defined as:

$$\text{Coefficient of Traction} = \frac{\text{Maximum Usable Pull}}{\text{Weight on Drivers}}$$

**(Equation 2.33)**

In an actual test, the coefficient of traction is determined by measuring the driving force of the vehicle just as slippage occurs. This force is the maximum useable pull. Since the weight on the drive wheels or tracks can also be measured, the coefficient of traction can easily be determined.

**Example 2-40:** Assuming the weight on the drive wheels of a vehicle is 18,000 pounds and that slippage occurs when the tractive force between the tires and road surface is 10,000 pounds, determine the coefficient of traction.

**Solution:** The coefficient of traction is:

$$\text{Coefficient of Traction} = \frac{10{,}000 \text{ lb.}}{18{,}000 \text{ lb.}} \qquad \text{(Eq. 2.33)}$$

$$= 0.56$$

The approximate values for coefficient of traction for typical travel surfaces are given in Table 2-6.

**Table 2-6: Coefficient of Traction Factors**

| Material | Traction Factors Rubber Tires | Beadless Tires | Tracks |
|---|---|---|---|
| Concrete | 0.90 | 0.45 | 0.45 |
| Clay loam, dry | 0.55 | 0.70 | 0.90 |
| Clay loam, wet | 0.45 | 0.55 | 0.70 |
| Rutted clay loam | 0.40 | 0.55 | 0.70 |
| Dry sand | 0.20 | 0.25 | 0.30 |
| Wet sand | 0.40 | 0.45 | 0.50 |
| Quarry pit | 0.65 | 0.70 | 0.55 |
| Gravel road (loose not hard) | 0.36 | 0.40 | 0.50 |
| Packed snow | 0.20 | 0.25 | 0.27 |
| Ice | 0.12 | 0.10 | 0.12 |
| Semi-skeleton shoes | | | |
| Firm earth | 0.55 | 0.75 | 0.90 |
| Loose earth | 0.45 | 0.50 | 0.60 |
| Coal, stockpiled | 0.45 | 0.50 | 0.60 |

Note: The elevated sprocket design track-type tractors (D11N, D10N, D9N and D8N), with their suspended undercarriage, provide up to 15% more efficient tractive effort than rigid-tracked track-type tractors.
(*Courtesy of Caterpillar Inc.*)

Since friction between the tire and road surface helps produce traction, reducing tire pressure will increase the ground contact area, which increases friction, resulting in greater traction. The coefficients of traction for rubber tires will vary, depending on the type of tire and tread design. Likewise, the coefficients for crawler tracks will vary, depending on the design of the shoe.

The standard track shoe consists of a flat steel plate with a single high cleat, or grouser. A *grouser* is a ridge or cleat across a track shoe which helps the track grip the ground. A standard track shoe gives good traction under most conditions, except over ice or frozen ground. It also tears the travel surface. Flat shoes reduce surface tearing but they don't provide much traction.

The best general-purpose track shoe has a triple grouser. It gives good traction and does the least damage to a paved road surface. A double grouser shoe offers more traction than the triple grouser shoe, and the single grouser offers maximum traction and mobility, especially in hilly terrain. On hard, smooth quarry floors that allow little or no grouser penetration, use narrow double grouser shoes.

A grouser shoe can be converted to a flat shoe by bolting a steel pad over the grouser shoe. Special types of flat shoes can be installed and covered with a rubber-faced pad. This type of shoe helps provide traction and reduces damage to the road surface; however, rubber pads are not as wear-resistant as steel shoes and the pushing ability is limited.

A semi-grouser is a flat shoe with two or three low-relief grousers. They do not provide as much traction as the standard track shoe, but they also have less tendency to tear up the working surface.

Shoes are available in differing widths. Wide shoes provide better floatation and traction, while narrow shoes allow easier steering and maneuverability. In severe underfoot conditions, narrow shoes impose lower forces on the undercarriage than do wide shoes, and narrow shoes normally have a longer life expectancy under these conditions. Track shoes have little effect on stability; therefore, machines working over rock should be equipped with the narrowest track shoes available that are adequate for the job. The pressure-bearing capabilities of the ground and machine weight will dictate the minimum width of track permitted.

Manufacturers' data should contain information regarding the weight distribution over the drive wheels. If this information is not available, you can estimate the weight on the drivers as follows:

1) For crawler tractors and 4-wheel-drive tractors, use the total tractor weight.
2) For four-wheel tractors towing scrapers, use 40% of the gross vehicle weight (including the weight of the scraper).
3) For two-wheel tractors towing scrapers, use 60% of the gross vehicle weight (including the weight of the scraper).

To be more precise, take your vehicle to a scale and weigh it, one axle at a time. Keep in mind that the weight of a load can shift toward the rear while climbing grades, which is advantageous for vehicles whose drive wheels are located at the rear.

**Example 2-41:** Determine the maximum usable drawbar pull of a 69,187-pound Caterpillar D8N crawler tractor while working over: a) firm earth, and b) loose earth.
**Solution:** a) On firm earth, the maximum usable drawbar pull is:

Maximum Usable Pull = 0.90 x 69,187 lb. (Eq. 2.32; Table 2-6)
= 62,268 pounds

b) On loose earth, the maximum usable drawbar pull is:

Maximum Usable Pull = 0.60 x 69,187 lb.
= 41,512 pounds

**Example 2-42:** Assuming a rolling resistance of 150 pounds per ton of gross vehicle weight, determine the steepest grade the crawler tractor given in the previous example can ascend while working over loose earth.
**Solution:** The gross vehicle weight is:

$$\text{G.V.W.} = \frac{69{,}187 \text{ lb.}}{2000 \text{ lb./Ton}}$$
$$= 34.6 \text{ tons}$$

The rolling resistance of the haul road in excess of 110 pounds per ton is:

$$\text{Excess Rolling Resistance (lb./Ton)} = 150 \text{ lb./Ton} - 110 \text{ lb./Ton}$$
$$= 40 \text{ pounds/ton}$$

The excess rolling resistance expressed in pounds is:

$$\text{Excess Rolling Resistance} = 40 \text{ lb./Ton} \times 34.6 \text{ Tons} \qquad \text{(Eq. 2.3)}$$
$$= 1384 \text{ pounds}$$

The maximum usable drawbar pull is:

$$\text{Maximum Usable Pull} = 41{,}512 \text{ pounds} \qquad \text{(Example 2-41)}$$

Therefore, the steepest grade the vehicle can ascend before slippage occurs is:

$$\text{Maximum Grade (\%)} = \frac{41{,}512 \text{ lb.} - 1384 \text{ lb.}}{69{,}187 \text{ lb.}} \qquad \text{(Eq. 2.26)}$$
$$= 0.58$$
$$= 58\%$$

Applying an 85% safety factor, the maximum grade is:

$$\text{Maximum Grade (\%)} = 0.85 \times 58\%$$
$$= 49\%$$

**Example 2-43:** Determine the usable rimpull of a Caterpillar 621E tractor-scraper traveling over loose dry sand while: a) empty and b) loaded with a 48,000-pound payload. Also, determine the steepest grade the unit can ascend in each scenario. Assume the following load data:

Operating weight (empty) = 67,195 lb.
Operating weight (loaded) = 115, 195 lb.
Weight distribution over drive wheels:
  Empty = 68%
  Loaded = 55%

**Solution:** In either case, the rolling resistance factor is 200 pounds per ton (Table 2-1) and the coefficient of traction is 0.20 (Table 2-6). The tractor-scraper weighs 67,195 pounds (33.6 tons) empty and 68% of this load is concentrated over the drive wheels. The unit weighs 115,195 pounds (57.6 tons) loaded and 55% of the load is distributed over the drive wheels.

a) While empty, the maximum usable pull is:

$$\text{Maximum Usable Pull (lb.)} = 0.20 \times 0.68 \times 67{,}195 \text{ lb.} \qquad \text{(Eq. 2.32; Table 2-6)}$$
$$= 9139 \text{ pounds}$$

The rolling resistance is:

$$\text{Rolling Resistance (lb.)} = 200 \text{ lb./Ton} \times 33.6 \text{ Tons} \qquad \text{(Eq. 2.3)}$$
$$= 6720 \text{ pounds}$$

The maximum grade that the vehicle can climb before slippage occurs is:

$$\text{Maximum Grade (\%)} = \frac{9139 \text{ lb.} - 6720 \text{ lb.}}{67{,}195 \text{ lb.}} \qquad \text{(Eq 2.26)}$$
$$= 0.036$$
$$= 3.6\%$$

b) While loaded, the maximum usable pull is:

$$\text{Maximum Usable Pull (lb.)} = 0.20 \times 0.55 \times 115{,}195 \text{ lb.} \qquad \text{(Eq. 2.32)}$$
$$= 12{,}671 \text{ pounds}$$

The rolling resistance is:

$$\text{Rolling Resistance (lb.)} = 200 \text{ lb./Ton} \times 57.6 \text{ Tons} \qquad \text{(Eq. 2.3)}$$
$$= 11{,}520 \text{ pounds}$$

The maximum grade the vehicle can climb is:

$$\text{Maximum Grade (\%)} = \frac{12{,}671 \text{ lb.} - 11{,}520 \text{ lb.}}{115{,}195 \text{ lb.}} \qquad \text{(Eq. 2.26)}$$
$$= 0.01$$
$$= 1\%$$

Note that altitude is not a consideration when determining maximum usable pull.

## LOAD-BEARING LIMITATIONS OF THE TRAVEL SURFACE

If the travel surface cannot support a vehicle, the vehicle will "bog down," regardless of the power produced by the engine. The pressure imposed on the ground surface is measured in pounds per square inch (psi), pounds per square foot (psf), or tons per square foot (tsf). The load-bearing capabilities (soil bearing powers) of typical ground surfaces are given in Table 2-7.

**Table 2-7: Soil Bearing Powers**

| Material | Bearing Power | |
|---|---|---|
| | lb./Sq. In. | U.S. Tons/Sq. Ft. |
| Rock (semi-shattered) | 70 | 5 |
| Rock (solid) | 350 | 24 |
| Clay, dry | 55 | 4 |
| medium dry | 27 | 2 |
| soft | 14 | 1 |
| Gravel, cemented | 110 | 8 |
| Sand, compact dry | 55 | 4 |
| clean dry | 27 | 2 |
| Quicksand and alluvial soil | 7 | 0.5 |

(*Courtesy of Caterpillar Inc.*)

The pressure imposed on a work surface can be determined by:

$$\text{Ground Contact Pressure (psi)} = \frac{\text{Load Over Wheel or Track (lb.)}}{\text{Ground Contact Area (Sq. In.)}}$$

**(Equation 2.34)**

Ground contact pressure given in terms of psi can be converted into terms of psf as follows:

$$\text{Ground Contact Pressure (psf)} = \text{Ground Contact Pressure (psi)} \times 144 \text{ Sq.In./Sq.Ft.}$$

**(Equation 2.35)**

Ground contact pressure can also be converted into terms of tons per square foot as follows:

$$\text{Ground Contact Pressure (tsf)} = \frac{\text{Ground Contact Pressure (psf)}}{\text{2000 lb./Ton}}$$

**(Equation 2.36)**

**Example 2-44:** Determine the ground contact pressure exerted by a 100,000-pound excavator equipped with a track shoe whose total ground contact area is 10,420 square inches.
**Solution:** The ground contact pressure is:

$$\text{Ground Contact Pressure (psi)} = \frac{100{,}000 \text{ lb.}}{10{,}420 \text{ Sq. In.}} \qquad \text{(Eq. 2.34)}$$

$$= 9.6 \text{ psi}$$

Referring to Table 2-7, all soils except quicksand and alluvial soil are capable of bearing the pressure exerted by the excavator.

The ground contact pressure of a wheel-mounted vehicle can be changed by varying the size of tire and the inflation pressure. When determining tire load, we can assume that 10% of the wheel load is carried by the sidewall of the tires and 90% is carried by the portion of the tire in contact with the travel surface. Thus, the ground contact area for a tire is:

$$\text{Ground Contact Area (Sq. In.)} = \frac{0.90 \text{ x Wheel Load (lb.)}}{\text{Inflation Pressure (psi)}} \qquad \textbf{(Equation 2.37)}$$

Before the ground can support a given load, the ground contact area of the tires must be greater than, or equal to the required bearing area of the ground. The required bearing area can be determined by:

$$\text{Required Bearing Area} = \frac{0.90 \text{ x Wheel Load}}{\text{Ground Bearing Capacity}} \qquad \textbf{(Equation 2.38)}$$

We can simplify the process of determining whether the ground surface can support a vehicle, by mathematically expressing the statement given in previous the paragraph. The ground will support a load, provided:

$$\text{Ground Contact Area} \geq \text{Required Bearing Area} \qquad \textbf{(Equation 2.39)}$$

where: $\geq$ means "greater than or equal to."

By applying this to Equations 2.37 and 2.38, we can say that the ground will support a wheel-mounted vehicle, provided:

$$\text{Inflation Pressure} \leq \text{Ground Bearing Capacity} \qquad \textbf{(Equation 2.40)}$$

where: $\leq$ means "less than or equal to."

**Example 2-45:** Determine the ground contact area and bearing area for the loaded Caterpillar 621E tractor-scraper given in Example 2-43. Assume that the travel surface is dry clay and that the tire inflation pressure is 50 psi.
**Solution:** The total gross vehicle weight is:

$$\text{G.V.W.} = 115{,}195 \text{ pounds} \qquad \text{(Example 2-43)}$$

The maximum wheel load will be exerted over the wheels carrying the heaviest portion of the load. The heaviest load is concentrated over the drive wheels, where the load imposed on each wheel is:

$$\text{Wheel Load} = \frac{0.55 \text{ x } 115{,}195 \text{ lb.}}{2} \qquad \text{(Example 2-43)}$$
$$= 31{,}679 \text{ pounds}$$

The ground contact area under each wheel is:

$$\text{Ground Contact Area} = \frac{0.90 \text{ x } 31{,}679 \text{ lb.}}{50 \text{ psi}} \qquad \text{(Eq. 2.37)}$$
$$= 570 \text{ square inches}$$

Referring to Table 2-7, the ground bearing capacity is 55 psi. Thus, the required bearing area is:

$$\text{Required Bearing Area} = \frac{0.90 \text{ x } 31{,}679 \text{ lb.}}{55 \text{ psi}} \qquad \text{(Eq. 2.38)}$$
$$= 518 \text{ square inches}$$

Since the ground contact area is greater than the required bearing area, the conditions given in Equation 2.39 are satisfied, and thus, the travel surface will support the scraper. This is no surprise since Equation 2.40 indicates that the given surface will support the vehicle as long as we do not inflate the tires beyond 55 psi.

## *Chapter 3*

# LOADERS

A loader is a tractor equipped with a bucket attached to the front of the machine. A loader is also referred to as a front-end loader, scoop loader, bucket loader or tractor shovel.

Loaders are very versatile machines. Although they are most often used for moving loose stockpiled material, they can also be used for levelling and stripping topsoil, cleanup work, backfilling (Photo 3-6), spreading fill (Photo 3-5), general excavating, demolition (Photo 3-1) and rock loading (Photo 3-2). They are also used for transporting heavy tools, construction materials and demolished debris (Photos 3-3 and 3-4).

**Photo 3-1. Track loader performing demolition.** (*Courtesy of Caterpillar Inc.*)

Spoil can be removed from trench areas and fill can be spread by back-dragging with the bucket (Photo 3-5). When working around a trench, always approach the trench at a 45-degree angle instead of parallel to the trench to avoid a cave-in (Photo 3-6). Back-dragging can also be used for finish grading. Partially loading the bucket will add weight to facilitate any back-dragging procedure; however, if the drive wheels of the loader are located at the rear, a loaded bucket transfers weight to the front of the machine, resulting in reduced traction.

Loaders are available with rubber tires, or crawler tracks. Wheel loaders can travel faster than track loaders and are best suited for projects where extensive travel is required to and from, and around the job site, or where track loaders are prohibited, such as over a paved road surface (Photo 3-7). However, tires require a firm travel surface and they are poorly suited for working over sharp-edged rock, especially if the travel surface is wet. You can help protect tires by covering them with protective mesh chains (Photo 1-2). Wheel loaders can travel along side slopes with grades up to 15%, and they can climb slopes with grades up to 30%.

(top photo)
**Photo 3-2. Rock loading.**
(*Courtesy of Caterpillar Inc.*)

(middle photo)
**Photo 3-3. Loader carrying tree stump.**
(*Photo by Dan Atcheson*)

(bottom photo)
**Photo 3-4. Loader transporting demolition debris.**
(*Photo by Dan Atcheson*)

**Photo 3-5. Back-dragging with a loader bucket.** (*Photo by Dan Atcheson*)

**Photo 3-6. Loader used for backfilling.** (*Photo by Dan Atcheson*)

**Photo 3-7. Articulated wheel loader.** (*Courtesy of Caterpillar Inc.*)

Track loaders are best suited for projects where good traction and low ground pressures are required, and where the working surface is abrasive enough to cause excessive tire wear (Photo 3-8). However, a track loader must be transported on a trailer if the travel distance is substantial, or if crawler tracks are prohibited over the travel surface. Track loaders can travel along side slopes with grades up to 35%, and they can climb slopes with grades up to 60%.

**Photo 3-8. Track loader loading a truck.** (*Courtesy of Caterpillar Inc.*)

As a general rule, loaders are most effective when the haul distance is less than 600 feet. This distance of loader operation is referred to as the *economic application zone* (Figure 3-1). The economic application zone is dependent upon underfoot conditions, grades, material type, operator skill and the size of loader. Provided that the material to be hauled is loose, a wheel loader is generally more efficient than either a wheel or track dozer when the haul distance exceeds 400 feet.

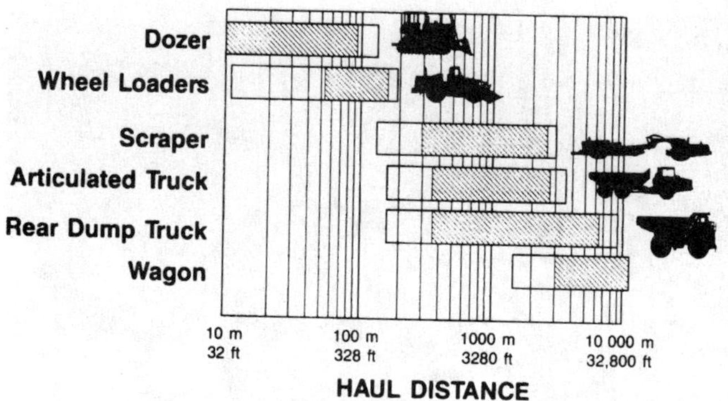


**Figure 3-1. General haul distances for mobile systems.**
(*Courtesy of Caterpillar Inc.*)

Loaders are available with a rigid frame or articulated frame. All track loaders are mounted on a rigid frame; however, some wheel loaders have a rigid frame and are steered from the rear axle. An articulated loader frame is hinged between the front and rear axles (Figure 3-2) (Photo 3-7). This allows greater maneuverability and provides a shorter turning radius than does a rigid frame.

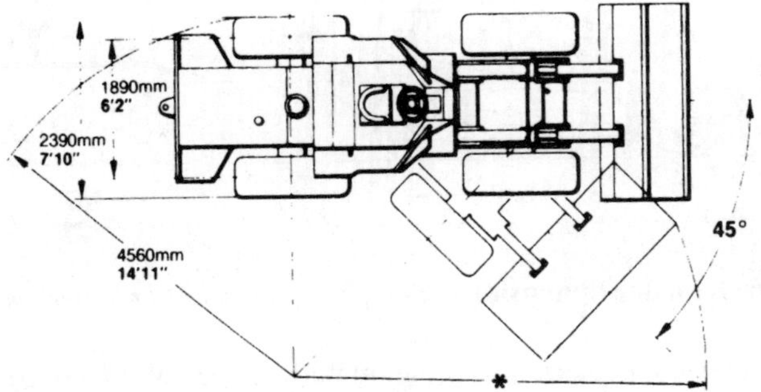


**Figure 3-2. Articulated wheel loader.**
*(Courtesy of Fiat-Allis North America, Inc.)*

Machine shipping dimensions, digging ranges and clearances are normally published with manufacturer's specifications for any given loader (Figures 3-3 and 3-4). Of the dimensions given, the most important with regard to job planning and loader selection include the maximum dump clearance (Dimension E in Figure 3-3), hinge pin height at full lift (F) and reach at full lift (N), because this dictates the size of truck that can be serviced by the loader. The maximum dump height is defined as the vertical distance from the ground to the lowest point of the cutting edge, with the bucket hinge pin at maximum height and the bucket at a 45-degree angle (Figures 3-3 and 3-4).

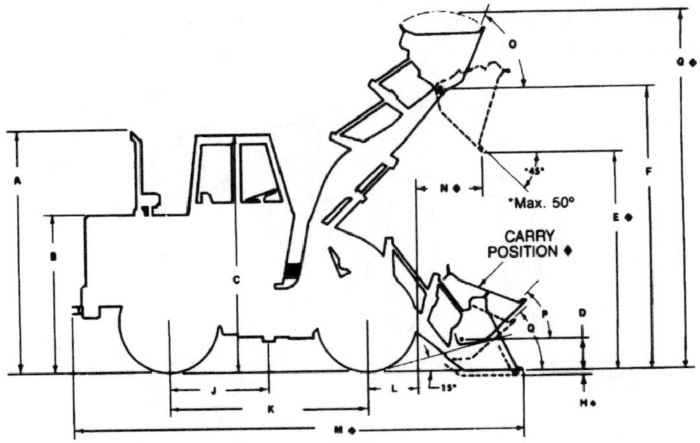


**Figure 3-3. Wheel loader dimensions.** *(Courtesy of Caterpillar Inc.)*

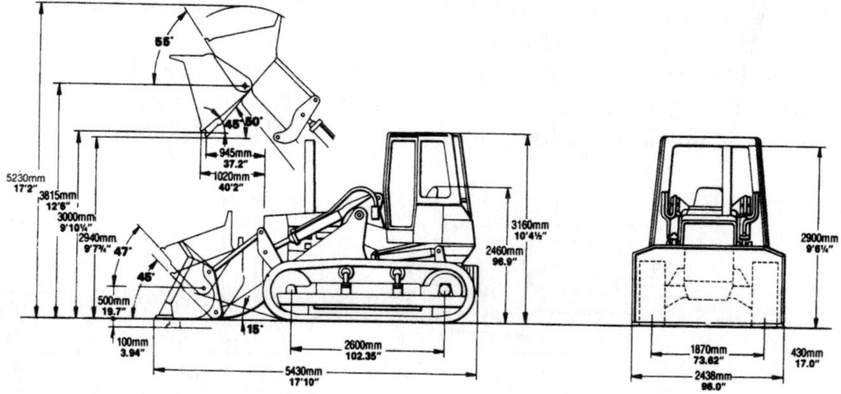


**Figure 3-4. Track loader dimensions.** (*Courtesy of Fiat-Allis North America, Inc.*)

Two basic loader buckets are available, including the solid (scoop) bucket (Figures 3-3 and 3-4) and the multi-segment (multi-purpose) bucket (Figure 3-5). The multi-segment bucket can be opened and the machine used for dozing as well as loading (Figure 3-5). Buckets are available with bolt-on edges, bolt-on teeth, or weld-on flush-mounted teeth. A reinforced spade nose bucket is also available for loading rock. Lightweight buckets are used for loading humus, saw-dust, or snow. Slatted buckets are used for handling loose rock or wood. The slats allow dirt to fall through the open slats. Side-dump buckets can dump forward, or to the side by using a cross ram (Photo 3-9). One end of the bucket is equipped with a chute to facilitate side dumping. An ejector bucket has a back wall that moves forward to completely clear the bucket as it unloads. A demolition (clamp) bucket resembles a clam shell and has an upper jaw that is controlled indepen-dently with hydraulic cylinders (Photo 3-10).

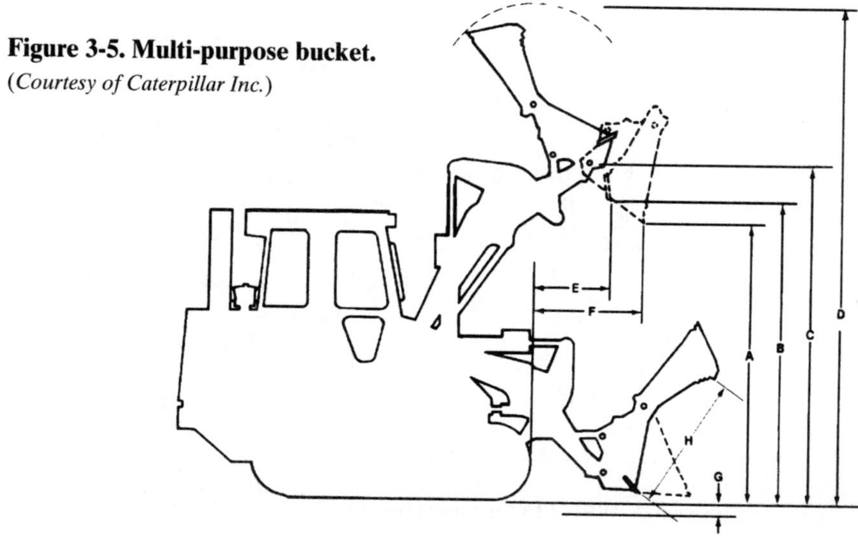


**Figure 3-5. Multi-purpose bucket.**
(*Courtesy of Caterpillar Inc.*)

**Photo 3-9.**
**Loader with side-dump bucket.**
(*Courtesy of Caterpillar Inc.*)

**Photo 3-10. Demolition bucket.**
(*Courtesy of Caterpillar Inc.*)

The *payload* of a loader bucket is rated in loose cubic yards of heaped capacity, which is the struck volume plus the volume above the strike-off plane parallel to the ground, having an angle of repose of 2 to 1 (Figure 3-6). Since different types of soil fill the bucket in varying amounts, the actual bucket payload is expressed as a percentage of heaped capacity. The percentage of heaped capacity is referred to as the *bucket fill factor,* or *bucket efficiency factor*. This factor varies, depending on the soil type, or the condition (size) of rock being loaded. Thus, the bucket payload can be determined by:

Bucket Payload (LCY) = Heaped Volume (LCY) x Bucket Fill Factor
**(Equation 3.1)**

where: Bucket fill factor is taken from Table 3-1.

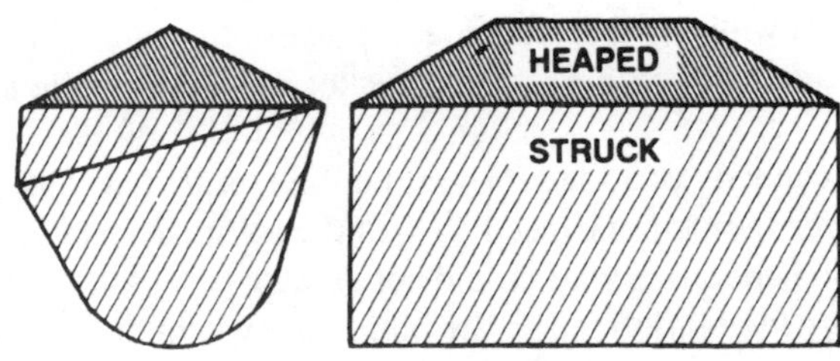


**Figure 3-6. Loader bucket ratings.** (*Courtesy of Caterpillar Inc.*)

Bucket fill factors are also affected by bucket penetration, breakout force, rollback angle, bucket profile and ground-engaging tools such as bucket teeth and cutting edges. Bucket penetration depends on the type of bucket, engine power, traction, soil type and operator skill. Bucket fill factors for loaders are given in Table 3-1.

**Table 3-1: Loader Bucket Fill Factors**

| Material | Fill Factor (% of Heaped Bucket Capacity) | |
|---|---|---|
| | Wheel Loader | Track Loader |
| Loose material | | |
| Mixed moist aggregates | 95-100 | 95-110 |
| Uniform aggregates | 95-100 | 95-110 |
| 1/8" - 3/8" | 90-95 | 90-110 |
| 1/2" - 3/4" | 85-90 | 90-110 |
| 1" and greater | 85-90 | 90-110 |
| Blasted rock | | |
| Well | 80-95 | 80-95 |
| Average | 75-90 | 75-90 |
| Poor | 60-75 | 60-75 |
| Other | | |
| Rock-dirt mixtures | 100-120 | 100-120 |
| Moist loam | 100-110 | 100-120 |
| Soil, boulders, roots | 80-100 | 80-100 |
| Cemented materials | 85-95 | 85-100 |

(*Courtesy of Caterpillar Inc.*)

When cleaning up a stockpile, as the size of the pile is reduced to the extent that the entire pile is moved by the bucket, the amount of material loaded into the bucket will continually decrease to a point where the final portion of material will require hand loading. Placing the stockpile against an immovable structure ("back-up") will help facilitate loading. Also, to ensure a full, even bucket load, be sure that both corners of the bucket contact the pile simultaneously (Figure 3-9).

The lifting capacity of a loader should not be exceeded. The operating load of a wheel loader should not exceed 50% of the full-turn static tipping load. For track loaders, the operating load should not exceed 35% of the static tipping load.

*Static tipping load* is defined as the minimum weight at the center of gravity of the load in the bucket which will lift the rear of the machine. Track loaders are considered to be lifted when the rear rollers are clear of the track. Wheel loaders are considered to be lifted when the rear wheels are clear of the ground while the machine is in full articulated position. Static tipping load is determined under the following conditions:

a) The loader is in a stationary position over a hard, level surface.
b) The loader is at standard operating weight while outfitted with standard equipment.
c) The bucket is tilted back.
d) The load is at the maximum forward position during the raising cycle.

Keep in mind that traveling down hill shifts more weight to the front of the loader and increases the chances for the machine to tip forward. When loading a truck, never take a downhill approach, and do not load a truck with the loader on a side slope with the bucket tilted with respect to the horizontal. Also, transport a load keeping the bucket as low as possible.

The tipping load of a loader can be increased by using counterweights, or by adding other attachments such as a cab, rippers, etc. to the rear of the loader. The tipping load can also be increased by adding water, or a calcium chloride solution ballast to the tires (hydroflating).

The operating load exerted on the loader can be determined by:

Payload Weight (lb.) = Bucket Payload (LCY) x Material Weight (lb./LCY)
= Heaped Capacity (LCY) x Bucket Fill Factor
x Material Weight (lb./LCY) **(Equation 3.2)**

The working ranges of various buckets can be determined by referring to a bucket selection charts such as those shown in Figures 3-7 and 3-8.

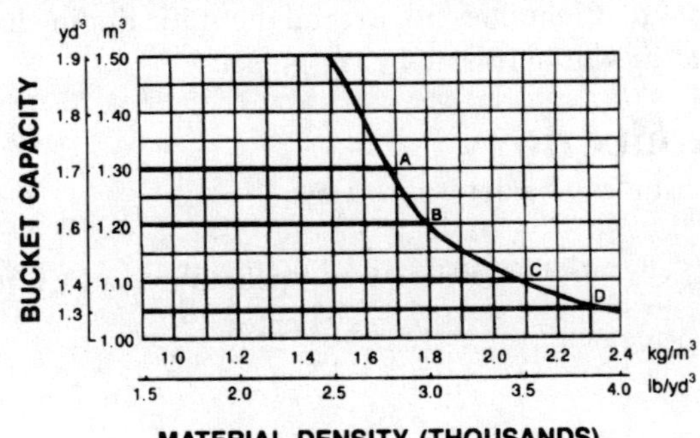


KEY

A — 1.3 $m^3$ (1.7 $yd^3$) General Purpose Bucket, bolt-on edge
B — 1.2 $m^3$ (1.6 $yd^3$) General Purpose Bucket, bolt-on teeth.
1.2 $m^3$ (1.6 $yd^3$) Penetration Bucket, weld-on, flush-mounted teeth.
C — 1.1 $m^3$ (1.4 $yd^3$) General Purpose Bucket, bolt-on teeth and segments.
1.1 $m^3$ (1.4 $yd^3$) General Bucket, bolt-on edge.
D — 1.0 $m^3$ (1.3 $yd^3$) General Purpose Bucket, bolt-on teeth.

**Figure 3-7. 910E bucket working ranges.**(*Courtesy of Caterpillar Inc.*)

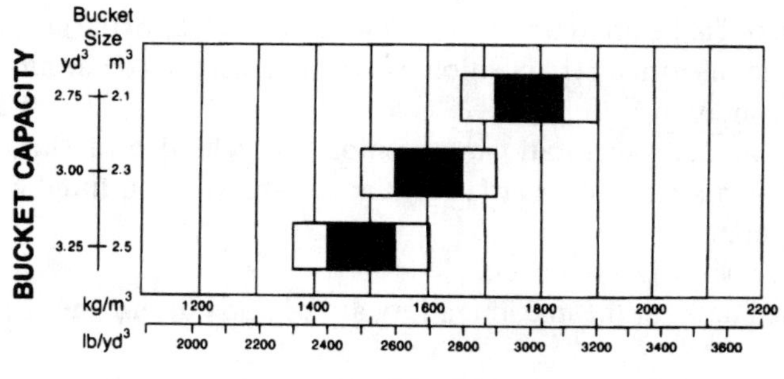


**Figure 3-8. 936F bucket working ranges.** *(Courtesy of Caterpillar Inc.)*

**Example 3-1:** Assuming a static tipping load of 17,319 pounds and the use of a 3-C.Y. heaped-capacity bucket, determine the weight of the operating load exerted on a wheel loader while loading moist loam. Also, determine if the loader is operating within its safe lifting capacity.

**Solution:** The payload weight is:

Payload Weight (lb.) = 3 LCY x 1.10 x 2100 lb./LCY (Tables 3-1 and 1-1)
= 6930 pounds

Fifty percent of the static tipping load is:

50% of Static Tipping Load = 0.50 x 17,319 lb.
= 8660 pounds

The given payload weight does not exceed the static tipping load; therefore, the loader is operating within its safe lifting capacity.

## LOADER PRODUCTION

Loader production can be determined by:

Production (LCY/Hr.) = Cycles/Hr. x Bucket Payload/Cycle x Efficiency Factor **(Equation 3.3)**

where:

$$\text{Cycles/Hour} = \frac{\text{3600 Sec./Hr.}}{\text{Cycle Time (Sec.)}}$$ **(Equation 3.4)**

or,

$$\text{Cycles/Hour} = \frac{\text{60 Min./Hr.}}{\text{Cycle Time (Sec.)}}$$ **(Equation 3.5)**

and,

Bucket Payload (LCY) = Heaped Volume (LCY) x Bucket Fill Factor (Eq. 3.1)

where: Bucket fill factor is taken from Table 3-1.

Job efficiency is taken from data given in Table 1-4, or is estimated as productive time per clock hour. However, do not use both methods simultaneously.

**WHEEL LOADER PRODUCTION**

Cycle time for wheel loaders consists of a basic cycle time, cycle time correction factors and travel time; i.e.,

Total Cycle Time = Basic Cycle Time + Cycle Correction Factors +Travel Time **(Equation 3.6)**

The *basic cycle time* includes time required for loading, dumping, four reversals of direction, turning and a 15-foot haul/return distance while at full throttle with a competent operator (Figures 3-9). All other things being equal, the basic cycle time is primarily dependent upon the size of the loader. The basic cycle times for various sized articulated loaders under normal conditions when loading trucks are shown in Table 3-2.

**Table 3-2: Basic Cycle Times**

| Struck Capacity (C.Y.) | Basic Cycle Time (Min.) |
|---|---|
| Up to 4 | 0.45 - 0.50 |
| 4 - 6.75 | 0.50 - 0.55 |
| 6.75 - 7.1 | 0.55 - 0.60 |
| 7.1 - 11.2 | 0.60 - 0.75 |

Material type, pile characteristics and miscellaneous factors will improve or reduce loader production. Cycle time correction factors from Table 3-3 must be added to, or deducted from the basic cycle time to obtain the corrected basic cycle time.

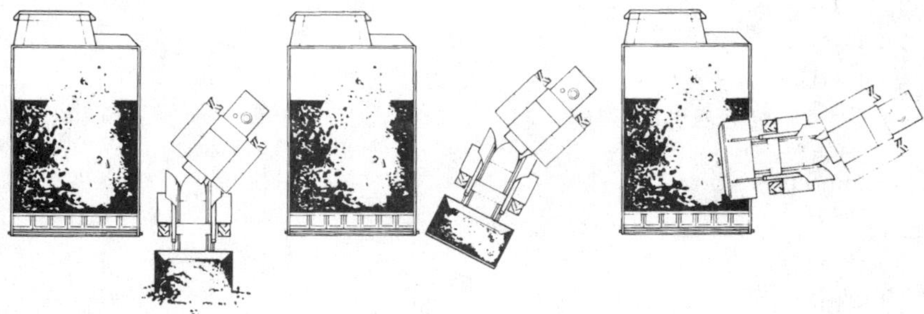

**Figure 3-9. Basic cycle time components.** (*Courtesy of Terex Corporation*)

**Table 3-3: Cycle Time Correction Factors**

| Job Conditions | Minutes added (+), or Subtracted (–) From Basic Cycle Time |
|---|---|
| Materials | |
| Mixed | + 0.02 |
| Up to 1/8" | + 0.02 |
| 1/8" to 3/4" | – 0.02 |
| 3/4" to 6" | 0.00 |
| 6" and over | + 0.03 and up |
| Bank, or broken | + 0.04 and up |
| Pile | |
| Conveyor-, or dozer-piled, 10' and up | 0.00 |
| Conveyor-, or dozer-piled, 10' or less | + 0.01 |
| Dumped by truck | + 0.02 |
| Miscellaneous | |
| Common ownership of trucks and loaders | Up to – 0.04 |
| Independently-owned trucks | Up to + 0.04 |
| Constant operation | Up to – 0.04 |
| Inconsistent operation | Up to + 0.04 |
| Small target | Up to + 0.04 |
| Fragile target | Up to + 0.05 |

(*Courtesy of Caterpillar Inc.*)

Truck-loading production can be increased by approaching the stockpile and the truck with the bucket at right angles, and not at an oblique angle (Figure 3-9). Also, take care to load the truck evenly, to prevent the truck from overturning while hauling the load. Because of the height and reach limitations of the loader, the pile will be higher on the side of the truck bed from which it is being loaded, unless the bucket pushes a portion of the pile across to the opposite side.

Rock or concrete loading time can be reduced by spreading out the stockpile to keep it low, and keeping the pieces small. Damage to the truck bed can be reduced by dumping soil into the bottom of the bed prior to loading rock or concrete.

When hauling is required, the loader travel time can be obtained from a travel time chart such as that shown in Figure 3-10. Keep in mind that maneuver, load and dump times must be added to the travel time to obtain the total cycle time. Travel time and spillage can be reduced by keeping the travel surface smooth. The loader can be used for travel-surface maintenance by lowering the bucket to the surface to cut high spots and fill low areas. The high spots can be cut during a forward pass, and low areas filled by back-dragging.

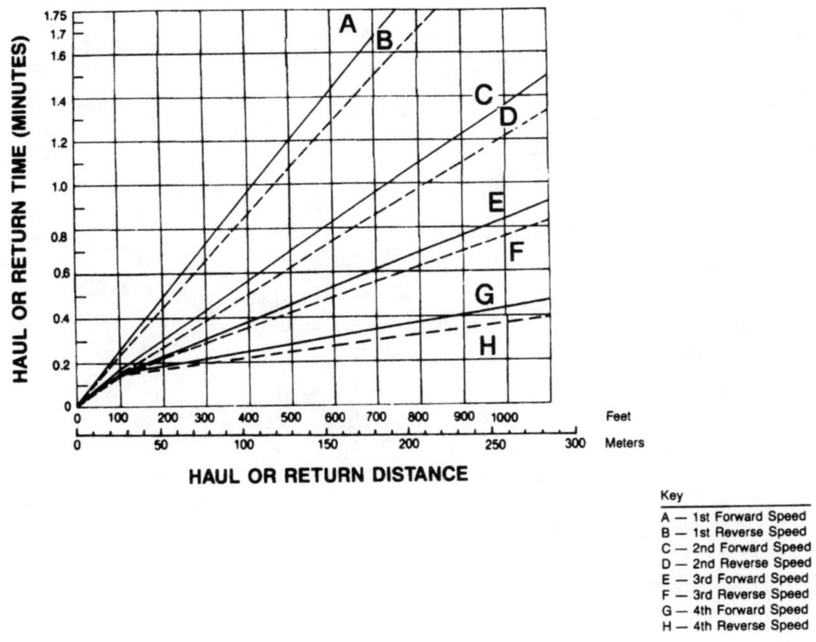


**Figure 3-10. 936F wheel loader travel time chart.**
*(Courtesy of Caterpillar Inc.)*

Referring to Table 3-2, we can see that the basic cycle times for loaders within given size ranges do not change appreciably; therefore, production can be optimized by using the largest loader available within a given cycle-time range. However, if a loader is dumping into trucks, the trucks should have a capacity of at least 4 times the loader bucket capacity. Also, try to use trucks that hold a whole number of bucket loads; otherwise, one of the bucket loads will be only partially loaded.

**Example 3-2:** Determine the hourly production rate of a Caterpillar 936F wheel loader working under the following conditions:

| | |
|---|---|
| Material | = 3/8" gravel |
| Pile | = Placed by dozer; pile height = 8 feet |
| Trucks | = Small; independently owned |
| Operation | = Constant |
| Travel | = 400 feet, each way (Assume 3rd gear forward on haul and return) |
| Efficiency | = 50-minute hour |
| Bucket | = 3 LCY (heaped capacity) |

**Solution:** The bucket payload is:

Bucket Payload (LCY) = 3 LCY x 0.95 (Eq. 3.1; Table 3-1)
= 2.85 LCY

The basic cycle time is:

Basic Cycle Time = 0.50 minutes (Table 3-2)

The cycle time correction factors are:

| | | |
|---|---|---|
| Material | = –0.02 Min. | (Table 3-3) |
| Pile | = +0.01 Min. | |
| Independently Owned Trucks | = +0.04 Min. | |
| Constant Operation | = –0.04 Min. | |
| Small Targets | = +0.04 Min. | |
| Total | = 0.03 Min. | |

The corrected basic cycle time is:

Corrected Basic Cycle Time = 0.50 Min. + 0.03 Min.
= 0.53 minutes

The round-trip travel time per cycle is:

Travel Time = 2 x 0.38 Min. (Figure 3-10)
= 0.76 minutes

Including travel time, the total cycle time is:

Total Cycle Time = 0.53 Min. + 0.76 Min. (Eq. 3.6)
= 1.29 minutes

At 100% efficiency, the number of cycles per hour is:

$$\text{Cycles/Hour} = \frac{\text{60 Min./Hr.}}{\text{1.29 Min./Cycle}} \qquad \text{(Eq. 3.5)}$$
$$= \text{47 cycles/hour}$$

Thus, the probable production rate is:

$$\text{Production (LCY/Hr.)} = \text{47 Cycles/Hr. x 2.85 LCY/Cycle x 50/60} \qquad \text{(Eq. 3.3)}$$
$$= \text{112 LCY/hour}$$

**Example 3-3:** Assuming 100% efficiency, work the previous example excluding travel time.
**Solution:** Excluding travel time, the corrected basic cycle time is:

Corrected Basic Cycle Time = 0.53 minutes (Example 3-2)

The bucket payload is:

Bucket Payload (LCY) = 2.85 LCY (Example 3-2)

At 100% efficiency, the number of cycles per hour is:

$$\text{Cycles/Hour} = \frac{\text{60 Min./Hr.}}{\text{0.53 Min./Cycle}} \qquad \text{(Eq. 3.5)}$$
$$= \text{113 cycles/hour}$$

At 100 % efficiency, the probable production rate is:

$$\text{Production (LCY/Hr.)} = \text{113 Cycles/Hr. x 2.85 LCY/Cycle x 1.00} \qquad \text{(Eq. 3.3)}$$
$$= \text{322 LCY/hour}$$

Some manufacturers express wheel loader production rates in terms of bank cubic yards per hour (Table 3-4); however, since a loader normally excavates loose material, we must use swell or shrinkage factors to interface bank-yard production with loose-yard production.

## Table 3-4: Wheel Loader Production (BCY/60-Minute Hour)

| Bucket Size (m³ or yd³) | | 1.0 | 1.5 | 2.0 | 2.5 | 3.0 | 3.5 | 4.0 | 4.5 | 5.0 | 5.5 | 6.0 | 6.5 | 7.0 | 7.5 | 8.0 | 8.5 | 9.0 | 9.5 | 10.0 |
|---|---|---|---|---|---|---|---|---|---|---|---|---|---|---|---|---|---|---|---|---|
| Cycle Time | Cycles Per Hr | Unshaded area indicates average production. | | | | | | | | | | | | | | | | | | |
| .35 | 171 | | | | | | | | | | | | | | | | | | | |
| .40 | 150 | 150 | 225 | 330 | 375 | 450 | 525 | | | | | | | | | | | | | |
| .45 | 133 | 133 | 200 | 268 | 332 | 400 | 466 | 530 | 600 | 665 | 730 | 800 | 865 | | | | | | | |
| .50 | 120 | 120 | 180 | 240 | 300 | 360 | 420 | 480 | 540 | 600 | 660 | 720 | 780 | 840 | 900 | 960 | 1003 | 1080 | 1140 | 1200 |
| .55 | 109 | 109 | 164 | 218 | 272 | 328 | 382 | 436 | 490 | 545 | 600 | 655 | 705 | 765 | 820 | 870 | 925 | 980 | 1008 | 1090 |
| .60 | 100 | 100 | 150 | 200 | 250 | 300 | 350 | 400 | 450 | 500 | 550 | 600 | 650 | 700 | 750 | 800 | 850 | 900 | 950 | 1000 |
| .65 | 92 | 92 | 138 | 184 | 230 | 276 | 322 | 368 | 416 | 460 | 505 | 555 | 600 | 645 | 690 | 735 | 780 | 830 | 875 | 920 |
| .70 | 86 | | | | | | | 342 | 386 | 430 | 474 | 515 | 560 | 600 | 645 | 690 | 730 | 775 | 815 | 860 |
| .75 | 80 | | | | | | | | | | | | | 560 | 600 | 640 | 680 | 720 | 760 | 800 |

*(Courtesy of Caterpillar Inc.)*

**Example 3-4:** Assuming 15% swell, and excluding travel time, use data from Table 3-4 to determine the production rate of the loader given in the previous example.
**Solution:** Excluding travel time, the corrected basic cycle time is:

Corrected Basic Cycle Time = 0.53 minutes (Example 3-3)

The bucket payload expressed in bank cubic yards is:

$$\text{Bucket Size} = \frac{2.85 \text{ LCY}}{1.15 \text{ LCY/BCY}} = 2.48 \text{ C.Y.} \qquad \text{(Example 3-3; 15\% swell)}$$

Referring to Table 3-4 while using a 2.5-C.Y. bucket and a 0.50- to 0.55-minute cycle time, we find that the probable production rate lies between 272 and 300 bank cubic yards per hour. Assuming a production rate of 282 bank cubic yards per hour, the loose-yard production rate is:

$$\text{Production (LCY/Hr.)} = 282 \text{ BCY/Hr.} \times 1.15 \text{ LCY/BCY} = 324 \text{ LCY/hour} \qquad \text{(15\% swell)}$$

**Example 3-5:** Work Example 3-2 assuming the loader is operating at an altitude of 15,000 feet above sea level.
**Solution:** Referring to Table 2-4, we can see that a 936F loader operates at only 83% efficiency at the given altitude; therefore, the probable production rate is:

$$\text{Production (15,000')} = 112 \text{ LCY/Hr.} \times 0.83 = 93 \text{ LCY/hour} \qquad \text{(Example 3-2; Table 2-4)}$$

## TRACK LOADER PRODUCTION

The production rates of track loaders is determined in a manner similar to that used for wheel loaders, except that specific cycle-time components are used in lieu of cycle correction factors. Excluding travel time, the *average basic cycle time* for a track loader ranges from 0.25 to 0.35 minutes.

The *load time* of a track loader is dependent primarily on the type of material being loaded, as shown in Table 3-5.

**Table 3-5: Load Time Components for Track Loaders**

| Material | Time (Min.) |
|---|---|
| Uniform Aggregates | 0.03 - 0.05 |
| Moist Mixed Aggregates | 0.03 - 0.06 |
| Moist Loam | 0.03 - 0.07 |
| Soil, Boulders, Roots | 0.04 - 0.20 |
| Cemented Materials | 0.05 - 0.20 |

(*Courtesy of Caterpillar Inc.*)

*Maneuver time* averages 0.20 minutes and includes time required for four reversals in direction, turning, and minimum travel while at full throttle with a competent operator (Figure 3-9).

*Dump time* varies from 0.02 to 0.10 minutes, depending on the size and strength of the dump target. Typical dump times into trucks vary from 0.04 to 0.07 minutes.

When hauling is required, the travel time can be obtained from a travel time chart such as that shown in Figure 3-11.

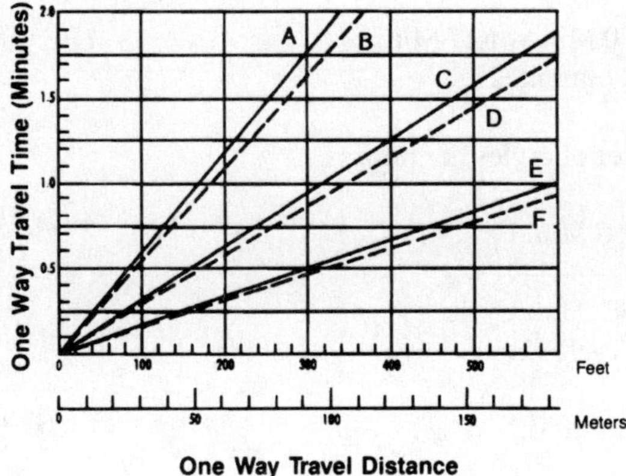


KEY
A — 1st forward speed
B — 1st reverse speed
C — 2nd forward speed
D — 2nd reverse speed
E — 3rd forward speed
F — 3rd reverse speed

**Figure 3-11.**
**935C Series II track loader travel time chart.**
(*Courtesy of Caterpillar Inc.*)

**Example 3-6:** Determine the hourly production rate of a Caterpillar 935C track loader working under the following conditions:

| | |
|---|---|
| Material | = Moist loam |
| Bucket Payload | = 3 LCY |
| Dump Targets | = Trucks |
| Travel | = 200 feet each way (Assume 2nd gear forward during haul and 2nd gear reverse during return) |
| Efficiency | = 55-minute hour |

**Solution:** The bucket payload is:

Bucket Payload (LCY) = 3 LCY x 1.10 (Eq. 3.1; Table 3-1)
= 3.3 LCY

The basic cycle time is:

| | | |
|---|---|---|
| Load time | = 0.05 Min. | (Table 3-5) |
| Maneuver time | = 0.20 Min. | (see text) |
| Dump time | = 0.05 Min. | (see text) |
| Total | = 0.30 Min. | |

The travel time is:

| | | |
|---|---|---|
| Haul time | = 0.62 Min. | (Figure 3-11) |
| Return time | = 0.55 Min. | |
| Total | = 1.17 Min. | |

Including travel time, the total cycle time is:

Total Cycle Time = 0.30 Min. + 1.17 Min. (Eq. 3.6)
= 1.47 minutes

At 100% efficiency, the number of cycles per hour is:

$$\text{Cycles/Hour} = \frac{60 \text{ Min./Hr.}}{1.47 \text{ Min./Cycle}} \quad \text{(Eq. 3.5)}$$

= 41 cycles/hour

Thus, the probable production rate is:

Production (LCY/Hr.) = 41 Cycles/Hr. x 3.3 LCY/Cycle x 55/60 (Eq. 3.3)
= 124 LCY/hour

**Example 3-7:** Assuming 100% efficiency, work the previous example excluding travel time.
**Solution:** Excluding travel time, the basic cycle time is:

Basic Cycle Time = 0.3 minutes (Example 3-6)

The bucket payload is:

Bucket Payload (LCY) = 3.3 LCY (Example 3-6)

At 100% efficiency, the number of cycles per hour is:

$$\text{Cycles/Hour} = \frac{\text{60 Min./Hr.}}{\text{0.3 Min./Cycle}} \quad \text{(Eq. 3.5)}$$
$$= \text{200 cycles/hour}$$

At 100% efficiency, the probable production rate is:

$$\text{Production (LCY/Hr.)} = \text{200 Cycles/Hr. x 3.3 LCY/Cycle x 1.00} \quad \text{(Eq. 3.3)}$$
$$= \text{660 LCY/hour}$$

Some manufacturers express track loader production rates in terms of bank cubic yards per hour (Table 3-6); however, since a loader normally excavates loose material, we must use swell or shrinkage factors to interface bank-yard production with loose-yard production.

## Table 3-6: Track Loader Production (BCY / 60-Minute Hour)

| Bucket Size ($m^3$ or $yd^3$ | | 1.0 | 1.5 | 2.0 | 2.5 | 3.0 | 3.5 | 4.0 | 4.5 | 5.0 |
|---|---|---|---|---|---|---|---|---|---|---|
| Cycle Time Hundredths of a minute | Cycles Per Hr | Unshaded area indicates average work range | | | | | | | | |
| 0.25 | 240 | 240 | 360 | 480 | 600 | 720 | 840 | 960 | | |
| 0.30 | 200 | 200 | 300 | 400 | 500 | 600 | 700 | 800 | | |
| 0.35 | 171 | 171 | 257 | 342 | 428 | 513 | 599 | 684 | 769 | |
| 0.40 | 150 | 150 | 225 | 300 | 375 | 450 | 525 | 600 | 675 | 750 |
| 0.45 | 133 | 133 | 200 | 268 | 332 | 400 | 466 | 530 | 600 | 665 |
| 0.50 | 120 | 120 | 180 | 240 | 300 | 360 | 420 | 480 | 540 | 600 |
| 0.55 | 109 | 109 | 164 | 218 | 272 | 328 | 382 | 436 | 490 | 545 |
| 0.60 | 100 | 100 | 150 | 200 | 250 | 300 | 350 | 400 | 450 | 600 |
| 0.65 | 92 | | | | | | | 368 | 416 | 460 |

*(Courtesy of Caterpillar Inc.)*

**Example 3-8:** Assuming 25% swell and excluding travel time, use data from Table 3-6 to determine the production rate of the loader given in the previous example.
**Solution:** Excluding travel time, the basic cycle time is:

Basic Cycle Time = 0.3 Minutes (Example 3-7)

Expressed in bank cubic yards, the bucket payload is:

$$\text{Bucket Payload (BCY)} = \frac{3.3\ \text{LCY}}{1.25\ \text{LCY/BCY}} \qquad \text{(Example 3-7; 25\% swell)}$$
$$= 2.64\ \text{BCY}$$

We will assume a bucket size of 2.5 cubic yards.

Referring to Table 3-6 while using a 2.5-C.Y. bucket and a 0.30-minute cycle time, we find that the probable production rate is 500 bank cubic yards per hour. Expressed in loose cubic yards per hour, the probable production rate is:

$$\text{Production (LCY/Hr.)} = 500\ \text{BCY/Hr.} \times 1.25\ \text{LCY/BCY} \qquad \text{(25\% swell)}$$
$$= 625\ \text{LCY/hour}$$

**Example 3-9:** Determine the time required for the track loader given in the previous example to excavate 10,000 loose cubic yards of soil.
**Solution:** The project duration is:

$$\text{Project Duration (Hr.)} = \frac{10{,}000\ \text{LCY}}{625\ \text{LCY/Hr.}} \qquad \text{(Eq. 1.17)}$$
$$= 16\ \text{hours}$$

**Example 3-10:** Assuming the track loader given in Example 3-8 costs $50.00 per hour, including operator's wages, determine the unit production cost.
**Solution:** The unit production cost is:

$$\text{Unit Production Cost} = \frac{\$50.00/\text{Hr.}}{625\ \text{LCY/Hr.}} \qquad \text{(Eq. 1.18)}$$
$$= \$0.08/\text{LCY}$$

## TRAVEL TIME FOR LARGE WHEEL LOADERS

The payloads of large wheel loaders often warrant their use over long haul distances. Travel times for some of the larger wheel loaders are determined from travel time charts such as those shown in Figures 3-12 and 3-13. Two travel time charts are given for the loader; one for the loader hauling its rated payload, and the other for the empty vehicle. Travel time charts take acceleration and deceleration of the vehicle into consideration at the loading and dumping areas.

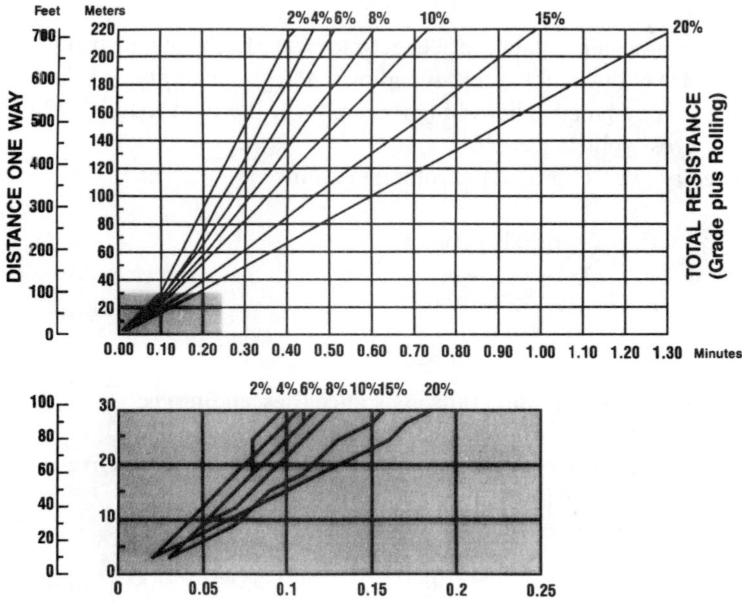


**Figure 3-12. 966F loader travel time chart (empty).** (*Courtesy of Caterpillar Inc.*)

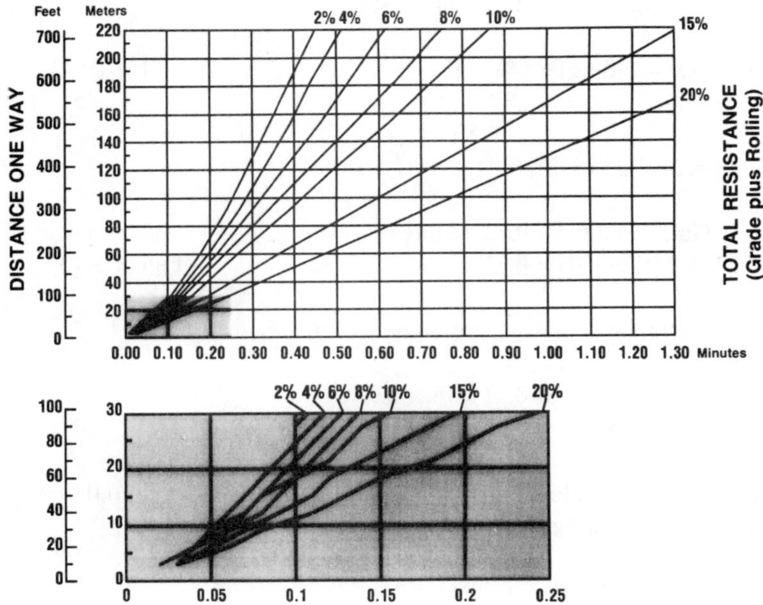


**Figure 3-13. 966F loader travel time chart (loaded).** (*Courtesy of Caterpillar Inc.*)

To use a travel time chart, enter the chart from the left and find the one-way travel distance. Project a horizontal line until it intersects the appropriate diagonal line representing total resistance (effective grade). From this intersection, drop vertically to obtain the travel time at the bottom of the chart.

**Example 3-11:** Assuming a haul distance of 500 feet each way, determine the travel time for a Caterpillar 966F wheel loader ascending a 4% grade empty and descending the same grade loaded. The rolling resistance of the haul road is 120 pounds per tons of gross vehicle weight.
**Solution:** Expressed as percent grade, the rolling resistance is:

$$\text{Rolling Resistance (\%)} = \frac{120 \text{ lb./Ton}}{20} \qquad \text{(Eq. 2.12)}$$
$$= 6\%$$

The grade resistance (ascending) and grade assistance (descending) is:

$$\text{Grade Resistance (Assistance)} = 4\% \qquad \text{(Given)}$$

While ascending, the effective grade (total resistance) is:

$$\text{Effective Grade (Ascending)} = 6\% + 4\% \qquad \text{(Eq. 2.13)}$$
$$= 10\%$$

While descending, the effective grade (total resistance) is:

$$\text{Effective Grade (Descending)} = 6\% - 4\% \qquad \text{(Eq. 2.14)}$$
$$= 2\%$$

Referring to Figures 3-12 and 3-13, the travel times are:

Travel Time (Empty) = 0.52 minutes (Figure 3-12)
Travel Time (Loaded) = 0.33 minutes (Figure 3-13)

Thus, the total travel time is:

$$\text{Total Travel Time} = 0.52 \text{ Min.} + 0.33 \text{ Min.}$$
$$= 0.85 \text{ minutes}$$

Travel time charts for track loaders do not include total resistance, and thus, are all similar to the travel time chart shown in Figure 3-11.

## MINIATURE (SKID-STEER) LOADERS

There are several miniature versions of earth-moving machines, including the skid-steer loader (Photo 3-11). These machines are designed for working in extremely confined areas. These machines can be very useful, provided that they are not overworked. They are four-wheel drive, but have relatively little power and traction, and are not as rugged as their larger counterparts; therefore, they should be confined to excavating and hauling loose material, only.

**Photo 3-11. Skid-steer loader.** (*Photo by Dan Atcheson*)

# Chapter 4

## TRUCKS AND WAGONS

Trucks (hauling units) and scrapers can both haul excavated materials, but since the scraper also excavates its hauled material, the soil must be conducive to the scraper's excavating and loading capabilities. Also, the excavation must be large enough for the scrapers to maneuver, and if a pit is being excavated, a ramp must be provided so that scrapers can enter and exit the pit. On the other hand, trucks work in conjunction with an excavator in areas inaccessible to scrapers, and trucks can haul almost any type of excavated material, including rock. Also, since the depth of cut of a scraper is limited, dissimilar layers of soil are better blended when an excavator/truck fleet is used. Larger scrapers are usually confined to off-highway travel and their payloads and top speeds are normally less than those of trucks. Thus, trucks are often used to carry heavy payloads at high speeds over great distances. Also, many trucks can be driven to a job site, while scrapers must be hauled on a trailer.

However, one disadvantage of using an excavator/haul-unit fleet, as compared to using scrapers, is that production will be slowed, if not halted, if the excavator breaks down. Also, a scraper can dump its payload in uniformly thick layers, while trucks cannot. This results in higher compaction production when scrapers are used. In addition, a scraper can perform other functions, such as haul-road maintenance, finish grading, general cleanup work and asphalt pavement removal.

### TYPES OF TRUCKS

Trucks are available in a wide variety of sizes and types, including standard and rock-type bodies, rigid and articulated bodies (Photos 4-2 and 4-3), and highway (Photo 4-2), or off-highway designs (Photo 4-1). They are manufactured with two-wheel, four-wheel, and six-wheel drives, and with one, two or three rear axles (Photos 4-1, 4-3 and 4-2). They are equipped with gasoline, diesel, butane or propane engines.

**Photo 4-1. Off-highway dump truck.**
(*Courtesy of Caterpillar Inc.*)

Trucks are manufactured with cable, or hydraulic dumping mechanisms. Trucks are normally designed to be dumped from the rear (rear-dump trucks) (Photos 4-2 and 4-3), and the rear dump body should be designed for the type of load carried. The body of a truck hauling wet, cohesive soil such as clay should not have any sharp angles or corners. Dry, non-cohesive materials such as dry sand or gravel will flow freely from any type of body. Rock should be loaded into a shallow body equipped with sloping sideboards.

**Photo 4-2. Rigid-body rear-dump truck.**
(*Photo by Dan Atcheson*)

**Photo 4-3. Articulated rear-dump truck.** (*Courtesy of Caterpillar Inc.*)

Truck shipping dimensions, including overhead clearance, are normally published with manufacturer's specifications for any given truck (Figure 4-1). Always watch for overhead obstructions such as highlines when raising the rear dump body.

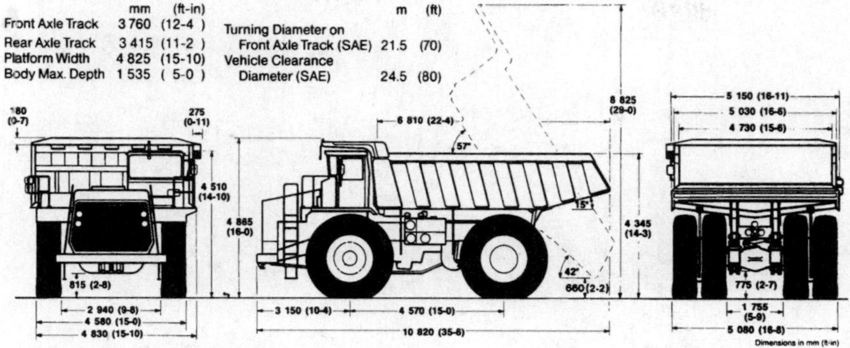


**Figure 4-1. Truck shipping dimensions.** (*Courtesy of Terex Corporation*)

Trailers pulled by tractors or truck-tractors are called *wagons* and are designed for hauling excavated material, only. Wagons are available with a variety of dumping mechanisms, including bottom-dump (Photos 4-6 and 4-7), rear-dump (Photo 4-4) and side-dump models (Photo 4-5).

Bottom-dump models are often equipped with longitudinal, or cross-flow dumping gates (switch gates). The longitudinal gates can be used when windrowing (placing longitudinal soil piles), or stockpiling is desired (Photo 4-6). Cross-flow gates are used for spreading layers of base materials and aggregates (Photo 4-7). Bottom-dump models can discharge their loads while moving, giving them a shorter dumping time than rear-dump trucks; however, since the doors have limited openings, bottom-dump units often experience difficulty when discharging wet, cohesive materials.

As a general rule, trucks are most effective when the haul distance ranges from 575 feet to 2 miles (articulated trucks), 575 feet to 6 miles (rear-dump trucks), and 4900 feet to 7 miles (wagons). These haul distances are referred to as *economic application zones* (Figure 3-1). The economic application zone is dependent upon the underfoot conditions, grades, material type, operator skill and the size and type of truck.

**Photo 4-4. Rear-dump trailer.** (*Courtesy of Athey Products Corporation*)

**Photo 4-5. Side-dump trailer.** (*Courtesy of Marathon LeTourneau Company*)

**Photo 4-6. Bottom-dump trailer with longitudinal dump gates.**
(*Courtesy of Athey Products Corporation*)

**Photo 4-7. Bottom-dump trailer with cross-flow dump gates.**
(*Courtesy of CMI Corporation*)

## PAYLOAD

*Payload* is the load that a vehicle can carry, exclusive of the vehicle weight. Payload can be expressed as weight capacity in pounds or tons, struck volume capacity in loose cubic yards, or heaped volume capacity in loose cubic yards. Regardless of the volume capacity, the rated *weight capacity* should never be exceeded. To do so will increase operating costs due to higher fuel consumption, reduced tire life, frequent failure of mechanical parts and higher maintenance costs. All of these items also equate to lost time and production while repairing and maintaining the vehicle.

*Struck capacity* of a truck is the volume of material that can be filled to the top of the sides of the rear body; therefore, the struck capacity remains constant for any given truck or wagon. Struck capacity is of value when the truck is hauling fluid materials such as concrete, or when there is a limit on the amount of spillage allowed.

*Heaped capacity* will vary, depending on the height that the soil can be extended above the sides of the rear body. The heaped capacity can be increased with the use of sideboards, provided the weight limitations are not exceeded. Heaped capacity is normally based on the heap having an angle of repose or 26.57 degrees, or a slope (run:rise) of 2 to 1 (2:1). Angles of repose for common materials are given in Table 4-1.

**Table 4-1: Slope and Angle of Repose of Various Materials**

| Material | Slope (Run:Rise) | Degrees |
|---|---|---|
| Coal, industrial | 1.4:1 - 1.3:1 | 35-38 |
| Common earth, dry | 2.8:1 - 1.0:1 | 20-45 |
| moist | 2.1:1 - 1.0:0 | 25-45 |
| wet | 2.1:1 - 1.7:1 | 25-30 |
| Gravel, round to angular | 1.7:1 - 0.9:1 | 30-50 |
| Sand, dry | 2.8:1 - 1.7:1 | 20-30 |
| moist | 1.8:1 - 1.0:1 | 30-45 |
| wet | 2.8:1 - 1.0:1 | 20-45 |

(*Courtesy of Caterpillar Inc.*)

The angle of repose for most soils under normal conditions ranges from 20 to 45 degrees; however, the actual angle of repose of a soil pile can be extremely variable. To simplify the concept of soil-pile behavior, we will turn our attention to coarse-grained aggregates; i.e., gravels and sands.

There are many forces at work when aggregate is dropped; however, the primary force which causes a soil pile to spread is gravity. The primary force responsible for preventing soil-pile spreading is friction between individual aggregates. Friction is created by surface-to-surface contact of the aggregate, and the finer the aggregate, the more surface area there will be per unit volume of aggregate. With these concepts in mind, we are ready to make some general rules regarding the behavior of soil piles:

1) Assuming similar-sized aggregate, larger aggregate will spread more than smaller aggregate, and thus, will have a lower angle of repose.
2) Assuming similar-sized aggregate, heavier aggregate will spread more than lighter aggregate.
3) Assuming similar-sized aggregate, well-rounded aggregate will spread more than angular-shaped aggregate.
4) The higher the water content of any aggregate, the more the soil pile will spread.

5) Well-graded aggregate will have a steeper angle of repose than poorly-graded aggregate.

If the angle of repose of a soil is not 2 to 1, adjustments must be made in order to determine the actual heaped capacity of a truck. Heaping occurs only above the sides of the rear body, so we can make volume adjustments in the soil above the struck volume, since the struck volume remains constant for any given truck. The heaped soil volume can be adjusted by applying the proper volume adjustment factor given in Table 4-2.

**Table 4-2: Heaped Volume Adjustment Factors**

| Slope (Run:Rise) | Volume Adjustment Factor |
|---|---|
| 3.0:1 | 0.69 |
| 2.8:1 | 0.74 |
| 2.6:1 | 0.79 |
| 2.4:1 | 0.85 |
| 2.2:1 | 0.92 |
| 2.0:1 | 1.00 |
| 1.8:1 | 1.09 |
| 1.6:1 | 1.21 |
| 1.4:1 | 1.34 |
| 1.2:1 | 1.50 |
| 1.0:1 | 1.69 |

**Example 4-1:** The struck capacity of a truck is 96 loose cubic yards (LCY) and its heaped capacity (based on a 2:1 slope) is 136.8 loose cubic yards. Determine the heaped capacity of the truck if the slope of the heap is: a) 3:1, or b) 1:1.
**Solution:** The volume of soil affected by the angle of repose is:

$$\text{Volume Above Sides (2:1 Slope)} = 136.8 \text{ LCY} - 96.0 \text{ LCY}$$
$$= 40.8 \text{ LCY}$$

a) With a 3:1 slope, this volume will be reduced to:

$$\text{Volume Above Sides (3:1 Slope)} = 0.69 \times 40.8 \text{ LCY} \qquad \text{(Table 4-2)}$$
$$= 28.2 \text{ LCY}$$

Thus, the total heaped capacity is:

$$\text{Heaped Volume (3:1 Slope)} = 96.0 \text{ LCY} + 28.2 \text{ LCY}$$
$$= 124.2 \text{ LCY}$$

b) With a 1:1 slope, the volume above the sides will be increased to:

$$\text{Volume Above Sides (1:1 Slope)} = 1.69 \times 40.8 \text{ LCY} = 69.0 \text{ LCY} \qquad \text{(Table 4-2)}$$

Thus, the total heaped capacity is:

$$\text{Heaped Volume (1:1 Slope)} = 96 \text{ LCY} + 69 \text{ LCY} = 165 \text{ LCY}$$

As a rule of thumb, some contractors estimate the heaped capacity under normal conditions as being midway between the manufacturer's struck and heaped capacity ratings. For example, the estimated heaped capacity of a truck with a rated 96-LCY struck capacity and 136-LCY heaped capacity would be:

$$\text{Average Heaped Capacity} = \frac{96 \text{ LCY} + 136 \text{ LCY}}{2} = 116 \text{ LCY}$$

As mentioned earlier, we must be careful not to exceed the weight limitations of the payload. Weights of various materials are given in Table 1-1. The maximum payload limit expressed in cubic yards can be determined by:

$$\text{Maximum Load Limit (C.Y.)} = \frac{\text{Maximum Weight Limit}}{\text{Weight/C.Y.}} \qquad \textbf{(Equation 4.1)}$$

**Example 4-2:** Assuming hauling damp sand weighing 2850 pounds per loose cubic yard, determine if you are hauling within the weight limitations of the truck given in the previous example if the maximum weight capacity is 195 tons and the angle of repose is 1:1.

**Solution:** The payload weighs:

$$\text{Payload (Tons)} = \frac{165 \text{ LCY} \times 2850 \text{ lb./LCY}}{2000 \text{ lb./Ton}} = 235 \text{ tons} \qquad \text{(Example 4-1)}$$

This exceeds the maximum allowable weight capacity by:

$$\text{Overload (Tons)} = 235 \text{ Tons} - 195 \text{ Tons} = 40 \text{ tons} \qquad \text{(Given)}$$

This is equivalent to an overload volume of:

$$\text{Overload (LCY)} = \frac{40 \text{ Tons x } 2000 \text{ lb./Ton}}{2850 \text{ lb./LCY}} = 28 \text{ LCY}$$

Thus, the maximum volume of damp sand we are allowed to haul is:

$$\text{Maximum Load (LCY)} = 165 \text{ LCY} - 28 \text{ LCY} = 137 \text{ LCY} \qquad \text{(Example 4-1)}$$

We could have also solved this problem by determining the maximum payload volume as follows:

$$\text{Maximum Load Limit (LCY)} = \frac{195 \text{ Tons x } 2000 \text{ lb./Ton}}{2850 \text{ lb./LCY}} = 137 \text{ LCY} \qquad \text{(Eq. 4.1)}$$

## INTERFACING PRODUCTION RATES OF EXCAVATORS AND HAULING UNITS

The payload of a dump truck is usually expressed in loose cubic yards. This is because of the fact that even if the soil being loaded is excavated from an in-place, natural condition (bank cubic yards, or BCY), by the time the soil is dropped into the truck body, it is in a disturbed, loose condition (Figure 7-20).

The production rate of an excavator loading a truck is expressed in loose, or bank cubic yards per hour, depending upon the physical condition of the soil being excavated. For example, if a dragline is excavating in-place material, its production rate is expressed in bank cubic yards per hour, but if the dragline is loading loose, stockpiled material, the production rate is expressed in loose cubic yards per hour.

As a general rule, the production rate of a loader is expressed in loose cubic yards per hour, whereas, the production rate of other equipment normally used for loading a truck (draglines, backhoes, shovels, mass excavators, clamshells, etc.) is expressed in bank, or loose cubic yards per hour.

Since the production rate of an excavator/haul-unit system is dependent upon the production rates of both the excavator and the trucks, we must interface the production rates of both units. Since production rate data tables for numerous excavators are based on bank-cubic-yard production, it is usually easier to express the truck payload in bank cubic yards.

The number of dipper cycles required to fill a truck can be determined by:

$$\text{Dipper Cycles} = \frac{\text{Truck Capacity}}{\text{Excavator Capacity}} \qquad \textbf{(Equation 4.2)}$$

**Example 4-3:** Assuming 30% swell, determine the number of dipper cycles required for a 2-BCY power shovel to fill a 14-LCY dump truck.
**Solution:** Expressed in bank cubic yards, the truck capacity is:

$$\text{Payload (BCY)} = \frac{14\text{ LCY}}{1.3\text{ LCY/BCY}} \qquad \text{(30\% swell)}$$
$$= 10.8\text{ BCY}$$

The number of dipper cycles required to fill the truck is:

$$\text{Dipper Cycles} = \frac{10.8\text{ BCY}}{2\text{ BCY/Cycle}} \qquad \text{(Eq. 4.2)}$$
$$= 5.4$$
$$= 6\text{ cycles}$$

## NUMBER OF HAUL UNITS REQUIRED

The number of trucks required to service an excavator is:

$$\text{N (Number of Trucks Required)} = \frac{\text{Haul Unit Cycle Time}}{\text{Load Time}} \qquad \textbf{(Equation 4.3)}$$

It's a good idea to have at least one or two standby units available to replace hauling units that break down. If rented equipment is not readily available, a rule of thumb is to have one standby unit for every five hauling units in operation.

## HAUL UNIT CYCLE TIME

The *haul unit cycle time* is the sum of the load, haul, turning and dump, return, and spot times; i.e.,

$$\text{Haul Unit Cycle Time} = \text{Load} + \text{Haul} + \text{Turning \& Dump} + \text{Return} + \text{Spot Times}$$
$$= \text{Load} + \text{Travel} + \text{Turning \& Dump} + \text{Spot Times}$$
**(Equation 4.4)**

Turning & dump times and spot times can be combined into what is referred to as *fixed time*; therefore, the haul unit cycle time can be defined as:

$$\text{Haul Unit Cycle Time} = \text{Load} + \text{Travel} + \text{Fixed Times} \qquad \textbf{(Equation 4.5)}$$

## LOAD TIME

The load time depends on the truck capacity, bucket capacity and the excavator cycle time. Load time can be calculated in two ways:

$$\text{Load Time} = \frac{\text{Haul Unit Capacity}}{\text{Excavator Production at 100\% Efficiency}} \qquad \textbf{(Equation 4.6)}$$

or,

$$\text{Load Time} = \text{Dipper Cycles x Excavator Cycle Time} \qquad \textbf{(Equation 4.7)}$$

Load time can be reduced by properly sizing haul units for the excavator used. The larger the target, the easier it is to hit. Using trucks too small for the excavator will increase excavator cycle time and will also cause excess spillage. When loading with a shovel, use trucks with a capacity of at least 4 times the bucket capacity, and when loading with a dragline, use trucks with a capacity of at least 5 times the bucket capacity. Also, try to use trucks that hold a whole number of bucket loads; otherwise, one of the dippers will be only partially loaded. Figure 1-6 can be used to help interface the excavator with hauling units.

You can also reduce load time by reducing the angle of swing of the excavator by spotting trucks in front of the excavator close to the excavation (Figure 1-5). However, soil conditions must be considered, since trucks spotted too close to an excavation consisting of weak soil could cause a cave-in.

Load time also depends on whether or not the trucks are owned, or leased. A leased truck is often abused by loading it rapidly. The soil is often dumped from the bucket higher above the truck than is recommended, and the bucket often hits the sides of the truck. Also, a leased truck is often overloaded.

**Example 4-4:** Assuming a power shovel producing 350 bank cubic yards per hour is loading a 20-LCY truck, determine the time required to load the truck if the soil swells 25%. Express the answer in seconds, minutes and hours.
**Solution:** The truck payload is:

$$\text{Payload (BCY)} = \frac{\text{20 LCY}}{\text{1.25 LCY/BCY}} \qquad \text{(25\% swell)}$$
$$= \text{16 BCY}$$

The load time is:

$$\text{Load Time} = \frac{\text{16 BCY}}{\text{350 BCY/Hr.}} \qquad \text{(Eq. 4.6)}$$
$$= \text{0.046 hours x 60 min./hr.}$$
$$= \text{2.76 minutes x 60 sec./min.}$$
$$= \text{166 seconds}$$

**Example 4-5:** Assuming a haul unit cycle time of 0.5 hours, determine the number of trucks required to service the power shovel under the conditions given in the previous example.
**Solution:** The number of hauling units required is:

$$N = \frac{0.5 \text{ Hr.}}{0.046 \text{ Hr.}} \qquad \text{(Example 4-4; Eq. 4.3)}$$
$$= 10.9$$
$$= 11 \text{ trucks}$$

**Example 4-6:** Assuming a 2-BCY power shovel has a dipper cycle time of 20 seconds, determine the time required to load a 20-LCY truck with a 16-BCY capacity. Express the answer in seconds, minutes and hours.
**Solution:** The number of dipper cycles required to load the truck is:

$$\text{Dipper Cycles} = \frac{16 \text{ BCY}}{2 \text{ BCY/Cycle}} \qquad \text{(Eq. 4.2)}$$
$$= 8 \text{ cycles}$$

The load time is:

$$\text{Load Time} = 8 \text{ Cycles x } 20 \text{ sec./Cycle} \qquad \text{(Eq. 4.7)}$$
$$= 160 \text{ seconds} \div 60 \text{ sec./min.}$$
$$= 2.67 \text{ minutes} \div 60 \text{ min./hr.}$$
$$= 0.045 \text{ hours}$$

**Example 4-7:** Assuming a haul unit cycle time of 1.0 hour, determine the number of trucks required to service the power shovel under the conditions given in the previous example.
**Solution:** The number of hauling units required is:

$$N = \frac{1.0 \text{ Hr.}}{0.045 \text{ Hr.}} \qquad \text{(Example 4-6; Eq. 4.3)}$$
$$= 22.2$$
$$= 23 \text{ trucks}$$

**Example 4-8:** Assuming a haul unit cycle time of 0.2 hours (excluding load time), determine the number of 20-LCY trucks with 16-BCY capacities required to service the dragline working under the conditions given in Example 1-12.
**Solution:** At 100% efficiency, the dragline production rate is:

Dragline Production (BCY/Hr.) = 207 BCY/hour (Example 1-12)

The load time is:

$$\text{Load Time (Hr.)} = \frac{16\ \text{BCY}}{207\ \text{BCY/Hr.}} \qquad \text{(Eq. 4.6)}$$
$$= 0.077\ \text{hours}$$

The number of hauling units required is:

$$N = \frac{0.2\ \text{Hr.} + 0.077\ \text{Hr.}}{0.077\ \text{Hr.}} \qquad \text{(Eq. 4.3)}$$
$$= 3.6$$
$$= 4\ \text{trucks}$$

**Example 4-9:** Assuming a haul unit cycle time of 0.5 hours (excluding load time), determine the number of 16-LCY trucks required to service the loader working under the conditions given in Example 3-7.
**Solution:** At 100% efficiency, the loader production rate is:

Loader Production (LCY/Hr.) = 660 LCY/hour (Example 3-7)

The load time is:

$$\text{Load Time (Hr.)} = \frac{16\ \text{LCY}}{660\ \text{LCY/Hr.}} \qquad \text{(Eq. 4.6)}$$
$$= 0.024\ \text{hours}$$

The number of hauling units required is:

$$N = \frac{0.5\ \text{Hr.} + 0.024\ \text{Hr.}}{0.024\ \text{Hr.}} \qquad \text{(Eq. 4.3)}$$
$$= 21.8$$
$$= 22\ \text{trucks}$$

Note in the two previous examples that the excavator production is based on 100% efficiency, as per Equation 4.6. The reason for this is that we want to be conservative (or safe) when determining the number of hauling units required to service the excavator. Referring to Equation 4.3, load time is in the denominator of the fraction. With any given numerator, the smaller the denominator, the larger will be the result after division is performed. In this case, the result is the number of hauling units (N) required. For instance, had we used an efficiency factor of 0.75 for the dragline production given in Example 4-8, the dragline production rate would have been:

Dragline Production (BCY/Hr.) = 207 BCY/Hr. x 0.75 (Example 4-8)
= 155 BCY/hour

The load time would have been:

$$\text{Load Time (Hr.)} = \frac{16 \text{ BCY}}{155 \text{ BCY./Hr.}} \qquad \text{(Example 4-8; Eq. 4.6)}$$
$$= 0.103 \text{ hours}$$

Thus, the number of hauling units required would have been:

$$N = \frac{0.2 \text{ Hr.} + 0.103 \text{ Hr.}}{0.103 \text{ Hr.}} \qquad \text{(Example 4-8; Eq. 4.3)}$$
$$= 2.9$$
$$= 3 \text{ trucks}$$

This is one hauling unit less than that determined in Example 4-8, and using 3 hauling units instead of 4 might be "stretching our luck," especially if one of the hauling units breaks down.

**TURNING & DUMP TIME**

Turning and dump time is the time required to maneuver and off-load at the dump area. Dump time is dependent upon the size, power and type of hauling unit, soil conditions, total resistance, traction, operator skill, and turning room and traffic at the fill, including trucks dumping simultaneously, bulldozers, graders and compactors. Some contractors allow trucks to drive over previously dumped soil to help compact it. This will increase dump time, but will reduce compaction time. Dump time can be reduced by providing an adequate number of graders and compactors to keep the fill smooth and compacted, and by giving trucks the right-of-way. In the absence of field timing tests and previous job data, you can use Table 4-3 to estimate turning and dump time. Turning & dump time is one of the components of fixed time in Equation 4.5.

**Table 4-3: Hauler Turning & Dump Times (Min.)**

| Operating Conditions | Turning & Dump Times (Min.) | |
|---|---|---|
| | Haulers | Bottom Dumps |
| Favorable | 1.0 | 0.3 |
| Average | 1.3 | 0.6 |
| Unfavorable | 1.5 - 2.0 | 1.5 |

(*Courtesy of Terex Corporation*)

## SPOT TIME

Spot time is the time required for a hauling unit to maneuver and get into the proper position at the loading area. Spot time is dependent upon the type of hauling unit, operator skill, total resistance, traffic at the loading site, turnaround room, and whether the trucks have a "straight shot" through the site, or have to back up to be loaded. When trucks are required to back into the loading zone, spot time can be reduced by installing spotting bumpers to help the driver properly position the truck. Production can also be increased by spotting trucks on each side of the excavator, which enables one truck to be loaded while another is being spotted (Figure 1-5). Depending on the spotting difficulty, it sometimes pays to have a spotting supervisor at the loading zone. In the absence of field timing tests and previous job data, you can use Table 4-4 to estimate spot time. Spot time is another component of fixed time in Equation 4.5.

**Table 4-4: Hauler Spot Times (Min.)**

| Operating Conditions | Spotting Time at Loading Machine (Min.) | |
|---|---|---|
| | Haulers | Bottom Dumps |
| Favorable | 0.15 | 0.15 |
| Average | 0.30 | 0.50 |
| Unfavorable | 0.50 | 1.00 |

(*Courtesy of Terex Corporation*)

## TRAVEL TIME

Travel time depends primarily on engine power, payload, total resistance and round-trip distance. Travel time also depends on the truck size, since larger trucks normally travel slower than smaller trucks. Travel times can be quickly estimated by using travel time charts such as those shown in Figures 4-2 and 4-3. Two travel time charts are shown for the hauling unit; one for the truck hauling its rated payload, and the other for the empty vehicle. Travel time charts take acceleration and deceleration of the truck into consideration at the loading and dumping areas.

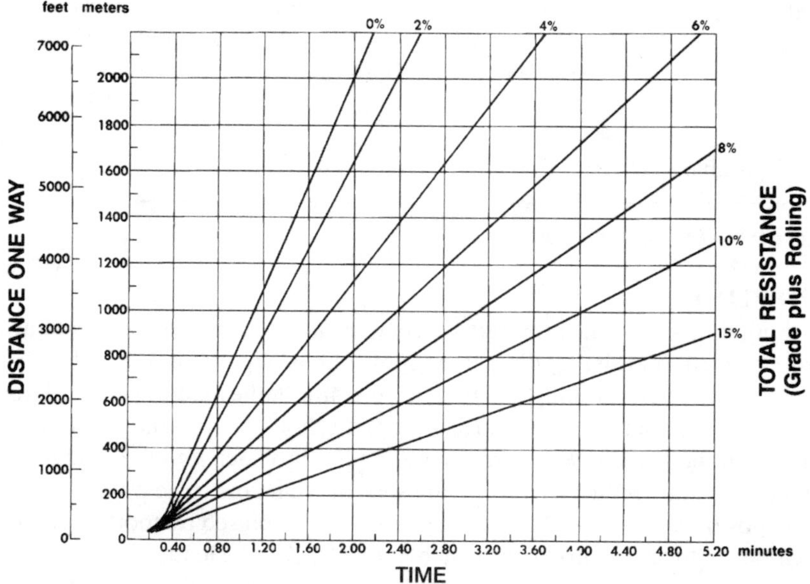


**Figure 4-2. 769C off-highway truck travel time chart (loaded).**
(*Courtesy of Caterpillar Inc.*)

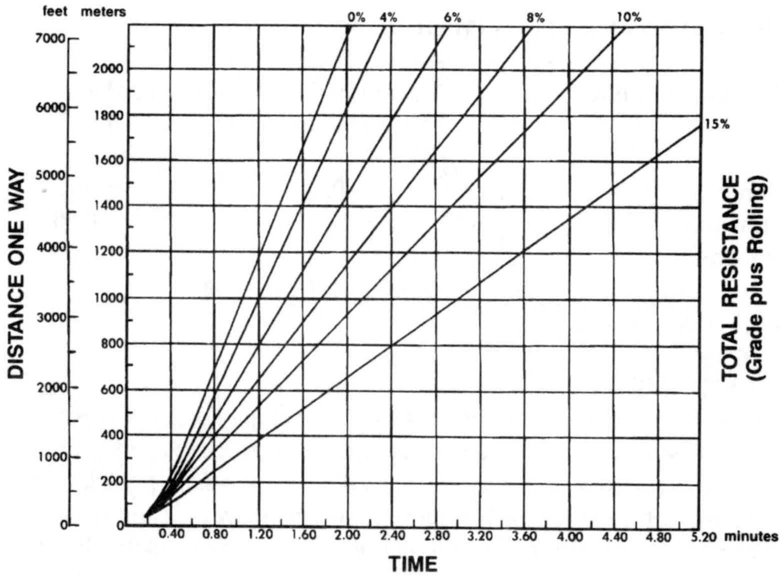


**Figure 4-3. 769C off-highway truck travel time chart (empty).**
(*Courtesy of Caterpillar Inc.*)

To use a travel time chart, enter the chart from the left and find the one-way travel distance. Project a horizontal line until it intersects the appropriate diagonal line representing total resistance (effective grade). From this intersection, drop vertically to obtain the travel time at the bottom of the chart.

When a truck is traveling down grade, the total resistance can be negative. This can cause the truck to accelerate, requiring the use of the retarder, a lower gear, and/or brakes. Under these conditions, travel time charts cannot be used. Instead, a brake-retarder performance curve such as that shown in Figure 4-4 must be used to determine the maximum safe speed during descent.

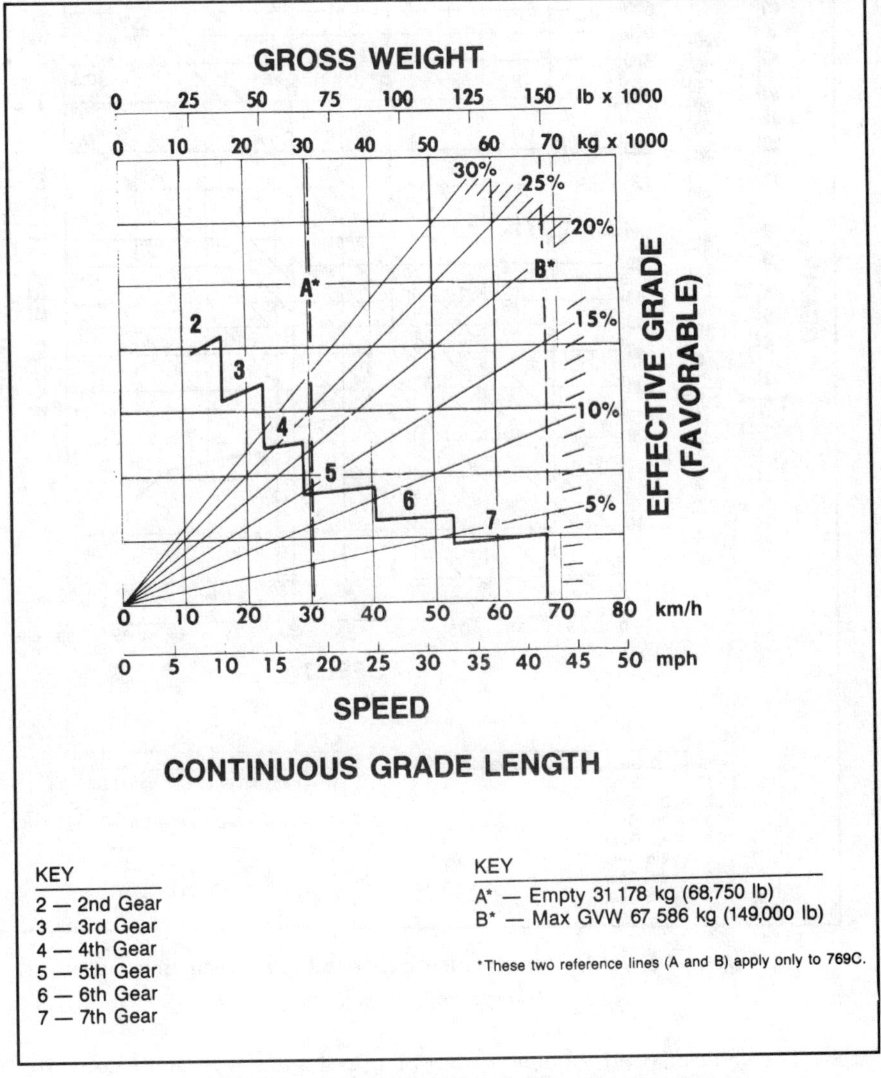


**Figure 4-4. 769C off-highway truck retarder curve.**

(*Courtesy of Caterpillar Inc.*)

If the truck is not carrying its rated payload, the travel time chart (loaded) cannot be used. Instead, a rimpull-speed-gradeability performance curve such as that shown in Figure 4-5 must be used to determine the maximum speed.

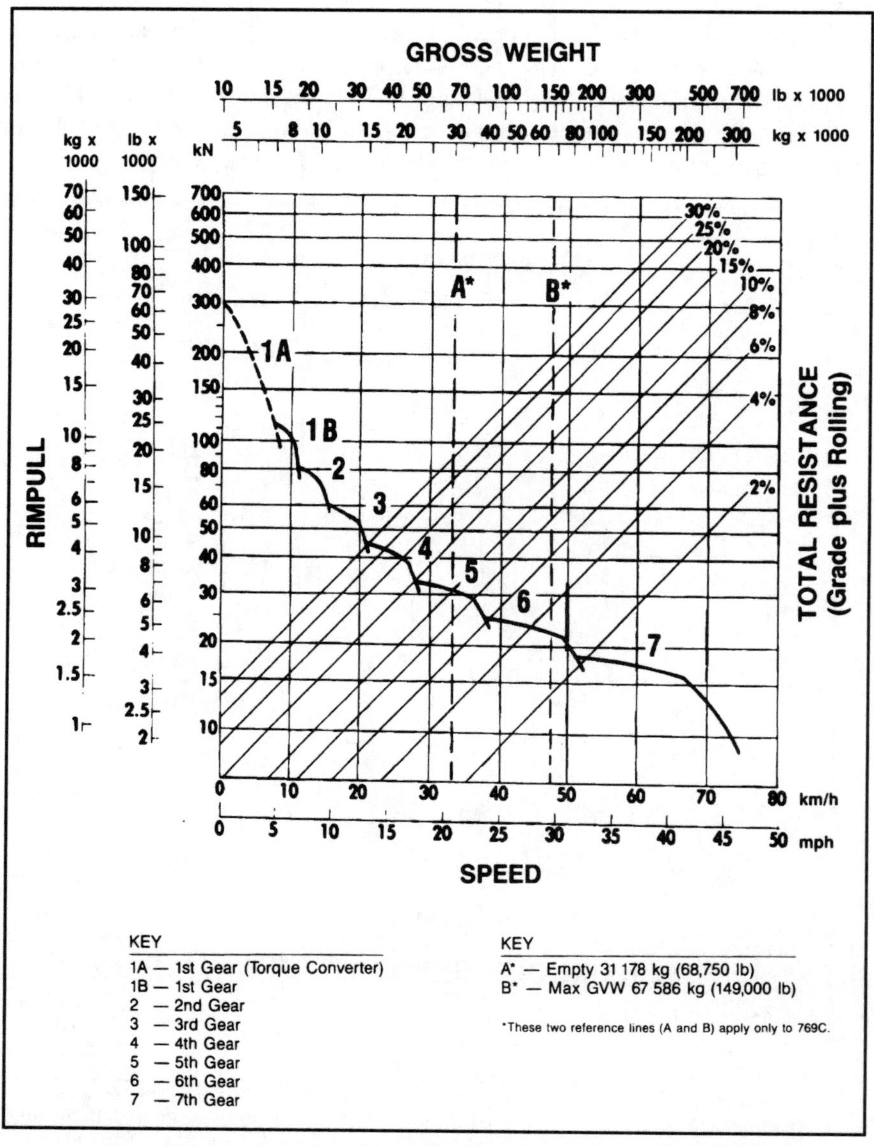


**Figure 4-5. 769C rimpull-speed-gradeability performance curve.**
(*Courtesy of Caterpillar Inc.*)

The vehicle speeds obtained from Figures 4-4 and 4-5 can be converted to travel time as follows:

$$\text{Travel Time (Min.)} = \frac{\text{Travel Distance (Ft.)}}{\text{88 Ft./Min. x Ave. Speed (mph)}} \qquad \textbf{(Equation 4.8)}$$

While travel time is one of the most important considerations for determining a fleet production rate, it is probably the most difficult time component to estimate. Even when travel time charts are available, their use is somewhat limited. Also, retarder curves and rimpull-speed-gradeability performance curves do not take acceleration, deceleration and time required to shift gears into consideration. To compensate for acceleration, deceleration and shifting gears, multiply the maximum haul unit speed by the appropriate factor in Table 4-5 to obtain an average haul unit speed. This is referred to as the *Average Speed Method.* To account for the deceleration of the loaded haul unit along the final short section of the loaded haul route (if any), use the factor under the column labelled "Level Haul Unit Starting From 0 MPH," within the table section labelled "Under 300 lb./hp.".

**Table 4-5: Average Speed Factors**

| Haul Road Length in Feet | Level Haul Unit Starting from 0 MPH | Unit in Motion When Entering Haul Road Section: Level | Unit in Motion When Entering Haul Road Section: Downhill Grade | Uphill Grade Factor |
|---|---|---|---|---|
| **under 300 lbs./hp** | | | | |
| 0-200 | 0-.40 | 0-.65 | 0-.67 | 1.00 |
| 201-400 | .40-.51 | .65-.70 | .67-.72 | |
| 401-600 | .51-.56 | .70-.75 | .72-.77 | (Entrance speed |
| 601-1000 | .56-.67 | .75-.81 | .77-.83 | greater than |
| 1001-1500 | .67-.75 | .81-.88 | .83-.90 | maximum |
| 1501-2000 | .75-.80 | .88-.91 | .90-.93 | attainable speed |
| 2001-2500 | .80-.84 | .91-.93 | .93-.95 | on section) |
| 2501-3500 | .84-.87 | .93-.95 | .95-.97 | |
| 3501 & up | .87-.94 | .95- | .97- | |
| **300-380 lbs./hp** | | | | |
| 0-200 | 0-.39 | 0-.62 | 0-.64 | 1.00 |
| 201-400 | .39-.48 | .62-.67 | .64-.68 | |
| 401-600 | .48-.54 | .67-.70 | .68-.74 | (Entrance speed |
| 601-1000 | .54-.61 | .70-.75 | .74-.83 | greater than |
| 1001-1500 | .61-.68 | .75-.79 | .83-.88 | maximum |
| 1501-2000 | .68-.74 | .79-.84 | .88-.91 | attainable speed |
| 2001-2500 | .74-.78 | .84-.87 | .91-.93 | on section) |
| 2501-3500 | .78-.84 | .87-.90 | .93-.95 | |
| 3501 & up | .84-.92 | .90-.93 | .95-.97 | |
| **380 & up lbs./hp** | | | | |
| 0-200 | 0-.33 | 0-.55 | 0-.56 | 1.00 |
| 201-400 | .33-.41 | .55-.58 | .56-.64 | |
| 401-600 | .41-.46 | .58-.65 | .64-.70 | (Entrance speed |
| 601-1000 | .46-.53 | .65-.75 | .70-.78 | greater than |
| 1001-1500 | .53-.59 | .75-.77 | .78-.84 | maximum |
| 1501-2000 | .59-.62 | .77-.83 | .84-.88 | attainable speed |
| 2001-2500 | .62-.65 | .83-.86 | .88-.90 | on section) |
| 2501-3500 | .65-.70 | .86-.90 | .90-.92 | |
| 3501 & up | .70-.75 | .90-.93 | .92-.95 | |

(*Courtesy of Terex Corporation*)

Grades and road surface conditions can vary along the haul road. Trucks may have to slow down for curves, not only for safety, but to preserve tire life, especially in hot weather. Truck traffic on narrow roads can become bottlenecked, especially when graders, dozers and water trucks are maintaining the haul road. Also, one slow truck can dictate overall fleet production when passing is dangerous, or impossible. Travel time in "civilian areas" is extremely difficult to predict.

Bad weather conditions such as wind, rain and blowing dust can affect visibility and slow production. Rain can change a relatively hard-surfaced dirt road into a quagmire, and adverse haul-road conditions will require that trucks haul less than their rated payloads.

The driver's skill can affect fleet production rates. Also, a leased truck will normally be run harder than an owned one, resulting in increased acceleration, faster gear-changing time and faster turning and hauling speed.

**Example 4-10:** Determine the number of Caterpillar 769C off-highway trucks required to service a dragline given in Example 1-12 under the following conditions:

| | |
|---|---|
| Haul Road | = Level rutted dirt with 2" tire penetration |
| Payload | = 32 LCY |
| Gross Vehicle Weight (Empty) | = 68,750 pounds |
| Maximum Gross Vehicle Weight (Loaded) | = 149,000 pounds |
| Maximum Load | = 80,250 pounds |
| Soil | = Common earth; 2500 lb./LCY |
| Swell | = 21% |
| Haul Distance (One Way) | = 5000 feet |
| Job Conditions | = Average |

**Solution:** We must make sure that we are not overloading the truck. The payload weighs:

$$\text{Payload (lb.)} = 32 \text{ LCY} \times 2500 \text{ lb./LCY} = 80{,}000 \text{ pounds} \qquad \text{(Given)}$$

The maximum payload limit is:

$$\text{Maximum Load Limit (LCY)} = \frac{80{,}250 \text{ lb.}}{2500 \text{ lb./LCY}} = 32.1 \text{ LCY} \qquad \text{(Eq. 4.1)}$$

Since the payload is only 32 loose cubic yards weighing 80,000 pounds, we are not exceeding the weight limitations of the payload.

The rolling resistance is:

$$\text{Rolling Resistance (lb./Ton)} = 100 \text{ lb./Ton} \qquad \text{(Table 2-1)}$$

We have no grade resistance since the haul road is level; therefore, the total resistance equals rolling resistance, and the effective grade is:

$$\text{Effective Grade (\%)} = \frac{100 \text{ lb./Ton}}{20} \qquad \text{(Eq. 2.12)}$$

$$= 5\%$$

Expressed in bank cubic yards, the capacity of the truck is:

$$\text{Payload (BCY)} = \frac{32 \text{ LCY}}{1.21 \text{ LCY/BCY}} \qquad \text{(21\% swell)}$$

$$= 26.4 \text{ BCY}$$

At 100% efficiency, the dragline production rate is 207 bank cubic yards per hour (Example 1-12); therefore, the load time is:

$$\text{Load Time (Min.)} = \frac{26.4 \text{ BCY}}{207 \text{ BCY/Hr.}} \qquad \text{(Eq. 4.6)}$$

$$= 0.128 \text{ hours x } 60 \text{ Min./Hr.}$$

$$= 7.68 \text{ minutes}$$

Referring to Figures 4-2 and 4-3, the travel times are:

Travel Time (Loaded) = 3.05 minutes (Figure 4-2)

and,

Travel Time (Empty) = 1.92 minutes (Figure 4-3)

The total travel time is:

$$\text{Total Travel Time (Min.)} = 3.05 \text{ Min.} + 1.92 \text{ Min.}$$

$$= 4.97 \text{ minutes}$$

The turning and dump time is:

Turning & Dump Time (Min.) = 1.3 minutes (Table 4-3)

The spot time is:

Spot Time (Min.) = 0.3 minutes (Table 4-4)

The total fixed time is:

$$\text{Fixed Time (Min.)} = 1.3 \text{ Min.} + 0.3 \text{ Min.}$$
$$= 1.6 \text{ minutes}$$

The haul unit cycle time is:

$$\text{Haul Unit Cycle Time (Min.)} = 7.68 \text{ Min.} + 4.97 \text{ Min.} + 1.6 \text{ Min.} \qquad \text{(Eq. 4.5)}$$
$$= 14.25 \text{ minutes}$$

The number of hauling units required is:

$$N = \frac{14.25 \text{ Min.}}{7.68 \text{ Min.}} \qquad \text{(Eq. 4.3)}$$
$$= 1.86$$
$$= 2 \text{ trucks}$$

**Example 4-11:** Assuming that the two trucks adequately service the dragline given in the previous example, determine the probable production rate of the excavator/haul-unit system.
**Solution:** Regardless of whether we are using 2 or 2000 trucks, the system production rate will not exceed the output of the excavator, since the excavator cannot adequately service any additional trucks; therefore, the probable production rate of the system is 172 bank cubic yards per hour, as per Example 1-12.

**Example 4-12:** Work the Example 4-10 assuming a haul route of 8 miles, each way.
**Solution:** The travel distance each way is:

$$\text{Distance (One Way)} = 8 \text{ Mi.} \times 5280 \text{ Ft./Mi.}$$
$$= 42{,}240 \text{ feet}$$

Figures 4-2 and 4-3 do not give travel time data for such a great distance; however, this data can be extrapolated by constructing a proportion equation for a loaded truck as follows:

$$\frac{3.05 \text{ Min.}}{5000 \text{ Ft.}} = \frac{x}{42{,}240 \text{ Ft.}} \qquad \text{(Example 4-10)}$$

Solving for x, we have a travel time (loaded) of::

$$x = \text{Travel Time (Loaded)} = \frac{3.05 \text{ x } 42{,}240}{5000}$$

$$= 25.77 \text{ minutes}$$

For an empty truck, the proportion equation is:

$$\frac{1.92 \text{ Min.}}{5000 \text{ Ft.}} = \frac{y}{42{,}240 \text{ Ft.}} \qquad \text{(Example 4-10)}$$

Solving for y, we have a travel time (empty) of:

$$y = \text{Travel Time (Empty)} = \frac{1.92 \text{ x } 42{,}240}{5000}$$

$$= 16.22 \text{ minutes}$$

The total travel time is:

$$\text{Total Travel Time (Min.)} = 25.77 \text{ Min.} + 16.22 \text{ Min.}$$

$$= 42 \text{ minutes}$$

The haul unit cycle time is:

$$\text{Haul Unit Cycle Time (Min.)} = 7.68 \text{ Min.} + 42 \text{ Min.} + 1.6 \text{ Min.} \qquad \text{(Example 4-10; Eq. 4.5)}$$

$$= 51.28 \text{ minutes}$$

The number of hauling units required is:

$$N = \frac{51.28 \text{ Min.}}{7.68 \text{ Min.}} \qquad \text{(Eq. 4.3)}$$

$$= 6.68$$

$$= 7 \text{ trucks}$$

**Example 4-13:** Work Example 4-10 assuming the trucks are working at an altitude of 13,000 feet above sea level. Assume that the dragline loading the trucks is derated to 70% of its rated power.

**Solution:** The dragline production rate will be reduced to:

Dragline Production (BCY/Hr.) = 0.70 x 207 BCY/Hr. (Given; Example 4-10)
= 145 BCY/hr.

The load time is:

$$\text{Load Time (Min.)} = \frac{26.4 \text{ BCY}}{145 \text{ BCY/Hr.}} \qquad \text{(Example 4-10; Eq. 4.6)}$$
= 0.182 hours x 60 Min./Hr.
= 10.92 minutes

Referring to Table 2-4, the 769C trucks operate at only 82% efficiency at the given altitude; therefore, the total travel time is:

$$\text{Total Travel Time (Min.)} = \frac{4.97 \text{ Min.}}{0.82} \qquad \text{(Example 4-10; Table 2-4)}$$
= 6.06 minutes

The haul unit cycle time is:

Haul Unit Cycle Time (Min.) = 10.92 Min. + 6.06 Min. + 1.6 Min.
(Example 4-10; Eq. 4.5)
= 18.58 minutes

The number of hauling units required is:

$$N = \frac{18.58 \text{ Min.}}{10.92 \text{ Min.}} \qquad \text{(Eq. 4.3)}$$
= 1.70
= 2 trucks

**Example 4-14:** Disregarding time required for acceleration and deceleration, use Figures 4-4 and 4-5 to determine the haul unit cycle time and number of trucks required to service the dragline under the conditions given in Example 4-10 if the trucks ascend a 10% grade while empty and descend the same grade with an 80,000-pound payload.

**Solution:** Expressed as percent grade, the rolling resistance is:

Rolling Resistance (%) = 5% (Example 4-10)

The effective grade that the empty truck ascends is:

Effective Grade (Ascending) = 5% + 10% (Eq. 2.13)
= 15%

The gross weight of the empty truck is:

G.V.W. (Empty) = 68,750 pounds (Example 4-10)

To determine the travel speed of the empty vehicle, refer to Figure 4-5 and find the empty gross weight of 68,750 pounds at the top of the chart. From this point, follow the vertical line labelled "A" until it intersects the diagonal line corresponding to the total resistance (adverse effective grade) of 15%. From this intersection, project a line horizontally to the left until you intersect the highest gear of the performance curve. From this point, drop vertically to obtain the maximum vehicle speed. In this instance, we can travel at 13 mph in 3rd gear.

The travel time of the empty vehicle is:

$$\text{Travel Time (Min.)} = \frac{5000 \text{ Ft.}}{88 \text{ Ft./Min. x } 13 \text{ mph}} \qquad \text{(Eq. 4.8)}$$
$$= 4.37 \text{ minutes}$$

The effective grade that the loaded truck descends is:

Effective Grade (Descending) = 5% – 10% (Eq. 2.14)
= –5%

The gross weight of the loaded truck is:

G.V.W, (Loaded) = 68,750 lb. + 80,000 lb. (Example 4-10)
= 148,750 pounds

To determine the travel speed of the loaded vehicle, refer to Figure 4-4 and find the loaded gross weight of 147,875 pounds at the top of the chart. From this point, project a vertical line until it intersects the diagonal line corresponding to a favorable effective grade (total resistance) of 5%. From this intersection, project a line horizontally to the left until you intersect the highest gear of the retarder curve. From this point, drop vertically to obtain the maximum *safe* vehicle speed. In this instance, we can travel at 33 mph in 6th gear.

The travel time of the loaded vehicle is:

$$\text{Travel Time (Min.)} = \frac{5000 \text{ Ft.}}{88 \text{ Ft./Min. x } 33 \text{ mph}} \qquad \text{(Eq. 4.8)}$$
$$= 1.72 \text{ minutes}$$

Excluding the time required for acceleration and deceleration, the total travel time is:

$$\text{Total Travel Time} = 4.37 \text{ Min.} + 1.72 \text{ Min.}$$
$$= 6.09 \text{ minutes}$$

The load time is:

$$\text{Load Time (Min.)} = 7.68 \text{ minutes} \qquad \text{(Example 4-10)}$$

The total fixed time is:

$$\text{Fixed Time (Min.)} = 1.6 \text{ minutes} \qquad \text{(Example 4-10)}$$

The haul unit cycle time is:

$$\text{Haul Unit Cycle Time (Min.)} = 7.68 \text{ Min.} + 6.09 \text{ Min.} + 1.6 \text{ Min.} \qquad \text{(Eq. 4.5)}$$
$$= 15.37 \text{ minutes}$$

The number of hauling units required is:

$$N = \frac{15.37 \text{ Min.}}{7.68 \text{ Min.}} \qquad \text{(Eq. 4.3)}$$
$$= 2 \text{ trucks}$$

**Example 4-15:** Use the Average Speed Method to determine the haul unit cycle time and the number of trucks required for the conditions given in the previous example if the truck engines are rated at 450 horsepower.
**Solution:** The ratio of weight to power is:

$$\text{Loaded} = \frac{148{,}750 \text{ lb.}}{450 \text{ hp}} \qquad \text{(Example 4-14)}$$
$$= 331 \text{ lb./hp}$$

$$\text{Empty} = \frac{68{,}750 \text{ lb.}}{450 \text{ hp}} \qquad \text{(Example 4-14)}$$
$$= 153 \text{ lb./hp}$$

The average speed of the truck descending grade is:

Average Speed (Loaded) = 33 mph x 0.92* (Example 4-14; Table 4-5)
= 30.4 mph

The average speed of the truck ascending grade is:

Average Speed (Empty) = 13 mph x 0.94* (Example 4-14; Table 4-5)
= 12.2 mph

The travel times are:

$$\text{Travel Time (Loaded)} = \frac{5000}{88 \times 30.4} \qquad \text{(Eq. 4.8)}$$
$$= 1.87 \text{ minutes}$$

$$\text{Travel Time (Empty)} = \frac{5000}{88 \times 12.2}$$
$$= 4.66 \text{ minutes}$$

The total travel time is:

Total Travel Time (Min.) = 1.87 Min. + 4.66 Min.
= 6.53 minutes

The load time is:

Load Time (Min.) = 7.68 minutes (Example 4-14)

The total fixed time is:

Fixed Time (Min.) = 1.6 minutes (Example 4-14)

* *We are assuming the value under the column labelled "Level Haul Unit Starting From 0 MPH."*

The haul unit cycle time is:

$$\text{Haul Unit Cycle Time (Min.)} = 7.68\ \text{Min.} + 6.53\ \text{Min.} + 1.6\ \text{Min.} = 15.8\ \text{minutes} \qquad \text{(Eq. 4.5)}$$

The number of haul units required is:

$$N = \frac{15.8\ \text{Min.}}{7.68\ \text{Min.}} = 2.05 = 2\ \text{trucks} \qquad \text{(Eq. 4.3)}$$

**COST PERFORMANCE**

By studying Equation 4.3, we can formulate some general rules regarding excavator/truck fleets:

1) With a given size excavator and truck, as the haul unit cycle time increases, the number of trucks required will also increase.
2) The greater the load time, the fewer trucks will be required. With a given size excavator and production rate, load time will increase with larger trucks. With a given size truck, load time will decrease as the excavator size and/or production is increased.

When determining the number of hauling units required, we have two major goals to consider: a) we must complete the project on schedule, and b) we want to move the soil at a minimum cost.

In reality, the size and type of excavator chosen and the size(s) and number of trucks will be dictated primarily by economics, equipment availability and other practical considerations. However, to simplify an already complex situation, let's assume that all needed equipment is available at a reasonable cost.

To complete a project on schedule requires that we choose an excavator that can excavate and load into trucks all of the soil to be moved within a specified time frame. The production rate required to complete a project on schedule can be determined by:

$$\text{Required Production} = \frac{\text{Total Quantity To Be Excavated}}{\text{Total Allowable Project Time}} \qquad \textbf{(Equation 4.9)}$$

To excavate and transport the soil at a minimum cost, we must determine the cost per cubic yard moved by using the following equation:

$$\text{Cost Performance (\$/C.Y.)} = \frac{\text{Fleet Cost/Hr. (Incl. Excavator)}}{\text{Fleet Production Rate/Hr.}}$$

**(Equation 4.10)**

When N trucks are used, the total fleet production rate will be that of the excavator, including job efficiency (*normal production*). However, when less than N trucks are used, the probable production rate will be:

$$\text{Production (Less Than N Units)} = \frac{\text{N (Actual)}}{\text{N (Required)}} \times \text{Normal Production}$$

**(Equation 4.11)**

where: N (Required) is *not* rounded off.

**Example 4-16:** Assuming the following information, determine the optimum number and size of trucks required to service the dragline given in Example 1-12.

Dragline Cost = \$75.00/hr. (incl. operator)
Spot Time = 0.8 min.
Turning & Dump Time = 1.0 min.

**Truck Options:**

| Capacity | Travel Time | Cost/Hr. (Incl. Operator) |
|---|---|---|
| 20 BCY | 0.57 hr. = 34.2 min. | 58.00 |
| 24 BCY | 0.66 hr. = 39.6 min. | 65.00 |
| 28 BCY | 0.71 hr. = 42.6 min. | 70.00 |

**Solution:** The actual (normal) production rate of the dragline is:

Normal Dragline Production (BCY/Hr.) = 172 BCY/hr. (Example 1-12)

As per Example 4-11, this production rate cannot be increased by adding more than N trucks to the system, since the excavator cannot adequately service them. At 100% efficiency, the dragline production rate is:

Dragline Production (BCY/Hr.) = 207 BCY/hr. (Example 1-12)

The load time required for a 20-BCY truck is:

$$\text{Load Time (20 BCY)} = \frac{\text{20 BCY}}{\text{207 BCY/Hr.}} \qquad \text{(Eq. 4.6)}$$

$$= 0.097 \text{ hours x 60 min./hr.}$$

$$= 5.82 \text{ minutes}$$

The total fixed time is:

$$\text{Fixed Time (Min.)} = 1.0 \text{ Min.} + 0.8 \text{ Min.} \qquad \text{(Given)}$$
$$= 1.8 \text{ minutes}$$

The number of 20-BCY hauling units required is:

$$N \text{ (20 BCY)} = \frac{5.82 \text{ Min.} + 34.2 \text{ Min.} + 1.8 \text{ Min}}{5.82 \text{ Min.}} \qquad \text{(Eq. 4.3)}$$
$$= 7.19$$
$$= 8 \text{ trucks}$$

The load time required for a 24-BCY truck is:

$$\text{Load Time (24 BCY)} = \frac{24 \text{ BCY}}{207 \text{ BCY/Hr.}}$$
$$= 0.116 \text{ hours x } 60 \text{ min./hr.}$$
$$= 6.96 \text{ minutes}$$

The number of 24-BCY hauling units required is:

$$N \text{ (24 BCY)} = \frac{6.96 \text{ Min.} + 39.6 \text{ Min.} + 1.8 \text{ Min.}}{6.96 \text{ Min.}}$$
$$= 6.95$$
$$= 7 \text{ trucks}$$

The load time required for a 28-BCY truck is:

$$\text{Load Time (28 BCY)} = \frac{28 \text{ BCY}}{207 \text{ BCY/Hr.}}$$
$$= 0.135 \text{ hours x } 60 \text{ min/hr.}$$
$$= 8.10 \text{ minutes}$$

The number of 28-BCY hauling units required is:

$$N \text{ (28 BCY)} = \frac{8.10 \text{ Min.} + 42.6 \text{ Min.} + 1.8 \text{ Min.}}{8.10 \text{ Min.}}$$
$$= 6.48$$
$$= 7 \text{ trucks}$$

When using N trucks of any size, the normal production rate will be 172 BCY per hour, but if we reduce the number of hauling units, the fleet production rate will also be reduced.Using a 7-unit 20-BCY truck fleet, the probable production rate is:

$$\text{Production (20 BCY)} = \frac{7}{7.19} \times 172 \text{ BCY/Hr.} \qquad \text{(Eq. 4.11)}$$
$$= 167 \text{ BCY/hour}$$

Using a 6-unit 24-BCY truck fleet, the probable production rate is:

$$\text{Production (24 BCY)} = \frac{6}{6.95} \times 172 \text{ BCY/Hr.}$$
$$= 148 \text{ BCY/hour}$$

Using a 6-unit 28-BCY truck fleet, the probable production rate is:

$$\text{Production (28 CY)} = \frac{6}{6.48} \times 172 \text{ BCY/Hr.}$$
$$= 159 \text{ BCY/hour}$$

The cost performance of an 8-unit 20-BCY truck fleet is:

$$\text{Cost/BCY} = \frac{\$75.00\text{/Hr.} + (8 \times \$58.00\text{/Hr.})}{172 \text{ BCY/Hr.}} \qquad \text{(Eq. 4.10)}$$
$$= \$3.13\text{/BCY}$$

The cost performance of a 7-unit 20-BCY truck fleet is:

$$\text{Cost/BCY} = \frac{\$75.00\text{/Hr.} + (7 \times \$58.00\text{/Hr.})}{167 \text{ BCY/Hr.}}$$
$$= \$2.88\text{/BCY}$$

Working the remainder of this problem in a similar fashion, we can tabulate our cost performance data as follows:

| Truck Capacity (BCY) | No. of Trucks | Fleet Cost/Hr. | Production (BCY/Hr.) | Cost Performance ($/BCY) |
|---|---|---|---|---|
| 20 | 8 | 539.00 | 172 | 3.13 |
| 20 | 7 | 481.00 | 167 | 2.88 |
| 24 | 7 | 530.00 | 172 | 3.08 |
| 24 | 6 | 465.00 | 148 | 3.14 |
| 28 | 7 | 565.00 | 172 | 3.28 |
| 28 | 6 | 495.00 | 159 | 3.11 |

Based on cost performance, the optimum choice is to use a 7-unit 20-BCY truck fleet at a unit production cost of $2.88/BCY. This assumes that the project can be completed on schedule, regardless of the decreased production rate.

**Example 4-17:** Assuming that the dragline-truck fleet in the previous example must excavate and transport 100,000 bank cubic yards of soil within 80 work days, determine the minimum production rate required when working 8-hour days.
**Solution:** The minimum required production is:

$$\text{Required Production (BCY/Hr.)} = \frac{100{,}000 \text{ BCY}}{80 \text{ Days x } 8 \text{ Hr./Day}} \qquad \text{(Eq. 4.9)}$$

$$= 157 \text{ BCY/hour}$$

(Conservatively rounded up)

Thus, the 7-unit 20-BCY truck fleet will produce adequately. Had the project required a production rate of 170 BCY/hour, the 7-unit 24-BCY truck fleet would be the most cost-efficient system.

## TRUCK SIZE

A cost performance study should always be made when determining the optimum number and size of trucks to be used with a given size excavator. This is the only way to determine the minimum fleet cost. However, there are some basic advantages and disadvantages to be considered when determining truck size.

Advantages of small trucks:

1) They are more maneuverable than large trucks.
2) Small trucks can travel through "civilian areas."
3) Small trucks are usually faster than large trucks.
4) Less production is lost when a small truck breaks down.
5) It is easier to "fine-tune" the balance of small trucks with the output of an excavator.
6) It takes less time to load a small truck.
7) Small trucks cause less damage to the haul road.
8) Repair parts are usually easy to find.

Advantages of large trucks:

1) Large trucks provide large targets for the excavator, so they are easier to load.
2) Fewer trucks (and drivers) are required.
3) Spotting time is reduced since fewer trucks are used.
4) There is less chance of trucks bunching up at any given location.
5) Fewer repair parts have to be stocked.
6) Engines of large trucks usually burn cheaper fuels.

*Chapter 5*

# TRACTORS AND DOZERS

Tractors are some of the most versatile and widely used equipment in the earthmoving industry. With appropriate modifications, tractors are manufactured as dozers, soil compactors (Chapter 7), landfill compactors (Photo 5-1), forklifts, log loaders (Photo 5-2), log skidders (Photo 5-3), agricultural tractors (Photo 5-7), scarifiers, rippers (Chapter 10) and loaders (Chapter 3). Some tractors are designed for landclearing and are equipped with clamp rakes, tree shredders, and grapple shears. Double-axle four-wheel tractors are often used for towing wagons (Chapter 4), compaction equipment and two-axle scrapers. Single-axle two-wheel tractors are manufactured specifically for towing single-axle scrapers (Chapter 6).

**Photo 5-1. Landfill compactor.** (*Courtesy of Caterpillar Inc.*)

**Photo 5-2. Log loader.** (*Courtesy of Caterpillar Inc.*)

**Photo 5-3. Log skidder.** (*Courtesy of Caterpillar Inc.*)

Dozers can be used to remove spoil from trench areas, and fill can be spread by back-dragging the blade. Dozers can also be used for backfilling. When working around a trench, always approach the trench at a 45-degree angle instead of parallel to the trench to avoid a cave-in. Back-dragging can also be used for finish grading.

Tractors are available with rubber tires, or crawler tracks. Wheel-mounted tractors can travel faster than crawler tractors and are best suited for projects where extensive travel is required to and from, and around the job site, or where crawler-tractors are prohibited, such as over a paved road surface (Photo 5-4). However, tires require a firm travel surface and they are poorly suited for working over sharp-edged rock, especially if the travel surface is wet. You can help protect tires by covering them protective mesh chains (Photo 1-2). Due to high ground contact pressure, a wheel tractor is useful for compacting soils. If the drive wheels of the dozer are located at the rear, a pushed load will tend to raise the front of the dozer and transfer weight to the back of the machine, resulting in greater traction.

**Photo 5-4. Wheel-mounted dozer with straight blade.** (*Courtesy of Caterpillar Inc.*)

Crawler tractors are best suited for projects where good traction and low ground pressures are required, and where the working surface is abrasive enough to cause excessive tire wear (Photos 5-5 and 5-6). However, a crawler-tractor must be transported on a trailer if the travel distance is substantial, or if crawler tracks are prohibited over the travel surface. Crawler tractors can operate on side slopes with grades up to 100%.

**Photo 5-5. Crawler dozer used where good traction and low ground pressure is required.**
(*Courtesy of Caterpillar Inc.*)

**Photo 5-6. Crawler dozer working over rock.** (*Courtesy of Caterpillar Inc.*)

Caterpillar Inc. manufacturers an agricultural tractor mounted on flexible rubber belts reinforced with steel cables bonded into the rubber (Photo 5-7). This tractor combines the mobility of wheels with the traction and floatation of tracks.

**Photo 5-7. Agricultural tractor with Mobil-Trac system.**
*(Courtesy of Caterpillar Inc.)*

## DOZERS

A tractor equipped with a front-mounted pusher blade which can be raised and lowered by hydraulic or cable control is called a *dozer,* or bulldozer. There are many blades available, and the size and type of blade used depends on the type of material to be moved, and the power and traction limitations of the tractor. Dozer blades consist of a cutting edge (or knife) mounted on the bottom of a concave *moldboard.* The cutting edge consists of a long center piece and two corners mounted at the ends of the blade.

Machine shipping dimensions, including maximum digging depth and blade tilt and pitch are normally published with manufacturer's specifications for any given dozer (Figure 5-1).

As a general rule, wheel or track dozers are most effective when the haul distance is less than 500 feet. This range of dozer operation is referred to as the *economic application zone*, or "power zone" (Figure 3-1). The economic application zone is dependent upon underfoot conditions, grades, material type, operator skill and the size and type of dozer. A track dozer is advantageous on short hauls over soft or muddy ground, and a wheel dozer is generally more effective on long hauls over firm ground.

Production can be increased by dozing down grade, or by using techniques known as slot dozing, blade-to-blade dozing, or side-by-side (S x S) dozing.

As a dozer makes its first pass over the ground, most of the soil will spill to the sides of the blade to form a *windrow* (longitudinal soil pile) on each side of the lane. As the dozer makes subsequent passes over the lane, a trench or slot will be formed, reducing or preventing further side spillage (Photo 5-8). Therefore, production will increase due to fuller blade loads being pushed. This technique is referred to as *slot dozing*. The dozer can doze subsequent slots spaced at intervals, leaving narrow uncut sections between the slots. Eventually, the uncut sections are cut away. Slot dozing can increase production rates by as much as 30% or more, as compared to dozing without slots.

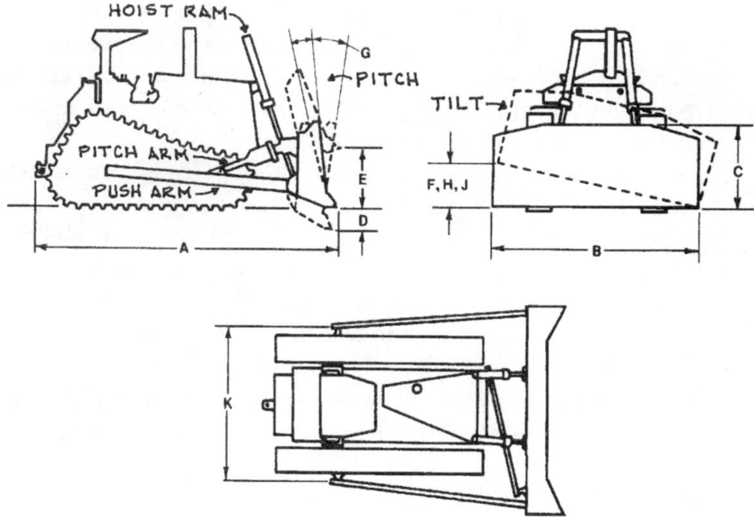


**Figure 5-1. Dozer dimensions.** (*Courtesy of Caterpillar Inc.*)

**Photo 5-8. Slot dozing.** (*Courtesy of Balderson Inc.*)

*Blade-to-blade* and *side-by-side* dozing are techniques using two dozers working adjacent to each other. The difference between the two techniques is that the blade-to-blade method uses two independent dozers operated in parallel by two operators with their blades almost touching, whereas, the side-by-side method uses two dozers mechanically coupled and operated by a single operator. The side-by-side method is more efficient and less expensive because less maneuvering time is required, and wages are paid to only one operator. When either method is used, the effective haul distance (or power zone) can be increased up to 1000 feet or more, and production rates will increase by at least 25%.

Most material is dozeable; however, dozer performance will vary with soil particle size and shape, the amount of voids, water content and soil temperature. Larger soil particles resist blade penetration and are harder to move than small particles. Also, angular-shaped particles are harder to doze than rounded, smooth particles due to increased friction. Well-graded soil lacking voids is a dense soil and is harder to move than poorly-graded soil full of voids.

Dry soil in the bank state usually increases the bond between soil particles and makes the soil difficult to move. However, a high moisture content increases the soil weight, also making it difficult to move. If dry soil freezes, the ability to move the soil is unaffected; however, frozen wet soil's bond is increased, making it difficult to move.

The weight and horsepower of a tractor determines its ability to doze, since the tractor cannot exert more pounds of push than its engine can develop. Also, with any given travel surface, the traction of the dozer is limited by the weight of the machine. Traction limitations and usable pull (push) were discussed in Chapter 2.

## DOZER BLADE PERFORMANCE

The potential performance of a dozer blade is expressed as horsepower per foot of cutting edge, or horsepower per loose cubic yard. The hp/foot rating indicates the blades ability to penetrate material and obtain a full blade load. The higher the horsepower-per-foot rating, the more aggressive the blade. The hp/LCY rating indicates the blade's ability to push material once the blade is loaded. The greater the horsepower-per-loose-cubic-yard rating, the faster the loose material can be pushed.

## DOZER BLADE ADJUSTMENTS

The blade is attached to the tractor by *push arms* which are normally mounted on the outside track frames (Figure 5-1). The blade can be pitched or tipped forward and back by *pitch arms* (pitch braces) which run diagonally from the push arms to the top of the blade (Figure 5-1). Both the straight and universal blades can be pitched. Pitching the blade varies the angle of attack of the blade against the ground and increases or decreases the blade penetration (Figure 5-1). Tipping the blade back reduces penetration, which is good for pushing loose soil over a firm, level surface (Figure 5-1).

The blade can be raised or lowered by a single or pair of hydraulic *hoist rams* which run from the top of the radiator guard to the bottom of the blade (Figure

5-1). Some dozers are equipped with cables instead of hoist rams.

With the exception of the cushion blade, earthmoving dozer blades can be tilted sideways (Figure 5-1). Tilting is useful for cutting ditches, penetrating soils with a hard crust, crowning roads and prying out underground obstacles. The blade can be tilted by adjustable pitch arms, or by a two-way hydraulic hoist ram.

The angle blade can be turned at an angle to the direction of travel of the tractor. This adjustment is especially useful for sidecasting or windrowing (Photo 5-9). The angle can be adjusted by manually changing the push arm length, or by adjusting a hydraulic control mounted on the push arm (Photo 5-10).

**Photo 5-9. Angle-dozing.** (*Photo by Dan Atcheson*)

**Photo 5-10. Blade angle controls.** (*Photo by Dan Atcheson*)

## TYPES OF DOZER BLADES

Dozer blades can be broadly classified as production dozing tools, or special application dozing tools. However, we will confine our discussion to earthmoving and landclearing blades.

### Universal Blade

The *universal* ("U") *blade* is manufactured with large wings on each end of the blade (Figure 5-2) (Photo 5-11). It has a relatively low hp/foot and hp/LCY rating; therefore, it is not used for high penetration, or moving heavy loads of material. This blade is used mainly for moving large loads of easily-dozed materials over long distances on projects such as land reclamation, stockpile work, coal handling, for charging hoppers, and trapping material for loaders. The universal blade can move tremendous loads. For example, a Caterpillar D11 dozer equipped with a 24-foot Balderson universal blade (BD11U-24) has a 94-LCY capacity. Depending on the type of tractor, this blade can be equipped with a single, or dual tilt cylinders, giving it the ability to ditch, pry out, level and perform many other utility tasks.

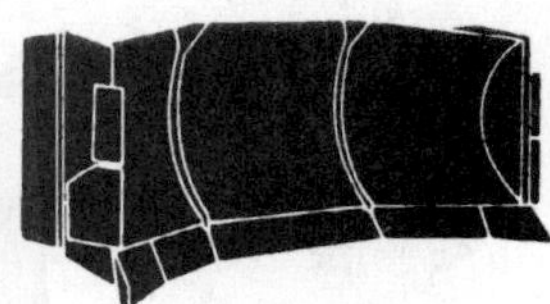

**Figure 5-2. Universal blade.** (*Courtesy of Caterpillar Inc.*)

**Photo 5-11. Universal blade.** (*Courtesy of Caterpillar Inc.*)

### Straight Blade

The *straight* ("S") *blade* is very maneuverable and versatile, and it can move a wide variety of materials (Figure 5-3) (Photo 5-4). In fact, it is considered to be the most versatile dozing blade. Although it is smaller than the "U" blade, it has a high hp/foot rating and is very aggressive in penetrating and obtaining a blade load. It has a high hp/LCY rating which enables the blade to easily move heavy material. This blade is best used for dozing material over a short to medium distance. It is also very useful for spreading materials in finish grading operations, and it can be equipped with a tilt cylinder to increase productivity and versatility. It can also be equipped with a *push plate* for pushing scrapers (Photo 5-12). A "bull" blade is similar to a straight blade, except that a bull blade cannot be tilted; therefore, a bull blade is seldom used anymore. Some straight blades are equipped with adjustable shanks at each end of the blade which can be extended up to 12 inches below the cutting edge, and can be used to rip tough or frozen soil, or to remove tree stumps.

**Figure 5-3. Straight blade.**
(*Courtesy of Caterpillar Inc.*)

**Photo 5-12. Straight blade used for pushing.** (*Courtesy of Rome Plow Company*)

**Semi-Universal Blade**

The *semi-universal* ("SU") *blade* combines characteristics of the straight and universal blades (Figure 5-4) (Photo 5-13). It has short wings on each end of the blade which help retain a load, and it can also penetrate and load materials quickly. It can be equipped with a tilt cylinder to increase productivity and versatility, or it can be equipped with a push plate for pushing scrapers. A *variable-radius* "SU" *blade* is also available with a variable-radius moldboard which causes the soil to move toward the center of the blade (Figure 5-5), and the extended side plates retain the load with less side spillage (Photo 5-14).

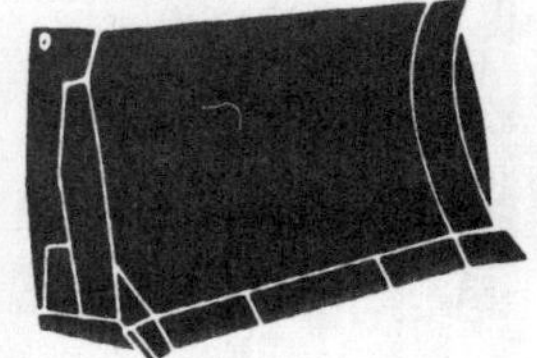

**Figure 5-4. Semi-universal blade.**
(*Courtesy of Caterpillar Inc.*)

**Photo 5-13. Dozer with semi-universal blade.** (*Courtesy of Caterpillar Inc.*)

**Figure 5-5. Variable-radius blade.**
(*Courtesy of Caterpillar Inc.*)

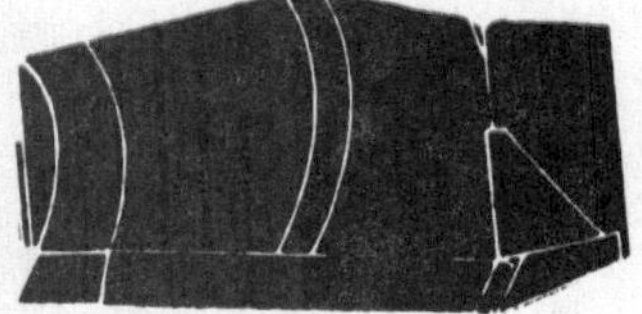

**Photo 5-14. Variable-radius semi-universal blade.** (*Courtesy of Balderson Inc.*)

**Angle Blade**

The *angle* ("A") *blade* is built so that it can be oriented straight, or at an angle of up to 25 degrees to either side of the straight position (Figure 5-6). Some angle blades can also be tilted, but not pitched. This blade can be used for pioneering roads, backfilling, cutting ditches and sidecasting (windrowing).

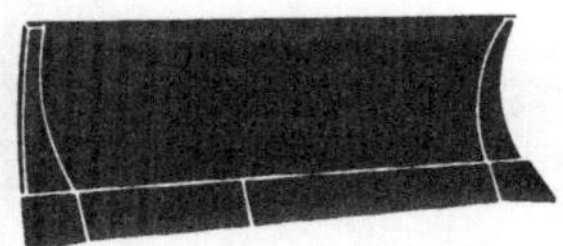

**Figure 5-6. Angle blade.**
(*Courtesy of Caterpillar Inc.*)

**Cushion Blade**

The *cushion* ("C") *blade* is installed on tractors used for push-loading scrapers (Figure 5-7) (Photo 5-15). This blade is equipped with rubber cushions to absorb the impact of contacting a scraper push block. This blade can also be used for cleanup work and other general dozing jobs. A *fixed plate* is sometimes used instead of a cushion blade (Photo 5-16). The fixed plate is relatively inexpensive; however, it does not always maintain proper contact with scrapers, especially when the vehicles are moving over a rough travel surface.

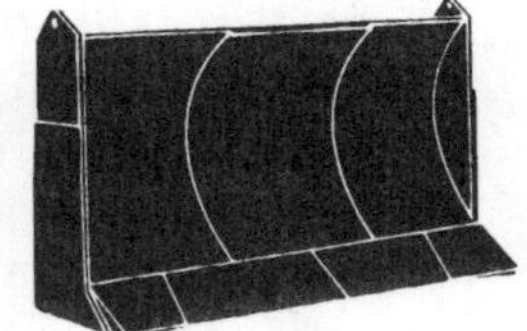

**Figure 5-7. Cushion blade.**
(*Courtesy of Caterpillar Inc.*)

**Photo 5-15. Dozer with cushion blade.** (*Courtesy of Caterpillar Inc.*)

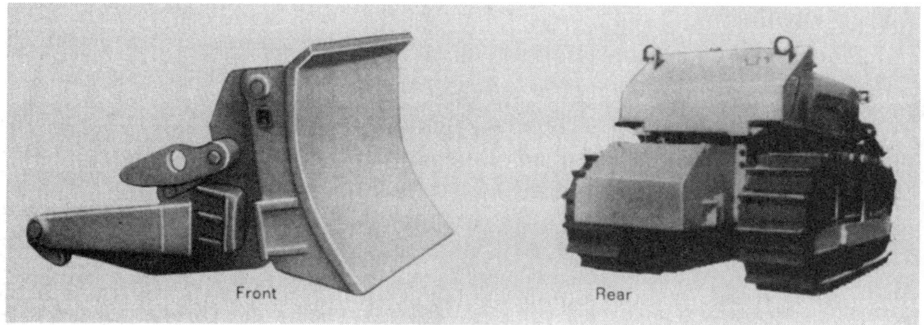


**Photo 5-16. Cushion pushblocks.** (*Courtesy of Fiat-Allis North America, Inc.*)

**Bowldozer Blade**

A *bowldozer* is a high-capacity blade used for moving lightweight material. It has a back wall and side walls reinforced by a rigid cross beam, but no bottom (Photo 5-17).

**Photo 5-17. Bowldozer blade.** (*Courtesy of Balderson Inc.*)

**Landfill Blade**

The *landfill blade* is designed to handle refuse. The blade is equipped with an open trash screen on the top of the blade which provides good visibility and protects the radiator (Photo 5-18).

**Photo 5-18. Landfill U-blade.** (*Courtesy of Balderson Inc.*)

**Stinger Blade**

A *stinger* (K/G) *blade* is an angle-blade tree cutter with a stinger projecting from the leading edge of the blade (Photo 5-19). This blade is used to weaken a tree by splitting the trunk. To further weaken the tree, the stinger can be inserted below grade as the dozer circles the tree while cutting the main horizontal roots of the tree. The blade can also be used to remove stumps, pile vegetation, cut drainage ditches and for general cleanup work. The angled blade allows debris to be pushed to one side for windrow piling.

**Photo 5-19.**
**Stinger blade.**
(*Courtesy of Rome Plow Company*)

**"V" Blade**

The *"V" blade* consists of two angled cutting blades with a stinger projecting from the leading center edge of the blade (Photo 5-20). This blade can shear trees and brush at ground level and cast the debris to each side, and it can be lowered below the ground surface to remove stumps. It can also be used to split tree trunks by ramming the tree with its stinger. This blade causes less wear and tear on the tractor, as compared to the stinger blade, and it requires less operator experience to maneuver.

**Photo 5-20.**
**"V" blade.**
(*Courtesy of Rome Plow Company*)

## Rake Blade

*Rake blades* are used for cleanup work behind the stinger and "V" blades, by removing and piling trees, rocks, roots, brush and small stumps (Photo 5-21). Granular material such as sand and gravel flows between the tines; therefore, vegetation can be transported and stacked for burning without excessive quantities of soil.

**Photo 5-21. Dozer with rake blade.** (*Courtesy of Balderson Inc.*)

## BLADE OPTIONS

The types of blades that can be installed depends on the type of tractor used. Options include blades built by the tractor manufacturer, and blades built by auxiliary equipment manufacturers (AEM blades). Table 5-1 indicates blade options available for various Caterpiller-built tractors.

### Table 5-1: Blade Options for Caterpillar-Built Machines

| | CATERPILLAR BLADES | | | | | | | SPECIAL ATTACHMENTS | | | | | | | | | | | | | | | |
|---|---|---|---|---|---|---|---|---|---|---|---|---|---|---|---|---|---|---|---|---|---|---|---|
| MODEL | S | U | SU | A | FS | LFS | P | RC | WC | CL | HU | LF | PAT | K/G | TP | VTC | RK | CU | CS | VR | WCS | SB | RCB |
| D8N | | ● | ● | ● | | | | ● | ● | ● | | ● | | ● | ● | ● | ● | ● | | ● | | ● | ● |
| D8N LGP | | | * | | | | | | | | | | | | | | | | | | | | |
| D9N | | ● | ● | | | | | ● | ● | ● | | ● | | ● | ● | ● | ● | ● | | ● | | | |
| D10N | | ● | ● | | | | | ● | ● | ● | | | | | | | | ● | | ● | | | |
| D11N | | ● | ● | | | | | ● | | ● | | | | | | | | ● | | ● | | | |

**CATERPILLAR SUPPLIED**
S — Straight
U — Universal
SU — Semi-Universal
A — Angling
FS — Fill Spreading
LFS — Landfill Spreading
P — Power Angle Tilt

**SPECIAL ATTACHMENTS SUPPLIED**
RC — Reclamation U
WC — Woodchips
CL — Coal
HU — Heavy U
LF — Landfill
PAT — Power Angle Tilt
K/G — KG Blade
VR — Variable Radius
TP — Tree Pusher
VTC — V-Tree Cutter
RK — Rake
CU — Cushion
CS — Coal Scoop
WCS — Wood Chip Scoop
SB — Slope Boards
RCB — Rolling Chopper Blade

**Note:** This chart suggests a range of blade options for Caterpillar built machines. It is not totally inclusive of all blades available. For additional information consult Caterpillar Special Attachments, Balderson.

(*Courtesy of Caterpillar Inc.*)

**BLADE CAPACITIES**

Blade capacity can be determined from the manufacturer's rating, or field-measured by creating a pile as follows:

1) Push a load onto a level area, then stop.
2) Move the dozer forward while raising the blade directly over the pile, forming a fairly symmetrical pile.
3) Reverse the dozer making sure the blade clears the pile.

The blade load, capacity, or volume can then be determined by:

$$\text{Blade Capacity (LCY)} = 0.0138 \times H \times W \times L \qquad \textbf{(Equation 5.1)}$$

where:

H = Average pile height (ft.) measured along the inside edge of each grouser mark (Figure 5-8)

$$= \frac{H_1 + H_2}{2}$$

W = Average pile width (ft.) measured along the inside edge of each grouser mark

$$= \frac{W_1 + W_2}{2}$$

and,

L = Greatest length of the pile (ft.)

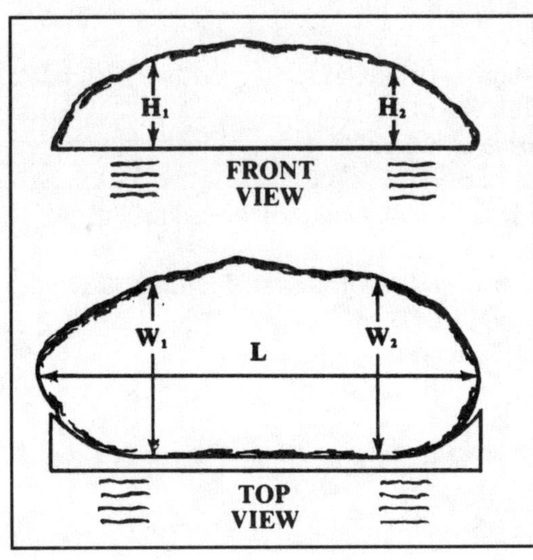


**Figure 5-8.**
**Measuring blade capacity.**
(*Courtesy of Terex Corporation*)

**Example 5-1:** Determine the blade capacity of a dozer that produces a pile with average dimensions of: L = 12 feet, W = 8 feet and H = 5 feet.
**Solution:** Expressed in loose cubic yards, the blade capacity is:

$$\text{Blade Capacity (LCY)} = 0.0138 \times 12' \times 8' \times 5' \quad \text{(Eq. 5.1)}$$
$$= 6.62 \text{ LCY}$$

**Example 5-2:** Assuming 25% swell, determine the blade capacity of the dozer given in the previous example in terms of bank cubic yards.
**Solution:** Expressed in bank cubic yards, the blade capacity is:

$$\text{Blade Capacity (BCY)} = \frac{6.62 \text{ LCY}}{1.25 \text{ LCY/BCY}} \quad \text{(25\% swell)}$$
$$= 5.30 \text{ BCY}$$

Blade capacity is also dependent on the grade and the direction of dozing. As a general rule, blade capacity increases about 5% for each percent grade while dozing down slope, with a maximum load occurring at about 20 to 25% favorable grade. For example, while dozing down a 10% grade, the anticipated load will be increased to:

$$\text{Capacity (\% of Normal)} = 100\% + (10\% \text{ Grade} \times 5\% \text{ Gain/\%-Grade})$$
$$= 100\% + 50\%$$
$$= 150\% \text{ of normal}$$

While dozing up slope, blade capacity decreases by about 2-1/2% for each percent grade up to about 20% grade, and further decreases at about 3/4% between 20% and 100% grade. For example, while dozing up a 15% grade, the anticipated load will be reduced to:

$$\text{Capacity (\% of Normal)} = 100\% - (15\% \text{ Grade} \times 2.5\% \text{ Loss/\%-Grade})$$
$$= 100\% - 38\%$$
$$= 62\% \text{ of normal}$$

While dozing up an 80% grade, the anticipated load will be reduced to:

$$\text{Capacity (\% of Normal)} = 100\% - (80\% \text{ Grade} \times 0.75\% \text{ Loss/\%-Grade})$$
$$= 100\% - 60\%$$
$$= 40\% \text{ of normal}$$

## DOZER PRODUCTION

Dozer production rates can be determined by:

$$\text{Production (LCY/Hr.)} = \frac{\text{Blade Capacity (LCY)}}{\text{Total Cycle Time (Hr.)}} \times \text{Efficiency Factor}$$

**(Equation 5.2)**

Total cycle time is the sum of the dozing, return and spot & start load times; i.e.,

$$\begin{aligned}\text{Total Cycle Time} &= \text{Dozing} + \text{Return} + \text{Spot \& Start Load Times}\\ &= \text{Dozing} + \text{Return} + \text{Fixed Time}\end{aligned}$$

**(Equation 5.3)**

Dozing speed can be determined for a given load by using a drawbar-pull performance curve (Figure 2-12). If a performance curve is not available, it is usually sufficiently accurate to assume the dozing speeds given in Table 5-2.

**Table 5-2: Average Dozing Speeds**

| Operating Conditions | Ave. Speed (mph) |
|---|---|
| Hard Materials; haul 100 ft. or less | 1.5 |
| Hard Materials; haul over 100 ft. | 2.0 |
| Loose Materials; haul 100 ft. or less | 2.0 |
| Loose Materials; haul over 100 ft. | 2.2 |

To return, the dozer is normally travels in reverse gear. Table 5-3 can be used to determine the reverse gear range of the dozer, then the reverse speed can be determined by referring to a travel speed table such as Table 5-4, or from field observation.

**Table 5-3: Maximum Return Speeds And Reverse Gear Ranges**

| Operating Conditions | Transmission Type | |
|---|---|---|
| | Direct Drive | Power Shift |
| Return 100 ft. or less | Max. reverse speed in gear used for dozing | Max. reverse speed in 2nd gear |
| Return over 100 ft. | Highest reverse speed | Max. reverse speed in 3rd gear |

**Table 5-4: D9N Dozer Speeds**

| Direction and Gear Range | Speed (mph) |
|---|---|
| Forward | |
| 1st | 2.5 |
| 2nd | 4.3 |
| 3rd | 7.5 |
| Reverse | |
| 1st | 3.0 |
| 2nd | 5.3 |
| 3rd | 9.3 |

(*Courtesy of Caterpillar Inc.*)

The dozing speeds obtained from Tables 5-2 and 5-4 can be converted to travel time as follows:

$$\text{Travel Time} = \frac{\text{Travel Distance (Ft.)}}{\text{88 Ft./Min. x Ave. Speed (mph)}} \qquad \textbf{(Equation 5.4)}$$

Spot & start load (*fixed*) time includes time required to shift gears and maneuver into the proper position to start loading. Fixed time can be determined from Table 5-5.

**Table 5-5: Spot & Start Load (Fixed) Time**

| Operating Conditions | Time (Min.) |
|---|---|
| Power Shift Transmission | 0.05 |
| Direct Drive Transmission | 0.10 |
| Hard Digging | 0.15 |

Job efficiency is taken from data given in Table 1-4, or estimated as productive time per clock hour. However, do not use both methods simultaneously.

**Example 5-3:** Assuming working a 50-minute hour, determine the production rate of a Caterpillar D9N bulldozer (power shift model) equipped with a 15.6-LCY-capacity blade digging hard material and pushing it 200 feet.
**Solution:** The dozing speed is:

$$\text{Dozing Speed} = 2.0 \text{ mph} \qquad \text{(Table 5-2)}$$

The dozing time is:

$$\text{Dozing Time (Min.)} = \frac{200 \text{ Ft.}}{88 \text{ Ft./Min. x } 2.0 \text{ mph}} \qquad \text{(Eq. 5.4)}$$

$$= 1.14 \text{ minutes}$$

Referring to Table 5-3, we can return at the maximum speed available in 3rd gear, reverse. Referring to Table 5-4, the maximum return speed is:

Return Speed = 9.3 mph

The return time is:

$$\text{Return Time (Min.)} = \frac{200}{88 \text{ x } 9.3} \qquad \text{(Eq. 5.4)}$$

$$= 0.24 \text{ minutes}$$

The fixed time is:

Fixed Time (Min.) = 0.15 minutes (Table 5-5)

The total cycle time is:

$$\text{Total Cycle Time (Min.)} = 1.14 \text{ Min.} + 0.24 \text{ Min.} + 0.15 \text{ Min.} \qquad \text{(Eq. 5.3)}$$

$$= 1.53 \text{ minutes}$$

In terms of hours, the total cycle time is:

$$\text{Total Cycle Time (Hr.)} = \frac{1.53 \text{ Min.}}{60 \text{ Min./Hr.}}$$

$$= 0.026 \text{ hours}$$

Including job efficiency, the probable hourly production rate is:

$$\text{Production (LCY/Hr.)} = \frac{15.6 \text{ LCY}}{0.026 \text{ Hr.}} \text{ x } 50/60 \qquad \text{(Eq. 5.2)}$$

$$= 500 \text{ LCY/hour}$$

## DOZER PRODUCTION CURVES

Dozer production rates can also be estimated by using production curves such as that shown in Figure 5-9. The production curve gives ideal production rates in loose cubic yards per hour for various types of Caterpillar dozers equipped with semi-universal blades.

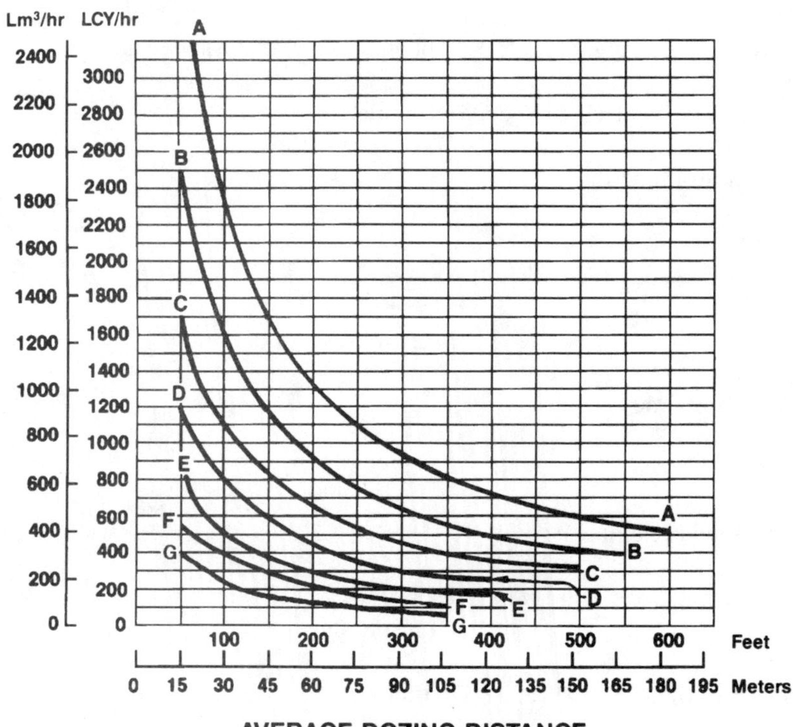


**KEY**

A — D11N-11SU
B — D10N-10SU
C — D9N-9SU
D — D8N-8SU
E — D7H-7SU
F — D6H-6SU
G — D5H XL-5SU XL

**Figure 5-9. Caterpillar dozing production with semi-universal blades (LCY/60-minute hour).**

(*Courtesy of Caterpillar Inc.*)

The production rates given in Figure 5-9 are based on the following assumptions:

| | |
|---|---|
| Efficiency | = 100% (60-minute hour) |
| Operator | = Excellent |
| Transmission | = Power shift with 0.05 minutes fixed time |
| Cut | = 50 feet prior to hauling |
| Dump time | = 0 seconds |
| Blade control | = Hydraulic |

Gear Ranges:

| | |
|---|---|
| Cut | = 1st gear forward |
| Carry | = 2nd gear forward |
| Return | = 2nd gear reverse |

Coefficient of traction:

| | |
|---|---|
| Track machines | = 0.5 or better |
| Wheel machines | = 0.4 or better |

Note: For wheel dozers, production rates fall by approximately 4% for each 1/100 decrease in the coefficient of traction below 0.40; i.e.,

$$\text{Loss of Production} = 4\% \times (0.40 - \text{Coef. of Traction}) \times 100 \qquad \textbf{(Equation 5.5)}$$

$$\text{Soil Weight} = 2300 \text{ lb./LCY}$$

Note: The soil weight correction factor can be obtained by:

$$\text{Soil Weight Correction Factor} = \frac{2300 \text{ lb./LCY}}{\text{Actual Weight/LCY}} \qquad \textbf{(Equation 5.6)}$$

The probable dozer production rate is obtained by tempering the ideal production rate given in Figure 5-9 with job condition correction factors obtained from Table 5-6 and Figure 5-10; i.e.,

$$\text{Actual Production} = \text{Ideal Production} \times \text{Correction Factors} \qquad \textbf{(Equation 5.7)}$$

**Table 5-6: Job Condition Correction Factors**

| Job Condition | Correction Factor | |
|---|---|---|
| | Track-Type Tractor | Wheel-Type Tractor |
| Operator | | |
| Excellent | 1.00 | 1.00 |
| Average | 0.75 | 0.60 |
| Poor | 0.60 | 0.50 |
| Material | | |
| Loose stockpile | 1.20 | 1.20 |
| Hard to cut; frozen- | | |
| with tilt cylinder | 0.80 | 0.75 |
| without tilt cylinder | 0.70 | – |
| cable-controlled blade | 0.60 | – |
| Hard-to-drift; "dead" (dry, non-cohesive material), or very sticky material | 0.80 | 0.80 |
| Rock, ripped or blasted | 0.60-0.80 | – |
| Dozing Method | | |
| Slot dozing | 1.20 | 1.20 |
| Side-by-side dozing | 1.15-1.25 | 1.15-1.25 |
| Visibility | | |
| Dust, rain, snow, fog or darkness | 0.80 | 0.70 |
| Job efficiency | | |
| 50 min./hr. | 0.83 | 0.83 |
| 40 min./hr. | 0.67 | 0.67 |
| Direct drive transmission | | |
| 0.1 min. fixed time | 0.80 | – |
| Bulldozer* | | |
| Adjust based on SAE capacity relative to the base blade used in the Estimated Dozing Production graphs. | | |
| Grades | (Refer to Figure 5-10) | |

* Angling blades and cushion blades are not considered production dozing tools. Depending on job conditions, the A-blade and C-blade will average 50-75% of straight blade production.

(*Courtesy of Caterpillar Inc.*)

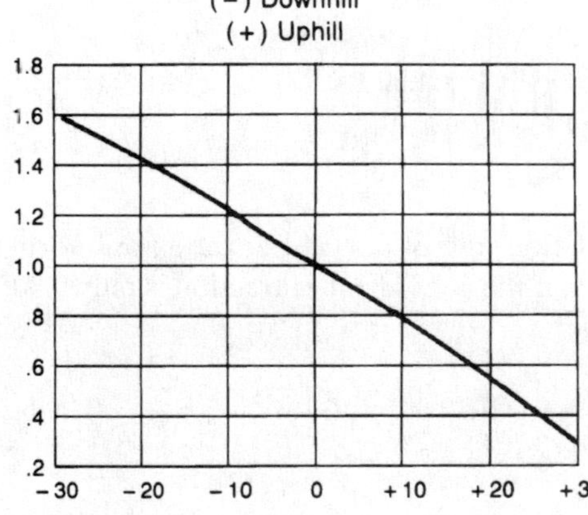


**Figure 5-10.**
**% grade versus dozing factor.**
*(Courtesy of Caterpillar Inc.)*

**Example 5-4:** Determine the hourly production rate of a Caterpillar D9N track dozer equipped with a semi-universal blade (with tilt cylinder) under the following conditions:

| | |
|---|---|
| Operator | = Average |
| Material | = Hard to cut |
| Soil Weight | = 2500 lb./LCY |
| Dozing Technique | = Slot dozing |
| Visibility | = Dusty |
| Transmission | = Power shift |
| Dozing | = Down 10% grade |
| Haul Distance | = 360 feet |
| Job efficiency | = 50-minute hour |

**Solution:** The ideal production rate is:

Ideal Production (LCY/Hr.) = 400 LCY/Hr. (Figure 5-9)

We will now temper the ideal production rate with the following job condition correction factors:

| Job Conditions | Correction Factor (Table 5-6) |
|---|---|
| Operator | 0.75 |
| Material | 0.80 |
| Slot Dozing | 1.20 |
| Visibility | 0.80 |
| Job Efficiency | 0.83 |
| Grade | 1.22 (Figure 5-10) |
| Soil Weight = 2300/2500 = | 0.92 (Eq. 5.6) |

Thus, the probable production rate will be:

$$\text{Production (LCY/Hr.)} = 400 \text{ LCY/Hr.} \times 0.75 \times 0.80 \times 1.20 \times 0.80 \times 0.83 \times 1.22 \times 0.92 = 215 \text{ LCY/hour} \quad \text{(Eq. 5.7)}$$

**Example 5-5:** Determine the production rate of a wheel dozer whose normal production rate is 300 LCY per hour, if the coefficient of traction is only 0.30.
**Solution:** The production loss will be:

$$\text{Production Loss} = 4\% \times (0.40 - 0.30) \times 100 = 40\% \quad \text{(Eq. 5.5)}$$

Since 40% of the production is lost, we have 60% production available; therefore, the probable production rate is:

$$\text{Actual Production (LCY/Hr.)} = 300 \text{ LCY/Hr.} \times 0.60 = 180 \text{ LCY/hour}$$

**Example 5-6:** Determine the production rate for the dozer given in Example 5-4 if the dozer is working at an altitude of 9000 feet above sea level.
**Solution:** Referring to Table 2-4, we can see that the D9N dozer operates at only 96% efficiency at the given altitude; therefore, the probable production rate is:

$$\text{Production (9000')} = 215 \text{ LCY/Hr.} \times 0.96 = 206 \text{ LCY/hour} \quad \text{(Table 2-4)}$$

**Example 5-7:** Determine the time required for the dozer given in the previous example to move 10,000 loose cubic yards of soil.
**Solution:** The project duration is:

$$\text{Project Duration (Hr.)} = \frac{10{,}000 \text{ LCY}}{206 \text{ LCY/Hr.}} = 48.5 \text{ hours} \quad \text{(Eq. 1.17)}$$

**Example 5-8:** Assuming the dozer given in Example 5-6 costs $180.00 per hour, including operator's wages, determine the unit production cost.
**Solution:** The unit production cost is:

$$\text{Unit Production Cost} = \frac{\$180.00\text{/Hr.}}{206 \text{ LCY/Hr.}} = \$0.87\text{/LCY} \quad \text{(Eq. 1.18)}$$

## *Chapter 6*

# SCRAPERS

With the exception of extremely large excavators, scrapers ("pans") are the most cost-efficient earthmovers available. Scrapers combine the ability to excavate and haul material economically over great distances. As a general rule, scrapers are most effective when the haul distance ranges from 500 feet to 2 miles. This range of scraper operation is referred to as the *economic application zone* (Figure 3-1). The economic application zone is dependent upon underfoot conditions, grades, material type, operator skill and the size of scraper. Scrapers are especially useful for alternating cuts and fills because they can excavate, haul and spread fill in a single cycle while also contributing to the compaction of the fill material.

One advantage of using scrapers instead of an excavator/haul-unit fleet is that production will be slowed, but not halted if one of the scrapers breaks down. However, if a single excavator is being used and breaks down, production comes to a standstill.

A scraper can dump its payload in uniformly thick layers, whereas, trucks cannot. As a result, compaction production is normally greater when scrapers are used. Also, scrapers are precision finishing machines that can help maintain haul roads by filling in low areas during the haul trip and cutting down high spots during the return trip.

Scrapers can clean up excess soil generated by motor graders and they can be used for finish grading (Photo 6-1). However, be sure that all tires are inflated with equal pressure because a low-pressure tire can cause the scraper to lean and gouge the ground surface on the side of the low tire. Also, keep an eye on the cutting edge of the blade since a badly worn blade is inefficient and will not cut properly or cleanly. It is important to control the moisture content of the soil being finished because soil that is too wet will stick to the blade and leave a rough surface.

**Photo 6-1. Elevating scraper picking up windrow left by grader.**
*(Photo by Dan Atcheson)*

Scrapers can be used to remove asphalt paving up to 4 inches thick. On thicker pavement, the scraper will require the assistance of a tractor pusher. Ripper teeth installed on the blade of the bowl will aid in removing asphalt, but don't rip the asphalt with other rippers prior to using scrapers, because this will cause the asphalt to stack up in front of the bowl. Also, ripped asphalt will preclude the use of an elevating scraper, since ripped asphalt will damage the paddles.

Before scrapers can be used, the site must be cleared of trees and boulders. Also, deep holes and gullies must be filled and sharp escarpments rounded off. Since the soil must be conducive to a scraper's excavating and loading capabilities, scraper production can be improved by preparing the soil before the scrapers enter the cut. For instance, hard, cohesive soils or rock should be ripped prior to scraper operations. On the other hand, loose, non-cohesive soils will load more easily if they have been pre-wetted. This will normally eliminate the need for water trucks at the fill area and will reduce traffic at the fill. Pre-wetted soil will also reduce the loss of traction due to a wet surface created by water trucks.

## TYPES OF SCRAPERS

Some scrapers are pulled by tractors *(trailer scrapers)* (Photos 6-2 and 6-3); however, large trailer scrapers require two axles to support the weight at the front of the scraper, because the hitch at the rear of the tractor cannot carry the load (Photo 6-3). Due to good traction and high drawbar pull available for loading, a scraper pulled by a crawler tractor is economical for short haul distances, especially on extremely steep grades, poorly-maintained haul roads, and over wet, slick underfoot conditions. For longer haul distances, scrapers pulled by wheel tractors, or wheel tractor-scrapers are more economical.

**Photo 6-2. Trailer scraper towed by a tractor.**

*(Courtesy of Miskin Scraper Works, Inc.)*

**Photo 6-3. Two-axle trailer scraper.** *(Courtesy of Rome Plow Company)*

A wheel *tractor-pulled scraper* is towed by a two-axle wheel tractor that can be used for purposes other than towing a scraper (Photo 6-2). This tractor-scraper combination is often referred to as a *three-axle scraper.* A *wheel tractor-scraper* is towed by a single-axle tractor designed expressly for towing a scraper (Figure 6-2). In this case, the scraper normally has only a single axle at the rear so that the weight of the front portion of the scraper is carried by the tractor. This increases the weight over the drive wheels of the tractor, resulting in increased traction. This tractor-scraper combination is often referred to as an *overhung* or *two-axle scraper* (Figure 6-2).

Some overhung scrapers are equipped with a single engine at the tractor and are referred to as *single-engine overhung scrapers,* or *standard scrapers* (Figure 6-2). Some single-engine scrapers are equipped with drive wheels on the scraper as well as the tractor and are referred to as *all-wheel-drive scrapers.* Other overhung scrapers have two engines, one at the tractor and the other at the rear of the scraper (Figure 6-3) (Photo 6-4). This type of tractor-scraper combination is an all-wheel-drive vehicle referred to as a *tandem-powered, twin-engine* or *four-wheel-drive scraper.* The second engine doubles the power, and more than doubles the traction as compared to a single-engine scraper. A *multi-bowl scraper* consists of a tractor towing scraper bowls coupled together (Photo 6-5).

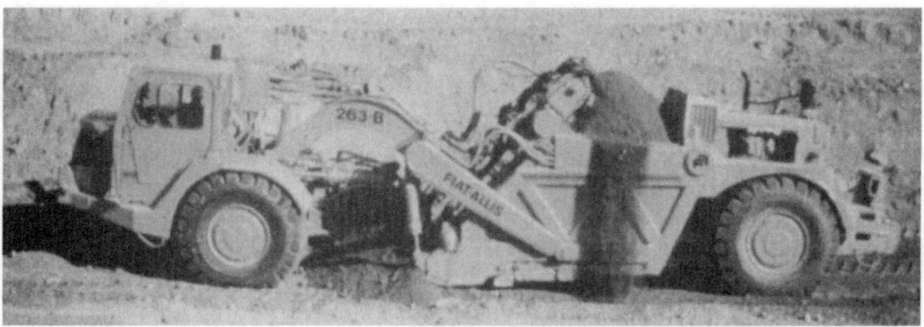


**Photo 6-4. Twin-engine overhung elevating scraper.**
*(Courtesy of Fiat-Allis North America, Inc.)*

**Photo 6-5. Multi-bowl scraper system.**
*(Courtesy of Miskin Scraper Works, Inc.)*

*Elevating* ("paddle wheel") *scrapers* are equipped with a ladder-type mechanism (elevator) that reduces loading resistance and assists during loading by lifting and distributing material into the bowl (Photo 6-6). The "ladder" consists of horizontal slates (flights or paddles) whose ends are connected to power-driven chains (Photo 6-7). The flights may cut in front of, or above the cutting edge of the bowl. The soil is thrown into the bowl as the flights travel upward. The movement of the flights can be reversed to dislodge obstacles and sticky soil, or to prevent material from entering the bowl during finish-grading operations. Some elevators have two speeds so that the elevator movement can be matched to the material conditions. Elevating scrapers usually require no assistance in loading and are referred to as *self-loading scrapers.* If a pusher is used to assist a self-loading unit, push cautiously to avoid damaging the elevator.

Elevating scrapers cannot be used to load rock, sticky soil, or extremely hard or large pieces of material, and they are not very useful if the cut area is short, because they can't cut deep enough to get a full load. In such instances, use an *open-bowl scraper* aided by a tractor pusher (Photos 6-8 and 6-10).

*Auger scrapers* are equipped with a power-driven auger located at the center of the scraper bowl. As the material flows over the scraper's cutting edge, it is lifted and distributed by the rotating auger. The auger scraper can usually load without pusher assistance. With either an elevating or auger scraper, tire life is often extended due to the assistance of the loading mechanisms.

Without a loading mechanism, a scraper must often be assisted during loading. *Push-pull scrapers* are equipped with hooks at the rear of the scraper and bails (hinged loops) at the front of the tractor so that they can couple together and assist one another while loading (Photo 6-9). In push-pull loading, the power of two scrapers is concentrated on loading one of them. The first scraper loads with the pushing assistance of the second scraper, then the second scraper loads with the pulling assistance of the first scraper. The scrapers then separate and haul their loads simultaneously. Push-pull loading eliminates the need for tractor pushers; however, since two loads are dumped almost simultaneously, you must have an adequate number of graders and compactors to handle the material at the fill. Also, this method of loading is not as efficient as using a tractor pusher.

**Photo 6-6. Elevating scraper.**
*(Courtesy of Caterpillar Inc.)*

**Photo 6-7. Scraper elevator and cutting edge.**
*(Photo by Dan Atcheson)*

**Photo 6-8. Open-bowl scraper.**
*(Courtesy of Caterpillar Inc.)*

**Photo 6-9. Push-pull scraper loading.**
*(Courtesy of Fiat-Allis North America, Inc.)*

Scrapers are built with a sturdy push block located at the rear of the scrapers which allows them to be assisted by a pusher tractor equipped with a reinforced cushion blade or push plate designed to absorb shock (Photo 6-3 and 6-10).

The type of scraper used depends on the size of the cut area, haul distance, soil type, size of rock, total resistance, traction, underfoot conditions and other job conditions. You can use Figure 6-1 to help in selecting the appropriate type of scraper for a particular project.

Machine shipping dimensions are normally published with manufacturer's specifications for any given scraper (Figure 6-2)

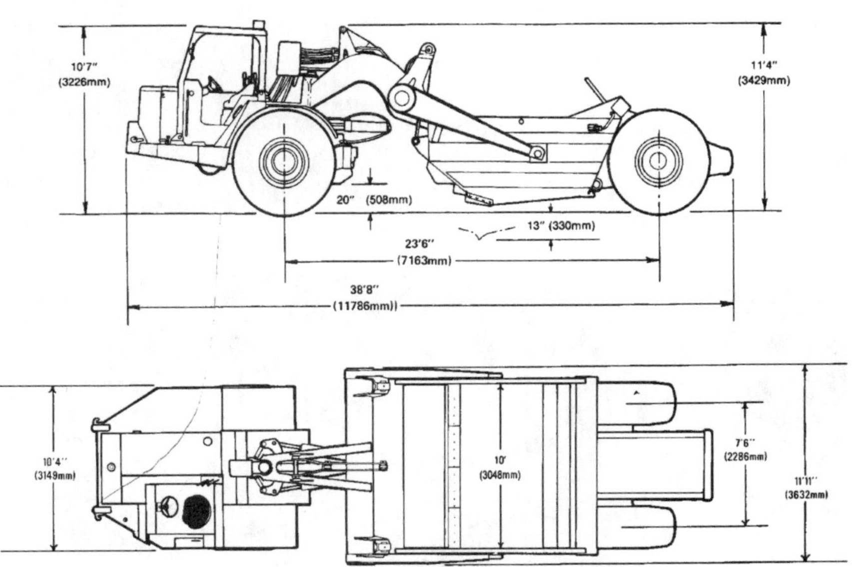


**Figure 6-2. Scraper dimensions.**
*(Courtesy of Fiat-Allis North America, Inc.)*

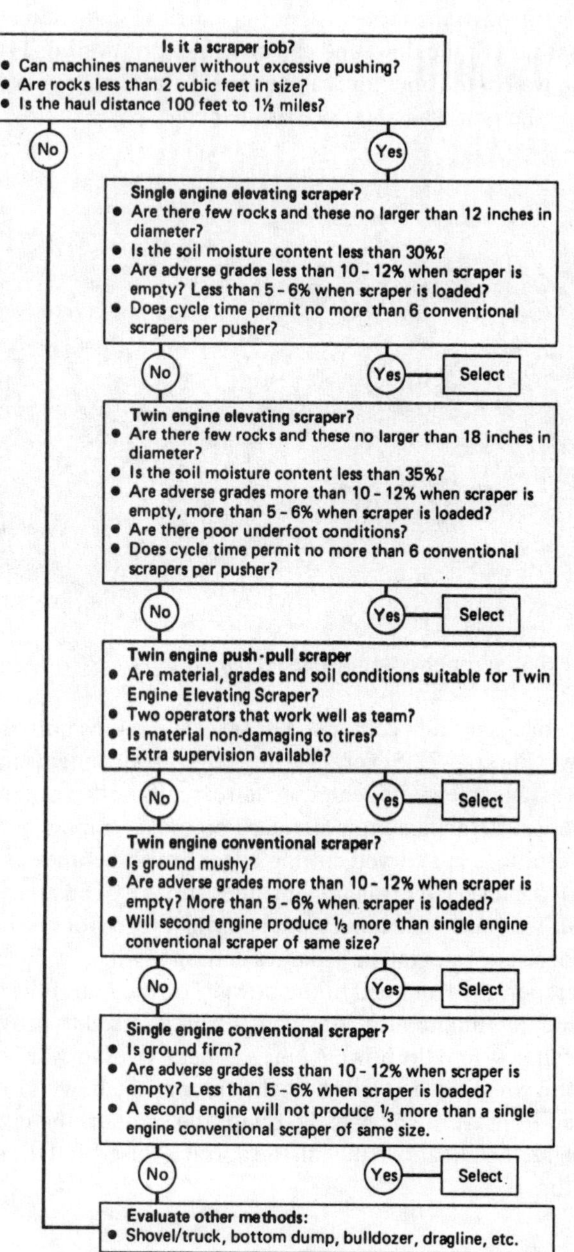


**Figure 6-1. Scraper selection guide.**

*(WABCO Construction and Mining Equipment Group)*

## SCRAPER COMPONENTS

The major components of a scraper are the bowl, apron and ejector (Figure 6-3). The *bowl* is the portion of the scraper that carries the loaded material. The *apron* is the front wall of the bowl and can be raised or lowered. The tailgate or *ejector* is the rear wall of the bowl and can be moved forward or back. Sometimes, the ejector forms the rear wall and floor of the bowl.

**Figure 6-3. Scraper components.** *(Courtesy of Terex Corporation)*

The cutting edge is usually comprised of several sections bolted to the front edge of the bowl (Photo 6-7). Several cutting edge arrangements are available. With the stinger arrangement, the center section extends farther out than the outer sections. The stinger arrangement is standard on most scrapers because it provides good penetration, but curved cutting edges are sometimes used in lieu of the stinger. With the level cut arrangement, all sections extend an equal distance beyond the bowl. The level cut arrangement is normally used for finish work. Some cutting edges consist of integral, or separate cutting teeth.

To load a scraper, the front end of the bowl is lowered until the cutting edge enters the ground. Simultaneously, the front apron is raised to provide an open slot so that soil can flow into the bowl. As the scraper moves forward, soil is forced into the bowl, and when the bowl is filled, the front of the bowl is raised and the apron is lowered to prevent spillage. To dump the scraper, the cutting edge is lowered to a desired height above the fill, the apron is raised and the soil is forced out by the ejector.

**PAYLOAD**

*Payload* is the load that a scraper can carry, exclusive of the vehicle weight. Payload can be expressed as weight capacity in pounds or tons, struck volume capacity in loose cubic yards, or heaped volume capacity in loose cubic yards. Regardless of the volume capacity, the rated weight capacity should never be exceeded. To do so will increase operating costs due to higher fuel consumption, reduced tire life, frequent failure of mechanical parts and higher maintenance costs. All of these items also equate to lost time and production while repairing and maintaining the vehicle.

*Struck capacity* of a scraper is the volume of material that can be filled to the top of the sides of the bowl; therefore, struck capacity remains constant for any given scraper. Struck capacity is of value when there is a limit on the amount of spillage allowed.

*Heaped capacity* will vary, depending on the height that the soil can be extended above the sides of the bowl. Heaped capacity is normally based on the heap having an angle of repose of 26.57 degrees, or a slope (run:rise) of 2 to 1 (2:1). Angles of repose for common materials are given in Table 6-1.

**Table 6-1: Angle of Repose of Various Materials**

| Material | Slope (Run:Rise) | Degrees |
|---|---|---|
| Coal, industrial | 1.4:1 - 1.3:1 | 35-38 |
| Common earth | | |
| dry | 2.8:1 - 1.0:1 | 20-45 |
| moist | 2.1:1 - 1.0:0 | 25-45 |
| wet | 2.1:1 - 1.7:1 | 25-30 |
| Gravel, round to angular | 1.7:1 - 0.9:1 | 30-50 |
| Sand | | |
| dry | 2.8:1 - 1.7:1 | 20-30 |
| moist | 1.8:1 - 1.0:1 | 30-45 |
| wet | 2.8:1 - 1.0:1 | 20-45 |

*(Courtesy of Caterpillar Inc.)*

The angle of repose for most soils under normal conditions range from 20 to 45 degrees; however, the actual angle of repose can be extremely variable, as was discussed in Chapter 4.

If the angle of repose of a soil is not 2 to 1, adjustments must be made in order to determine the actual heaped capacity of a scraper. Heaping occurs only above the sides of the scraper bowl, so we can make volume adjustments in the soil above the struck volume, since the struck volume remains constant for any given scraper. The heaped soil volume can be adjusted by applying the proper volume adjustment factor given in Table 6-2.

**Table 6-2: Heaped Volume Adjustment Factors**

| Slope (Run:Rise) | Volume Adjustment Factor |
|---|---|
| 3.0:1 | 0.69 |
| 2.8:1 | 0.74 |
| 2.6:1 | 0.79 |
| 2.4:1 | 0.85 |
| 2.2:1 | 0.92 |
| 2.0:1 | 1.00 |
| 1.8:1 | 1.09 |
| 1.6:1 | 1.21 |
| 1.4:1 | 1.34 |
| 1.2:1 | 1.50 |
| 1.0:1 | 1.69 |

**Example 6-1:** The struck capacity of a Caterpillar 621E scraper is 14 loose cubic yards (LCY) and its heaped capacity (based on a 2:1 slope) is 20 loose cubic yards. Determine the heaped capacity of the scraper if the slope of the heap is: a) 3:1, or b) 1:1.

**Solution:** The volume of soil affected by the angle of repose is:

Volume Above Sides (2:1 Slope) = 20 LCY – 14 LCY
= 6 LCY

a) With a 3:1 slope, this volume will be reduced to:

Volume Above Sides (3:1 Slope) = 0.69 x 6 LCY (Table 6-2)
= 4.1 LCY

Thus, the total heaped capacity is:

Heaped Volume (3:1 Slope) = 14 LCY + 4.1 LCY
= 18.1 LCY

b) With a 1:1 slope, the volume above the sides will be increased to:

Volume Above Sides (1:1 Slope) = 1.69 x 6 LCY (Table 6-2)
= 10.1 LCY

Thus, the total heaped capacity is:

Heaped Volume (1:1 Slope) = 14 LCY + 10.1 LCY
= 24.1 LCY

As a rule of thumb, some contractors estimate the heaped capacity under normal conditions as being midway between the manufacturer's struck and heaped capacity ratings. For example, the estimated heaped capacity of a scraper with a rated 14-LCY struck capacity and 20-LCY heaped capacity would be:

$$\text{Average Heaped Capacity} = \frac{14\ \text{LCY} + 20\ \text{LCY}}{2}$$

$$= 17\ \text{LCY}$$

As mentioned earlier, be careful not to exceed the weight limitations of the payload. Weights of various materials are given in Table 1-1. The maximum payload limit expressed in cubic yards can be determined by:

$$\text{Maximum Load Limit (C.Y.)} = \frac{\text{Maximum Weight Limit}}{\text{Weight/C.Y.}} \qquad \textbf{(Equation 6.1)}$$

**Example 6-2:** Assuming hauling damp sand weighing 2850 pounds per loose cubic yard, determine if you are hauling within the weight limitations of the scraper given in the previous example if the maximum capacity is 20 tons and the angle of repose is 3:1.
**Solution:** The payload weighs:

$$\text{Payload (Tons)} = \frac{18.1\ \text{LCY x } 2850\ \text{lb./LCY}}{2000\ \text{lb./Ton}} \qquad \text{(Example 6-1)}$$

$$= 25.8\ \text{tons}$$

This exceeds the maximum allowable weight capacity by:

$$\text{Overload (Tons)} = 25.8\ \text{Tons} - 20\ \text{Tons} \qquad \text{(Given)}$$

$$= 5.8\ \text{tons}$$

This is equivalent to an overload volume of:

$$\text{Overload (LCY)} = \frac{5.8\ \text{Tons x } 2000\ \text{lb./Ton}}{2850\ \text{lb./LCY}}$$

$$= 4.1\ \text{LCY}$$

Thus, the maximum volume of damp sand we are allowed to haul is:

$$\text{Maximum Load (LCY)} = 18.1\ \text{LCY} - 4.1\ \text{LCY} \qquad \text{(Example 6-1)}$$

$$= 14\ \text{LCY}$$

We could have also solved this problem by determining the maximum payload volume as follows:

$$\text{Maximum Load Limit (LCY)} = \frac{20 \text{ Tons x } 2000 \text{ lb./Ton}}{2850 \text{ lb./LCY}} \qquad \text{(Eq. 6.1)}$$
$$= 14 \text{ LCY}$$

As a general rule, the production rate of a scraper is normally expressed in bank cubic yards per hour; therefore, we must express the scraper payload in terms of bank cubic yards. However, the degree of compaction within the bowl will vary, depending on the type of scraper and loading method used. For example, an elevating or auger scraper will generally deposit the soil in a loose condition within the bowl. However, open-bowl scrapers will deposit the soil using "brute force" which increases the compactive effort exerted on the soil, especially if a tractor-pusher is used to assist the scraper through the cut. Therefore, the end result is that soil is deposited in the bowl somewhere between a loose and bank condition. Some tests have shown that the loose-yard payload can be increased by 10%.

**Example 6-3:** Assuming 30% swell, determine the payload of a 20-LCY elevating scraper in terms of bank cubic yards.
**Solution:** Expressed in bank cubic yards, the scraper payload is:

$$\text{Payload (BCY)} = \frac{20 \text{ LCY}}{1.30 \text{ LCY/BCY}} \qquad \text{(30\% swell)}$$
$$= 15.4 \text{ BCY}$$

**Example 6-4:** Work the previous example using a standard scraper assisted by a tractor-pusher.
**Solution:** Assuming that the compactive effort increases the payload by 10% during loading, the partially compacted payload is:

$$\text{Payload (BCY)} = 15.4 \text{ BCY x } 1.10$$
$$= 16.9 \text{ BCY}$$

## TOTAL SCRAPER CYCLE TIME

The total scraper cycle time is the sum of the load, haul, turning and dump, return, and spot and delay times; i.e.,

$$\begin{aligned}\text{Total Cycle Time} &= \text{Load + Haul + Turning \& Dump + Return} \\ &\quad + \text{Spot \& Delay Times} \\ &= \text{Load + Turning \& Dump + Spot \& Delay + Travel Times}\end{aligned}$$

**(Equation 6.2)**

Load times, turning & dump times and spot & delay times can be combined into what is referred to as *fixed time;* therefore, the total scraper cycle time can be defined as:

Total Cycle Time = Fixed Time + Travel Time **(Equation 6.3)**

**LOAD TIME**

The load time of a scraper is dependent upon the size, power and type of scraper, the amount of pusher power available, soil condition, total resistance, traction, traffic at the cut and operator skill. In the absence of field timing tests and previous job data, you can use Table 6-3 to estimate load time.

**Table 6-3: Scraper Loading Times (Min.)**

| Job Conditions | Push-Loaded Single Engine | Push-Loaded Twin Power | Twin-Hitched (Two Scrapers) | Self-Loaded Elevator | Self-Loaded Twin Power |
|---|---|---|---|---|---|
| Favorable | 0.5 | 0.4 | 0.7 | 0.8 | 0.6 |
| Average | 0.7 | 0.6 | 1.0 | 1.1 | 0.8 |
| Unfavorable | 1.0 | 0.9 | 1.4 | 1.5 | 1.0 |

*(Courtesy of Terex Corporation)*

Load time can be reduced by loading down grade, since this will reduce the total resistance and the amount of pusher power required. *Straddle loading* can also reduce load time. When the straddle-loading method is used, cuts are spaced at intervals leaving ridges between the cuts. Eventually, the ridges are cut away and removed. When a scraper is excavating a hill, start the cut at the top of the hill and make the cut as deep as possible, raising the bowl as the cut proceeds down the slope. This will flatten the hill and increase production.

Load at a rate so that the tires do not spin, since this will cause excessive and expensive tire wear. If the loading rate is too rapid, lift the bowl and make a lighter cut. Sometimes it is more productive to make a shallow cut in a faster, higher gear than to make a deep cut in a slower, lower gear.

Ripping hard, cohesive soil prior to scraper excavation will reduce load time. Raising and lowering the scraper bowl will also speed up loading when soft, non-cohesive soil is being loaded.

Using pusher assistance will increase the loading rate (Photo 6-10). Use pushers powerful and heavy enough to push the scrapers through the cut and accelerate (boost) the scrapers as they leave the cut. Adequate boosting will allow scrapers to spread the soil remaining in front of the bowl as it is lifted. Pushers are normally crawler tractors weighing at least 20 tons. Four-wheel-drive wheel-mounted tractor pushers will increase production, provided they have adequate traction, since the speed of both vehicles is limited by the speed of the pusher. The pusher normally pushes in first gear, limiting track-mounted pushers to about 1-1/2 mph and wheel-mounted pushers to 2-1/2 mph, or more.

**Photo 6-10. Push loading.** *(Courtesy of Caterpillar Inc.)*

Some pushers are equipped with a fixed push plate while others use a reinforced dozer blade (Photos 5-15 and 5-16). Since a dozer blade can be equipped with rippers that engage only when the dozer is backing up, the dozer can be used for back-ripping or cleanup work while waiting for a scraper.

Use an adequate number of pushers to service all of the scrapers, and if congestion occurs, reduce the load time of some of the scrapers so that they are more evenly spaced.

Pushers are often used in tandem to increase the loading rate. As a general rule, each pound of push (including the scraper's effort) loads one pound of soil into the bowl per minute of load time.

There are three basic methods used for push loading, including back-track loading, chain loading and shuttle loading (Figure 6-4). When *back-track loading,* the pusher assists a scraper, then returns and transfers to another scraper. When *chain loading,* the pusher transfers from one scraper to another without having to return to its original position, since the second scraper is located in front and to the side of the first scraper. When *shuttle loading,* the pusher transfers from one scraper to another moving in the opposite direction. You cannot shuttle load effectively unless the site is laid out so that the scrapers can leave the cut through two exits.

a. Back Track Loading

b. Chain Loading

c. Shuttle Loading

**Figure 6-4.**
**Push-loading methods.**
*(Courtesy of Terex Corporation)*

The pusher cycle time is:

Pusher Cycle Time = Load + Boost + Return + Transfer Times

**(Equation 6.4)**

Since return time is eliminated from the pusher cycle time when chain loading or shuttle loading is used, scraper load time will be reduced when either of these two push-loading methods is used.

## TURNING AND DUMP TIME

Turning and dump time is required to maneuver and off-load at the fill area. Turning and dump time is dependent upon the size, power and type of scraper, soil conditions, total resistance, traction, operator skill, and turning room and traffic at the fill, including scrapers dumping simultaneously, bulldozers, graders and compactors. Some contractors allow scrapers to drive over previously dumped soil to help compact it. This will increase dump time, but will reduce compaction time. In the absence of field timing tests and previous job data, you can use Table 6-4 to estimate turning and dump time.

**Table 6-4: Scraper Turning and Dump Times (Min.)**

| Job Conditions | Single Engine | Twin Power | Elevator | Twin Power in Coal |
|---|---|---|---|---|
| Favorable | 0.3 | 0.3 | 0.3 | 0.2 |
| Average | 0.5 | 0.5 | 0.5 | 0.3 |
| Unfavorable | 1.0 | 0.9 | 0.8 | 0.5 |

*(Courtesy of Terex Corporation)*

Turning and dump time can be reduced by dumping down grade. Keeping the cutting edge of the bowl high enough off the ground will help keep wet, cohesive soil from becoming clogged as it passes through the front of the bowl. Provide pushers if the scrapers require additional power to dump their loads and be boosted out of the fill. Dump time can be reduced by providing an adequate number of graders and compactors to keep the fill smooth and compacted, and by giving scrapers the right-of-way.

When the scraper dumps its load, start by placing the fill in the lowest spots. This will produce a more level surface and will increase production.

## SPOT AND DELAY TIME

Spot and delay time is the time required for the scraper to maneuver into the proper position and wait for pusher assistance at the cut. Spot and delay time is mainly dependent upon the type of scraper used, the number of pushers available and the type of push-loading method used. In the absence of field timing tests and previous job data, you can use Table 6-5 to estimate spot and delay time.

**Table 6-5: Scraper Spot and Delay Times (Min.)**

| Job Conditions | Push-Loaded Single Engine | Twin Power | Twin-Hitched (Two Scrapers) | Self-Loaded Elevator | Self-Loaded Twin Power in Coal |
|---|---|---|---|---|---|
| Favorable | 0.2 | 0.1 | Negligible | Negligible | Negligible |
| Average | 0.3 | 0.2 | 0.1 | Negligible | Negligible |
| Unfavorable | 0.5 | 0.5 | 0.3 | 0.2 | Negligible |

*(Courtesy of Terex Corporation)*

Spot and delay time can be reduced by using self-loading scrapers and by maneuvering down grade. When push loading, spot and delay time can be reduced by providing an adequate number of pushers to service the scrapers, and by using either the chain-, or shuttle-loading method.

## TRAVEL TIME

Travel time depends primarily on the engine power, payload, total resistance and round-trip distance. Travel times can be quickly estimated by using travel time charts such as those shown in Figures 6-5 and 6-6. Two travel time charts are shown for the scraper; one for the scraper hauling its rated payload, and the other for the empty vehicle. Travel time charts take acceleration and deceleration of the scraper into consideration at the loading and dumping areas.

To use a travel time chart, enter the chart from the left and find the one-way travel distance. Project a horizontal line until it intersects the appropriate diagonal line representing total resistance (effective grade). From this intersection, drop vertically to obtain the travel time at the bottom of the chart.

When a scraper is traveling down grade, the total resistance can be negative. This can cause the scraper to accelerate, requiring the use of the retarder, a lower gear, and/or brakes. Under these conditions, travel time charts cannot be used. Instead, a brake-retarder performance curve such as that shown in Figure 6-7 must be used to determine the maximum safe speed during descent.

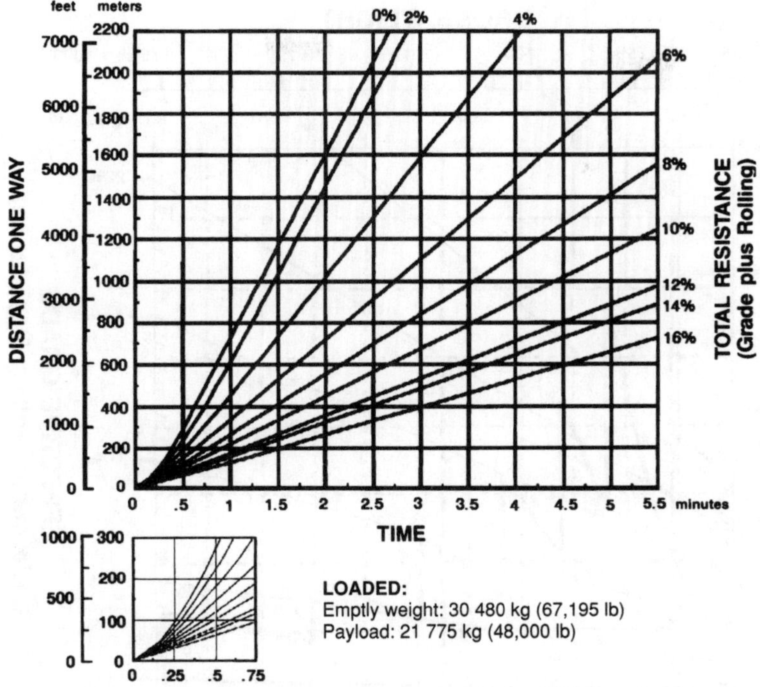


**Figure 6-5. 621E scraper travel time chart (loaded).** *(Courtesy of Caterpillar Inc.)*

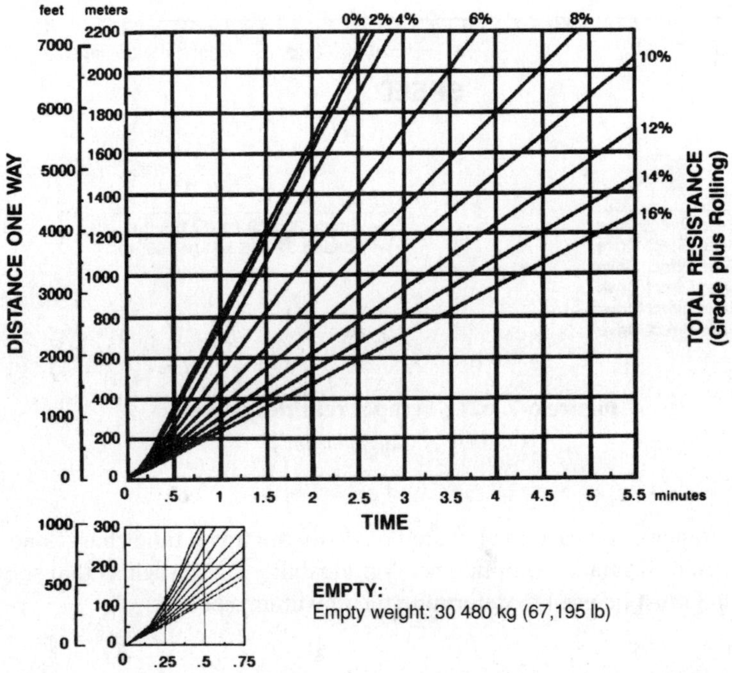


**Figure 6-6. 621E scraper travel time chart (empty).** *(Courtesy of Caterpillar Inc.)*

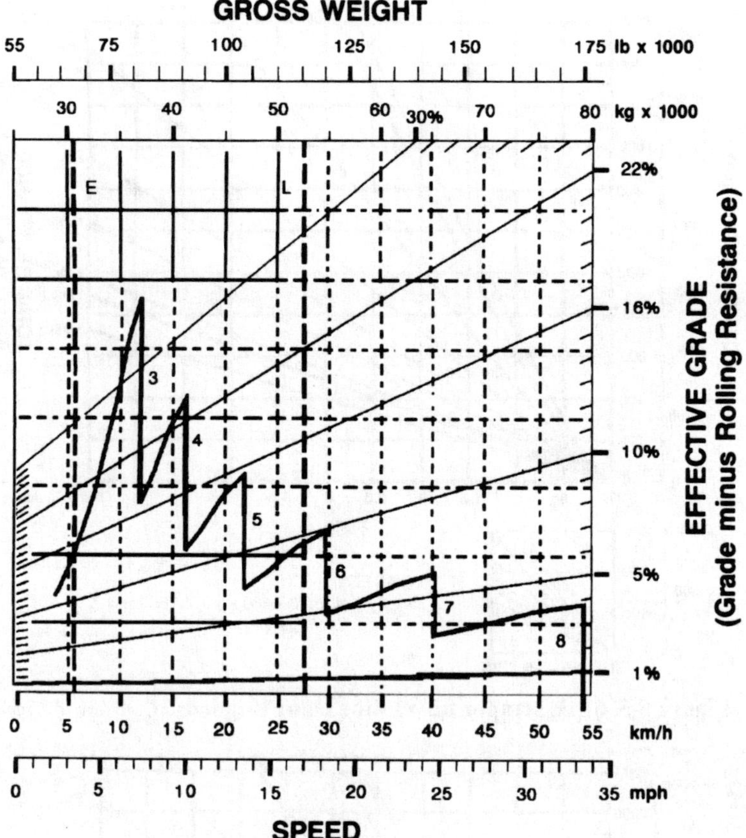


KEY
3 — 3rd Gear Direct Drive
4 — 4th Gear Direct Drive
5 — 5th Gear Direct Drive
6 — 6th Gear Direct Drive
7 — 7th Gear Direct Drive
8 — 8th Gear Direct Drive

KEY
E — Empty 30 480 kg (67,195 lb)
L — Loaded 52 255 kg (115,195 lb)

**Figure 6-7. 621E scraper retarder curve.**

*(Courtesy of Caterpillar Inc.)*

If the scraper is not carrying its rated payload, the travel time chart (loaded) cannot be used. Instead, a rimpull-speed-gradeability curve such as that shown in Figure 6-8 must be used to determine the maximum speed.

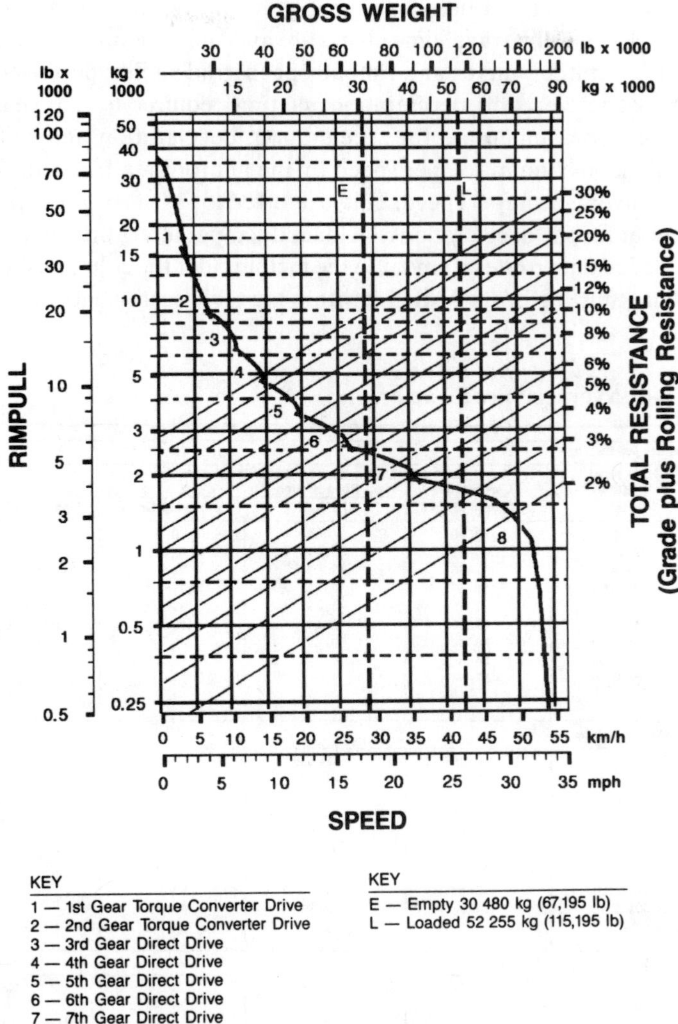


**Figure 6-8. 621E rimpull-speed-gradeability performance curve.**

*(Courtesy of Caterpillar Inc.)*

The vehicle speeds obtained from Figures 6-7 and 6-8 can be converted to travel time as follows:

$$\text{Travel Time (Min.)} = \frac{\text{Travel Distance (Ft.)}}{\text{88 Ft./Min. x Ave. Speed (mph)}} \qquad \textbf{(Equation 6.5)}$$

While travel time is one of the most important considerations for determining scraper production rates, it is probably the most difficult time component to estimate. Even when travel time charts are available, their use is somewhat limited. Also, retarder curves and rimpull-speed-gradeability performance curves do not take acceleration, deceleration and time required to shift gears into consideration. To compensate for acceleration, deceleration and shifting gears, multiply the maximum scraper speed by the appropriate factor in Table 6-6 to obtain an average scraper speed. This is referred to as the *Average Speed Method.* To account for the deceleration of a loaded scraper along the final short section of the loaded haul route (if any), use the factor under the column labelled "Level Haul Unit Starting From 0 MPH," within the table section labelled "Under 300 lb./hp."

**Table 6-6: Average Speed Factors**

| Haul Road Length in Feet | Level Haul Unit Starting from 0 MPH | Unit in Motion When Entering Haul Road Section | | |
|---|---|---|---|---|
| | | Level | Downhill Grade | Uphill Grade Factor |
| **under 300 lbs./hp** | | | | |
| 0-200 | 0-.40 | 0-.65 | 0-.67 | 1.00 |
| 201-400 | .40-.51 | .65-.70 | .67-.72 | |
| 401-600 | .51-.56 | .70-.75 | .72-.77 | (Entrance speed |
| 601-1000 | .56-.67 | .75-.81 | .77-.83 | greater than |
| 1001-1500 | .67-.75 | .81-.88 | .83-.90 | maximum |
| 1501-2000 | .75-.80 | .88-.91 | .90-.93 | attainable speed |
| 2001-2500 | .80-.84 | .91-.93 | .93-.95 | on section) |
| 2501-3500 | .84-.87 | .93-.95 | .95-.97 | |
| 3501 & up | .87-.94 | .95- | .97- | |
| **300-380 lbs./hp** | | | | |
| 0-200 | 0-.39 | 0-.62 | 0-.64 | 1.00 |
| 201-400 | .39-.48 | .62-.67 | .64-.68 | |
| 401-600 | .48-.54 | .67-.70 | .68-.74 | (Entrance speed |
| 601-1000 | .54-.61 | .70-.75 | .74-.83 | greater than |
| 1001-1500 | .61-.68 | .75-.79 | .83-.88 | maximum |
| 1501-2000 | .68-.74 | .79-.84 | .88-.91 | attainable speed |
| 2001-2500 | .74-.78 | .84-.87 | .91-.93 | on section) |
| 2501-3500 | .78-.84 | .87-.90 | .93-.95 | |
| 3501 & up | .84-.92 | .90-.93 | .95-.97 | |
| **380 & up lbs./hp** | | | | |
| 0-200 | 0-.33 | 0-.55 | 0-.56 | 1.00 |
| 201-400 | .33-.41 | .55-.58 | .56-.64 | |
| 401-600 | .41-.46 | .58-.65 | .64-.70 | (Entrance speed |
| 601-1000 | .46-.53 | .65-.75 | .70-.78 | greater than |
| 1001-1500 | .53-.59 | .75-.77 | .78-.84 | maximum |
| 1501-2000 | .59-.62 | .77-.83 | .84-.88 | attainable speed |
| 2001-2500 | .62-.65 | .83-.86 | .88-.90 | on section) |
| 2501-3500 | .65-.70 | .86-.90 | .90-.92 | |
| 3501 & up | .70-.75 | .90-.93 | .92-.95 | |

*(Courtesy of Terex Corporation)*

Grades and road surface conditions can vary along the haul road. Scrapers may have to slow down for curves, not only for safety, but to preserve tire life, especially in hot weather. Scraper traffic on narrow roads can become bottlenecked, especially when graders, dozers and water trucks are maintaining the haul road. Also, one slow scraper can dictate overall production when passing is dangerous, or impossible.

Bad weather conditions such as wind, rain and blowing dust can affect visibility and slow production. Rain can change a relatively hard-surfaced dirt road into a quagmire, and adverse haul-road conditions will require that scrapers haul less than their rated payloads.

The driver's skill can affect scraper production rates. Also, a leased scraper will normally be run harder than an owned one, resulting in increased acceleration, faster gear-changing time and increased turning and hauling speed.

Production can be increased by hauling from two cuts to a single fill (Figure 6-9), or by hauling from a single cut to two fills (Figure 6-10). Either procedure will eliminate one turn from each cycle as compared to the normal method of hauling from a single cut to a single fill, and returning (Figure 6-11).

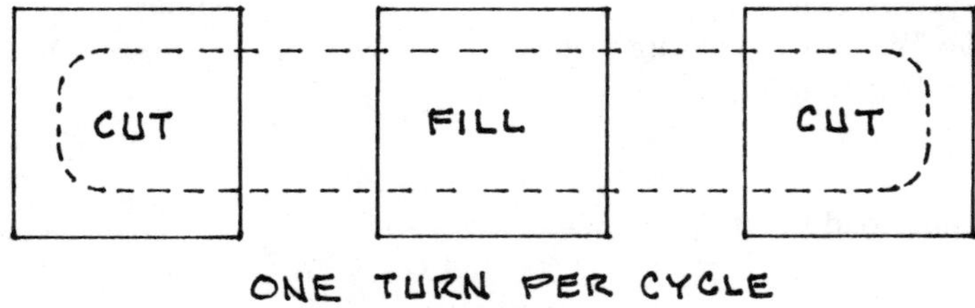


**Figure 6-9. Two cuts hauled to a single fill area.**

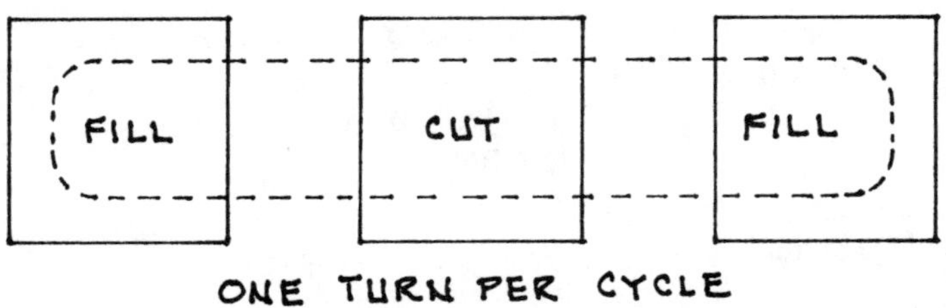


**Figure 6-10. One cut hauled to two fill areas.**

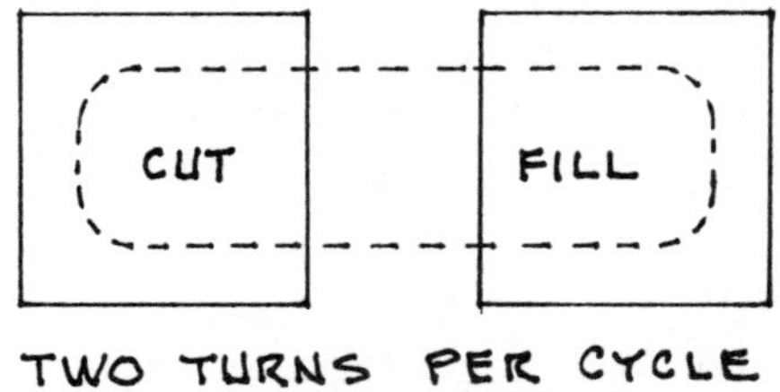


**Figure 6-11. Single cut hauled to a single fill.**

**Example 6-5:** Determine the total scraper cycle time and the hourly production rate for a 20-LCY single-engine Caterpillar 621E standard scraper working under the following conditions:

| | |
|---|---|
| Haul Road | = Level; loose earth |
| Rolling Resistance | = 80 pounds per ton of G.V.W. |
| Gross Vehicle Weight (Empty) | = 67,195 pounds |
| Maximum Gross Vehicle Weight (Loaded) | = 115,195 pounds |
| Maximum Load | = 48,000 pounds |
| Soil | = Common earth; 2500 lb./LCY |
| Swell | = 25% |
| Haul Distance (One Way) | = 3000 feet |
| Job Conditions | = Average |
| Loading Method | = Push-loading |
| Job Efficiency | = 50-minute hour |

**Solution:** We must make sure that we are not overloading the scraper. The maximum payload limit is:

$$\text{Maximum Load Limit (LCY)} = \frac{48{,}000 \text{ lb.}}{2500 \text{ lb./LCY}} = 19.2 \text{ LCY} \qquad \text{(Eq. 6.1)}$$

The maximum payload expressed in bank cubic yards is:

$$\text{Payload (BCY)} = \frac{19.2 \text{ LCY}}{1.25 \text{ LCY/BCY}} = 15.4 \text{ BCY} \qquad \text{(25\% swell)}$$

The fixed cycle times are:

Load Time = 0.7 minutes (Table 6-3)
Turning and Dump Time = 0.5 minutes (Table 6-4)
Spot and Delay Time = 0.3 minutes (Table 6-5)

The total fixed time is:

$$\text{Total Fixed Time (Min.)} = 0.7 \text{ Min.} + 0.5 \text{ Min.} + 0.3 \text{ Min.} = 1.5 \text{ minutes}$$

The rolling resistance is:

Rolling Resistance (lb./Ton) = 80 lb./ton (Given)

We have no grade resistance since the haul road is level; therefore, the total resistance (effective grade) equals rolling resistance expressed in %-grade; i.e.,

$$\text{Effective Grade (\%)} = \frac{80 \text{ lb./Ton}}{20} \qquad \text{(Eq. 2.16)}$$

$$= 4\%$$

Referring to Figures 6-5 and 6-6, the travel times are:

Travel Time (Loaded) = 1.8 minutes (Figure 6-5)

and,

Travel Time (Empty) = 1.3 minutes (Figure 6-6)

The total travel time is:

Total Travel Time (Min.) = 1.8 Min + 1.3 Min.
= 3.1 minutes

The total scraper cycle time is:

Total Scraper Cycle Time (Min.) = 1.5 Min. + 3.1 Min. (Eq. 6.3)
= 4.6 minutes

Working at 100% efficiency, the number of cycles per hour is:

$$\text{Cycles/Hr.} = \frac{60 \text{ Min./Hr.}}{4.6 \text{ Min./Cycle}} \qquad \text{(Eq. 1.6)}$$

$$= 13 \text{ cycles/hour}$$

Working at 100% efficiency, the hourly production rate of an individual scraper is:

Production (BCY/Hr.) = 15.4 BCY/Cycle x 13 Cycles/Hr.
= 200.2 BCY/hour

Working at a job efficiency of only 50 minutes per hour, the probable production rate is:

Production (BCY/Hr.) = 200.2 BCY/Hr. x 50/60
= 166.8 BCY/hour

**Example 6-6:** Determine the production rate of the scraper given in the previous example while working at 9000 feet above sea level.
**Solution:** Referring to Table 2-4, we can see that a 621E scraper operates at only 87% efficiency at the given altitude; therefore, the probable production rate is:

Production (9000') = 0.87 x 166.8 BCY/Hr. (Example 6-5; Table 2-4)
= 145.1 BCY/hour

**Example 6-7:** Use Figures 6-6 and 6-7 to determine the total cycle time and production rate of the scraper given in Example 6-5, assuming the following conditions:
Haul Route:
Section 1: Level loading area
Section 2: Down 10% grade; 4000 ft. (loaded)
Section 3: Level dumping area
Section 4: Level turnaround; 400 ft. (empty)
Section 5: Up 10% grade; 4000 ft. (empty)
Section 6: Level turnaround into loading area; 600 ft. (empty)

**Solution:** The total fixed cycle time for Sections 1, 3, and a portion of Section 6 is:

Total Fixed Cycle Time (Min.) = 1.5 minutes (Example 6-5)

Expressed in % grade, the rolling resistance (effective grade) along all level haul road sections (Sections 1, 3, 4 and 6) is:

Rolling Resistance (%) = 4% (Example 6-5)

The effective grade (total resistance) that the scraper descends (Section 2) is:

Effective Grade (Descending) = 4% – 10% (Eq. 2.14)
= –6%

The effective grade (total resistance) that the scraper ascends (Section 5) is:

Effective Grade (Ascending) = 4% + 10% (Eq. 2.13)
= 14%

The gross weight of the loaded scraper is:

G.V.W. (Loaded) = 115,195 pounds (Example 6-5)

To determine the travel speed of the loaded scraper as it descends along Section 2 of the haul road, refer to Figure 6-7 and find the loaded gross weight of 115,195 pounds at the top of the chart. From this point, follow the vertical line labelled

"L" until it intersects the diagonal line corresponding to a favorable effective grade (total resistance) of 6%. From this intersection, project a horizontal line to the right until you intersect the highest gear of the retarder curve. From this point, drop vertically to obtain the maximum safe vehicle speed at the bottom of the chart. In this instance, the maximum safe speed is approximately 24.8 mph in 7th gear.

The travel time of the loaded scraper along Section 2 of the haul road is:

$$\text{Travel Time (Section 2)} = \frac{\text{4000 Ft.}}{\text{88 Ft./Min. x 24.8 mph}} \quad \text{(Eq. 6.5)}$$

$$= \text{1.8 minutes}$$

To determine the travel time of the empty scraper along Section 4 of the haul road, refer to Figure 6-6 and find the one-way distance of 400 feet at the left side of the chart. Project a horizontal line to the right until it intersects the diagonal line corresponding to the total resistance (effective grade) of 4%. From this intersection, project a vertical line to the bottom of the chart to obtain the travel time. In this instance, we have a travel time of approximately 0.3 minutes. Continuing to refer to Figure 6-6, we can see that the travel times of the empty scraper along the remaining sections of the haul road are:

| Section | Effective Grade (Total Resistance) | Distance | Travel Time |
|---|---|---|---|
| 5 | 14% | 4000' | 4.5 Min. |
| 6 | 4% | 600' | 0.4 Min. |

The total travel time is:

$$\text{Total Travel Time (Min.)} = \text{1.8 Min + 0.3 Min. + 4.5 Min. + 0.4 Min}$$

$$= \text{7.0 minutes}$$

The total cycle time is:

$$\text{Total Cycle Time (Min.)} = \text{1.5 Min. + 7.0 Min.} \quad \text{(Example 6-5; Eq. 6.3)}$$

$$= \text{8.5 minutes}$$

Working a 50-minute hour, the number of cycles per hour is:

$$\text{Cycles/Hr.} = \frac{\text{60 Min./Hr.}}{\text{8.5 Min./Cycle}} \times 50/60 \quad \text{(Eq. 1.6)}$$

$$= \text{5.9 cycles/hour}$$

Therefore, the probable production rate is:

Production (BCY/Hr.) = 15.4 BCY/Cycle x 5.9 Cycles/Hr. (Example 6-5)
= 90.9 BCY/hour

**Example 6-8:** Use the Average Speed Method to determine the total cycle time and production rate for the scraper given in the previous example if the scraper engine is rated at 330 horsepower.
**Solution:** To use the Average Speed Method, we must determine the *maximum vehicle speeds* along each section of the haul road, without regard to acceleration, deceleration or gear-shift time. We used the retarder curve shown in Figure 6-7 to determine the scraper speed along Section 2 of the haul road in Example 6-7. A retarder curve does not consider acceleration or deceleration; therefore, the maximum speed along Section 2 of the haul road is approximately 24.8 mph. However, the travel time charts (Figures 6-5 and 6-6) used to work Example 6-7 take acceleration and deceleration into consideration, which means that average vehicle speed was used to compile the charts. Therefore, we must use Figure 6-8 to determine the maximum vehicle speeds along Sections 4, 5 and 6 of the haul road. This will allow us to use the Average Speed Method to determine the average speeds along these sections.

To determine the maximum travel speed of the empty scraper along Section 4 of the haul road, refer to Figure 6-8 and find the empty gross weight of 67,195 pounds at the top of the chart. From this point, follow the vertical dashed line labelled "E" until it intersects the diagonal line corresponding to the total resistance (effective grade) of 4%. Then project a horizontal line to the right until you intersect the highest gear of the performance curve. From this intersection, drop vertically to the bottom of the chart to obtain the maximum vehicle speed. In this instance, we can travel approximately 31 mph in 8th gear. Continuing to refer to Figure 6-8, we can see that the maximum speeds of the empty scraper along the remaining sections of the haul road are:

| Section | Effective Grade (Total Resistance) | Speed |
|---|---|---|
| 5 | 14% | 11 mph |
| 6 | 4% | 31 mph |

The ratio of weight to power is:

$$\text{Loaded} = \frac{115{,}195 \text{ lb.}}{330 \text{ HP}}$$
$$= 349.1 \text{ lb./hp}$$

$$\text{Empty} = \frac{67{,}195 \text{ lb.}}{330 \text{ HP}}$$
$$= 203.6 \text{ lb./hp}$$

The average speeds along various haul-road sections are:

| Section | Maximum Speed | x | Ave. Speed Factor* (Table 6-6) | = | Average Speed |
|---|---|---|---|---|---|
| 2 | 24.8 mph | x | 0.92 | = | 22.8 mph |
| 4 | 31.0 mph | x | 0.70 | = | 21.7 mph |
| 5 | 11.0 mph | x | 1.00 | = | 11.0 mph |
| 6 | 31.0 mph | x | 0.75 | = | 23.3 mph |

* *Maximum values from Table 6-6 were used in this problem.*

The travel times along various haul road sections are:

| Section | Distance | ÷ | (88 x mph) (Eq. 6.5) | = | Travel Time |
|---|---|---|---|---|---|
| 2 | 4000' | ÷ | (88 x 22.8) | = | 1.99 Min. |
| 4 | 400' | ÷ | (88 x 21.7) | = | 0.21 Min. |
| 5 | 4000' | ÷ | (88 x 11.0) | = | 4.13 Min. |
| 6 | 600' | ÷ | (88 x 23.3) | = | 0.29 Min. |
| Total Travel Time | | | | = | 6.62 Min. |

The total cycle time is:

Total Cycle Time = 1.50 Min. + 6.62 Min. (Example 6-7; Eq. 6.3)
= 8.12 minutes

Working a 50-minute hour, the number of cycles per hour is:

$$\text{Cycles/Hr.} = \frac{60 \text{ Min./Hr.}}{8.12 \text{ Min./Cycle}} \times 50/60 \qquad \text{(Eq. 1.6)}$$
$$= 6.2 \text{ cycles/hour}$$

Therefore, the probable production rate is:

Production (BCY/Hr.) = 15.4 BCY/Cycle x 6.2 Cycles/Hr. (Example 6-7)
= 95.5 BCY/hour

## SCRAPER PRODUCTION CURVES

Scraper production rates can also be estimated by using production curves such as that shown in Figure 6-12. The production curve shown gives production rates in bank cubic yards per 60-minute hour with a fixed time of 1.4 minutes while carrying a payload of 48,000 pounds.

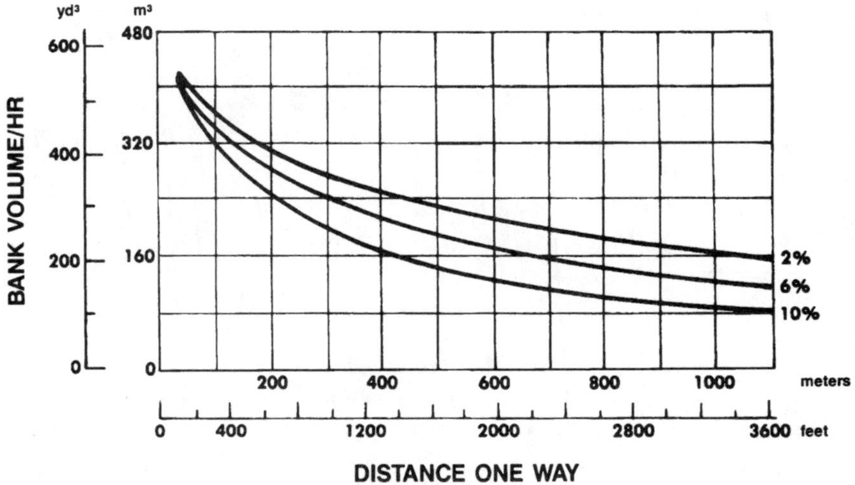


**Figure 6-12. 621E scraper production chart.** *(Courtesy of Caterpillar Inc.)*

**Example 6-9:** Use Figure 6-12 to determine the ideal production rate of the scraper working under the conditions given in Example 6-5.
**Solution:** Referring to Figure 6-12, find the one-way haul distance of 3000 feet at the bottom of the chart and project a vertical line to the curve corresponding to a total resistance (effective grade) of 4%, which lies midway between the 2% and 6% curves. From this intersection, project a horizontal line to the left side of the chart to obtain the hourly production rate. In this instance, the ideal production rate(100% efficiency) is approximately 200 bank cubic yards per hour.

## NUMBER OF PUSHERS REQUIRED

As we discussed earlier, pusher cycle time consists of the following time components:

$$\text{Pusher Cycle Time} = \text{Load} + \text{Boost} + \text{Return} + \text{Transfer Times} \quad \text{(Eq. 6.4)}$$

Field studies indicate that when using any loading method, the sum of the boost and transfer times averages about 0.25 minutes. When using the back-track loading method, return time is about 40% of the load time. Thus, when we use the back-track loading method, the pusher cycle time is:

Pusher Cycle Time (Min.) = 0.25 Min. + (1.4 x Load Time) **(Equation 6.6)**

You can also estimate pusher cycle time by multiplying the load time by the pusher factors given in Table 6-7. Tandem pushing means that two tractors are positioned behind the scraper so that both tractors are assisting.

**Table 6-7: Pusher Factors**

| Loading Method | Single Pusher | Tandem Pusher |
|---|---|---|
| Back-track loading | 1.5 | 2.0 |
| Chain loading | 1.3 | 1.5 |
| Shuttle loading | 1.3 | 1.5 |

*(Courtesy of Terex Corporation)*

The number of scrapers served by a single pusher can be determined as follows:

$$\text{No. Scrapers Served by One Pusher} = \frac{\text{Scraper Cycle Time}}{\text{Pusher Cycle Time}}$$ **(Equation 6.7)**

If we know the total number of scrapers to be used at a site, we can determine the number of pushers required to serve the scraper fleet as follows:

$$\text{No. Pushers Required} = \frac{\text{No. Scrapers}}{\text{No. Scrapers Served by One Pusher}}$$ **(Equation 6.8)**

**Example 6-10:** Assuming each scraper has a load time of 1 minute (back-track loading with single pusher) and a total cycle time of 5 minutes, use Table 6-7 to determine the number of pushers required to service a fleet of 12 scrapers.
**Solution:** The pusher cycle time is:

Pusher Cycle Time (Min.) = 1.5 x 1.0 Min. (Table 6-7)
= 1.5 minutes

The number of scrapers served by each pusher is:

$$\text{No. Scrapers Served by One Pusher} = \frac{5.0 \text{ Min.}}{1.5 \text{ Min.}}$$ (Eq. 6.7)

= 3.3 scrapers/pusher

The total number of pushers required to service the entire fleet of scrapers is:

$$\text{No. Pushers Required} = \frac{\text{12 Scrapers}}{\text{3.3 Scrapers/Pusher}} \qquad \text{(Eq. 6.8)}$$
$$= 3.6$$
$$= 4 \text{ pushers}$$

In the previous example, we were given the total number of scrapers used. Had we not been given this information, we could exercise the following options:

| No. Scrapers | No. Pushers | Effect |
|---|---|---|
| 3 | 1 | Pusher waiting |
| 4 | 1 | Scraper waiting |
| 6 | 2 | Pusher waiting |
| 7 | 2 | Scraper waiting |
| 9 | 3 | Pusher waiting |
| 10 | 3 | Scraper waiting |
| 13 | 4 | Pusher waiting |
| 14 | 4 | Scraper waiting |

Normally, it is be better for the pusher to be waiting for a scraper, since the pusher can be accomplishing back-ripping or cleanup work during the waiting period.

**Example 6-11:** Assuming 260,000 bank cubic yards must be excavated and transported within 30 working days (assume 8-hour days) under the conditions given in Example 6-5, determine the number of scrapers and pushers required. Assume that the back-track loading method is used, and use Equation 6.6 to determine pusher cycle time.
**Solution:** The minimum required production rate is:

$$\text{Required Production (BCY/Hr.)} = \frac{\text{260,000 BCY}}{\text{30 Days x 8 Hr./Day}} \qquad \text{(Eq. 1.13)}$$
$$= \text{1083 BCY/hour}$$

The number of scrapers required is:

$$\text{No. Scrapers} = \frac{1083 \text{ BCY/Hr.}}{166.8 \text{ BCY/Hr./Scraper}} \qquad \text{(Example 6-5)}$$
$$= 6.5$$
$$= 7 \text{ scrapers}$$

The load time is:

$$\text{Load Time (Min.)} = 0.7 \text{ minutes} \qquad \text{(Example 6-5)}$$

The pusher cycle time is:

$$\text{Pusher Cycle Time (Min.)} = 0.25 \text{ Min.} + (1.4 \times 0.7 \text{ Min.}) \qquad \text{(Eq. 6.6)}$$
$$= 1.23 \text{ minutes}$$

The total scraper cycle time is:

$$\text{Scraper Cycle Time (Min.)} = 4.6 \text{ minutes} \qquad \text{(Example 6-5)}$$

The number of scrapers served by each pusher is:

$$\text{No. Scrapers Served by One Pusher} = \frac{4.6 \text{ Min.}}{1.23 \text{ Min.}} \qquad \text{(Eq. 6.7)}$$
$$= 3.7 \text{ scrapers/pusher}$$

The number of pushers required to service the entire fleet of scrapers is:

$$\text{No. Pushers Required} = \frac{7 \text{ Scrapers}}{3.7 \text{ Scrapers/Pusher}} \qquad \text{(Eq. 6.8)}$$
$$= 1.9$$
$$= 2 \text{ pushers}$$

Since we require only 1.9 pushers but have 2 available, the pushers will have to wait for scrapers, so we can use the pushers intermittently for ripping and cleanup work.

**Example 6-12:** Assuming each scraper costs $175.00 per hour and each pusher-tractor costs $180.00 per hour, including operators' wages, determine the unit production cost for the fleet given in the previous example.
**Solution:** Using 7 scrapers, we have an hourly production rate of:

Production (BCY/Hr.) = 7 Scrapers x 166.8 BCY/Hr./Scraper (Example 6-11)
= 1168 BCY/hour

The unit production cost is:

$$\text{Unit Production Cost} = \frac{(7 \times \$175.00) + (2 \times \$180.00)}{1168 \text{ BCY/Hr.}} \qquad \text{(Eq. 1.18)}$$

= $1.36/BCY

## OPTIMUM LOAD TIME

Loading a scraper to maximum capacity does not necessarily ensure optimum production rates. Caterpillar Inc. has run field tests where scrapers are loaded for controlled periods of time and the resulting payloads weighed. The results are then tabulated (Table 6-8) and the data plotted to form a load-growth curve such as that shown in Figure 6-13. We can see that the rate of loading decreases as load time increases; therefore, the law of diminishing returns comes into play at some point during the loading period. Optimum load time can be determined mathematically, or graphically.

**Table 6-8: Development of a Load Growth Curve**
**(631C Scraper and 1-D9G Pusher)**

| Load Time (Min.) | Payload (BCY) |
|---|---|
| 0.2 | 10.4 |
| 0.4 | 17.0 |
| 0.6 | 20.6 |
| 0.8 | 22.5 |
| 1.0 | 23.7 |
| 1.2 | 24.4 |

*(Courtesy of Caterpillar Inc.)*

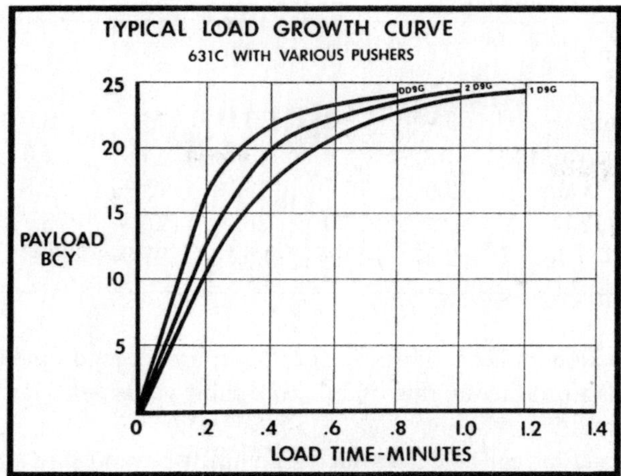


**Figure 6-13. Typical load-growth curves.** *(Courtesy of Caterpillar Inc.)*

**Example 6-13:** Assuming a cycle time (excluding load time) of 3 minutes, determine the optimum load time of the scraper whose load-growth curve is shown in Figure 6-13. Assume using one D9G pusher-tractor, and working a 50-minute hour.
**Solution:** Using a load time of 0.2 minutes, the total cycle time is:

$$\text{Total Cycle Time (Min.)} = 3.0 \text{ Min.} + 0.2 \text{ Min.} = 3.2 \text{ minutes}$$

The number of haul trips (cycles) made each 50-minute hour is:

$$\text{Cycles/Hr.} = \frac{60 \text{ Min./Hr.}}{3.2 \text{ Min./Cycle}} \times 50/60 = 15.6 \text{ cycles/hour} \qquad \text{(Eq. 1.6)}$$

The payload hauled during each cycle is:

$$\text{Payload/Cycle (BCY)} = 10.4 \text{ BCY/cycle} \qquad \text{(Table 6-8)}$$

The probable production rate is:

$$\text{Production (BCY/Hr.)} = 10.4 \text{ BCY/Cycle} \times 15.6 \text{ Cycles/Hr.} = 162.2 \text{ BCY/hour}$$

Continuing in this fashion while using other load-time and payload data from Figure 6-13 and Table 6-8, we can tabulate the production-rate performance data as follows:

| Load time (min.) | 0.2 | 0.4 | 0.6 | 0.8 | 1.0 | 1.2 |
|---|---|---|---|---|---|---|
| Cycle time less load time (min.) | 3.0 | 3.0 | 3.0 | 3.0 | 3.0 | 3.0 |
| Total cycle time (min.) | 3.2 | 3.4 | 3.6 | 3.8 | 4.0 | 4.2 |
| Cycles/50-min. hour | 15.6 | 14.7 | 13.9 | 13.2 | 12.5 | 11.9 |
| Load/cycle (BCY) | 10.4 | 17.0 | 20.6 | 22.5 | 23.7 | 24.4 |
| Production (BCY/hr.) | 162.2 | 249.9 | 286.3 | 297.0 | 296.3 | 290.4 |

Based on production and cost performance, the optimum load time is 0.8 minutes which yields a production rate of 297 bank cubic yards per hour.

Optimum load time can be determined graphically by plotting a load-growth curve starting at the origin as shown in Figure 6-14. After plotting the load-growth curve, determine the location of Point A to the left of the origin as follows:

Point A = Total Cycle Time – Load Time **(Equation 6.9)**
= Turning & Dump + Spot & Delay + Travel Times

Draw a line from Point A tangent to (touching) the load-growth curve (Point B), then drop a vertical line from Point B to the horizontal axis (Point C). The load time corresponding to Point C is the optimum load time. In this instance, the optimum load time is approximately 0.75 minutes.

Referring to Figure 6-14, it is obvious that as the total cycle time increases (the longer the haul trip), the optimum load time also increases. A graphical representation of this concept is shown in Figure 6-15.

## DRAG SCRAPERS

Drag scrapers are towed behind a tractor or truck and are used for leveling and smoothing work on small-scale projects (Photo 6-11).

**Photo 6-11. Drag scraper.** *(Courtesy of Miskin Scraper Works, Inc.)*

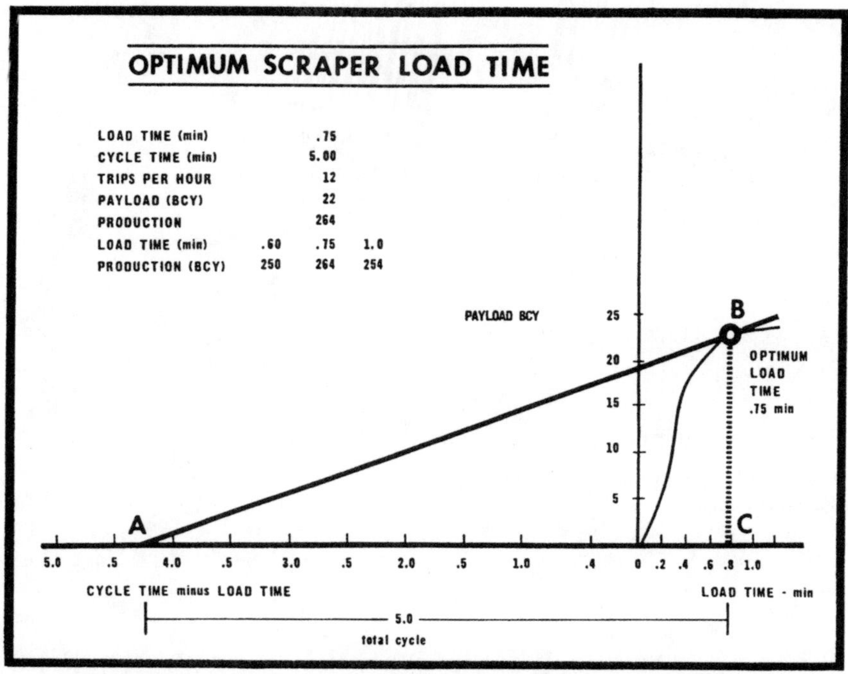


**Figure 6-14. Graphical method for determining optimum load time.**

*(Courtesy of Caterpillar Inc.)*

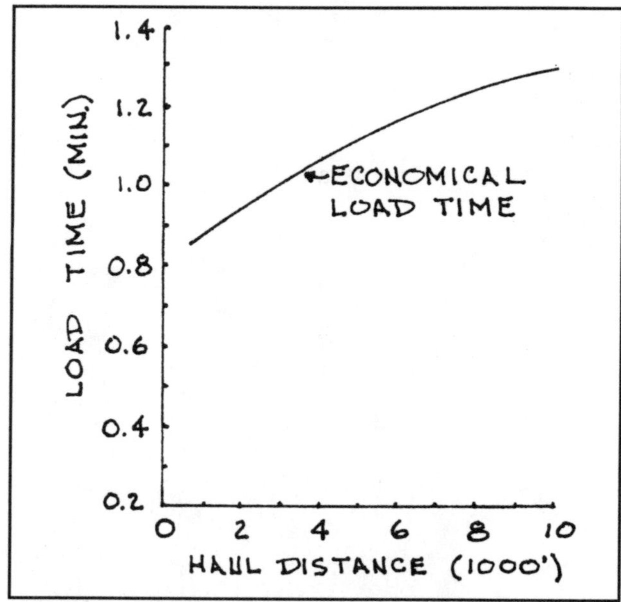


**Figure 6-15. The effect of haul distance on the economical loading time of a scraper.**

*Chapter 7*

# SOIL COMPACTION

The need for good compaction beneath structures was realized and practiced by the road builders of the Roman Empire, and some of these roads are still in existence.

There are many benefits that result from good compaction. Compaction squeezes out air voids in the soil and replaces them with soil. As a result, the load-bearing capability and stability of the soil is increased (Figures 7-1 and 7-2).

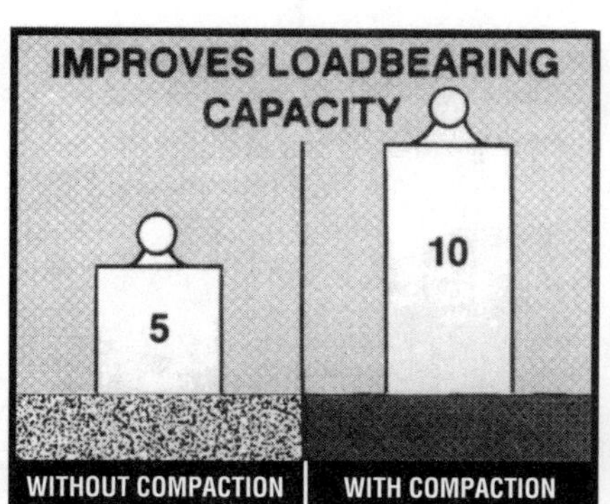


**Figure 7-1. Compaction increases load-bearing capacity.**
*(Courtesy of Wacker Corporation)*

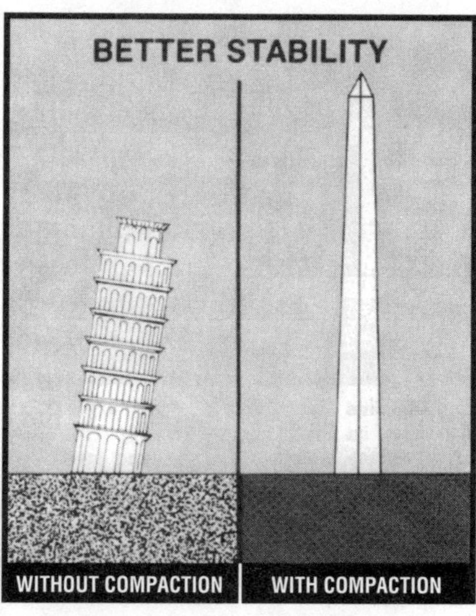


**Figure 7-2. Compaction increases stability.**
*(Courtesy of Wacker Corporation)*

Compacted soil reduces water penetration so that water flow and drainage is better controlled (Figure 7-3). Without compaction, water fills air voids, resulting in the soil's swelling, followed by shrinkage as the soil dries (Figure 7-4). The problem is compounded if the water freezes.

Poorly-compacted soil will eventually settle due to vibration and water seepage, whereas, a well-compacted soil will not settle (Figure 7-5). Settlement of a foundation or slab will eventually causes walls, ceilings and floors to crack, doors to shut out-of-plumb, floor finishes to wear unevenly, and even structural failure (Photo 7-1). Sometimes, the only remedy is to demolish the entire building, remove and replace the fill, and re-build. Under these extreme circumstances, the money and time required to defend a lawsuit (including attorney fees, expert witness fees and court costs), plus the cost of making restitution can wreck the finances and reputation of a previously lucrative earthmoving business that took a lifetime of hard work to build.

**Figure 7-3. Compaction decreases water seepage.**
*(Courtesy of Wacker Corporation)*

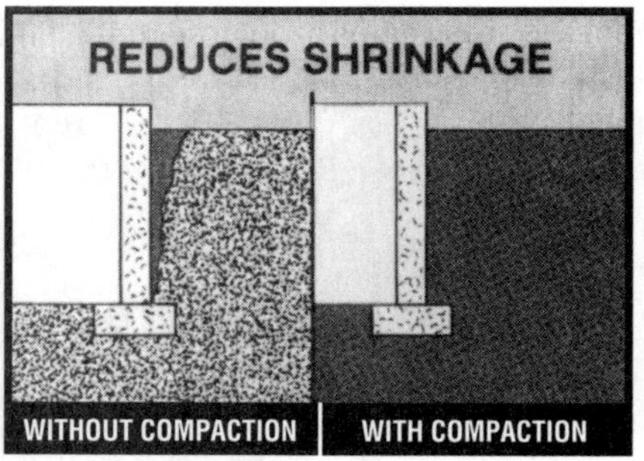


**Figure 7-4. Compaction reduces shrinkage.**
*(Courtesy of Wacker Corporation)*

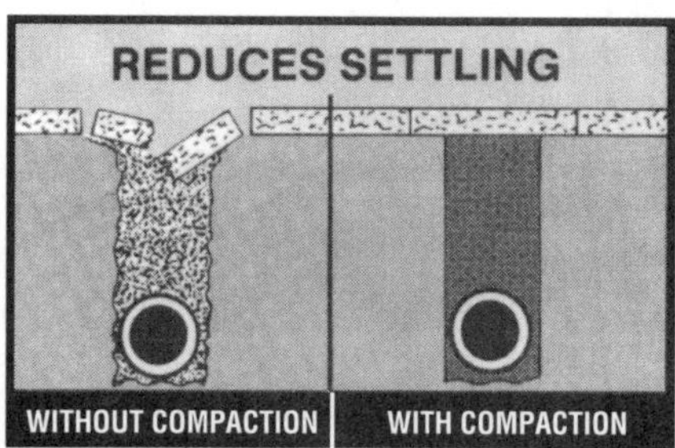


**Figure 7-5. Compaction reduces settling.**
*(Courtesy of Wacker Corporation)*

**Photo 7-1. Cracks resulting from settlement.**
*(Photo by Dan Atcheson)*

## COMPACTION FUNDAMENTALS

Compaction (consolidation or compression) is used to strengthen soil by increasing the density of the soil. Increased soil density is obtained by expelling excess air and water from the soil while forcing soil particles closer together and filling the voids between the particles. The main factors that affect compaction are material gradation, moisture content and compaction effort.

## MATERIAL GRADATION

*Material gradation* refers to the distribution of different particle sizes within the soil. The distribution is normally expressed as a percentage (by weight) of a given soil sample. A sample that contains a good, even distribution of particle sizes is described as *well-graded.* If the soil composed predominantly of one size particle, or if it lacks a given particle size, it is said to be *poorly-graded* (Figure 7-6).

A well-graded soil can be compacted to a greater density than a poorly- graded soil because in a well-graded soil, the smaller particles will fill voids between the larger particles, leaving fewer voids after compaction (Figure 7-6).

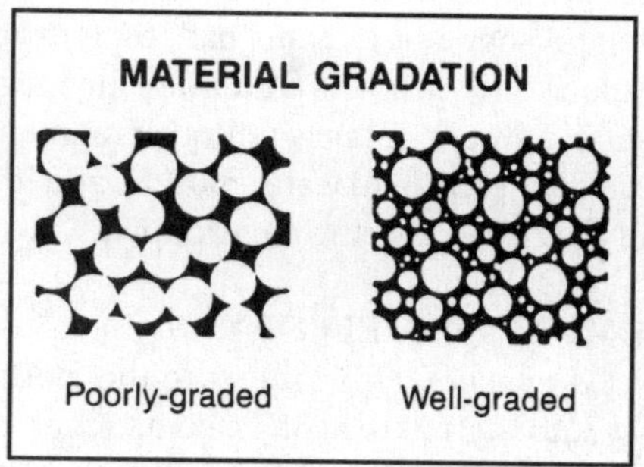


**Figure 7-6. Material gradation.**
*(Courtesy of Caterpillar Inc.)*

## MOISTURE CONTENT

The *moisture content* of the soil determines the extent to which soil can be compacted. Water lubricates soil so that soil particles can slide into a more compacted condition. Water also fills voids which otherwise would contain air, only. However, too much water will cause soil particles to float out of the voids, resulting in decreased density. Therefore, too little or too much water makes it impossible to achieve proper compaction.

The moisture content at which maximum soil density can be obtained with a given compaction effort is called the *optimum moisture content.* Figure 7-7 shows the relationship between moisture content and dry density. The curve shown in Figure 7-7 is referred to as a compaction curve, control curve, *moisture-density curve* or Proctor curve.

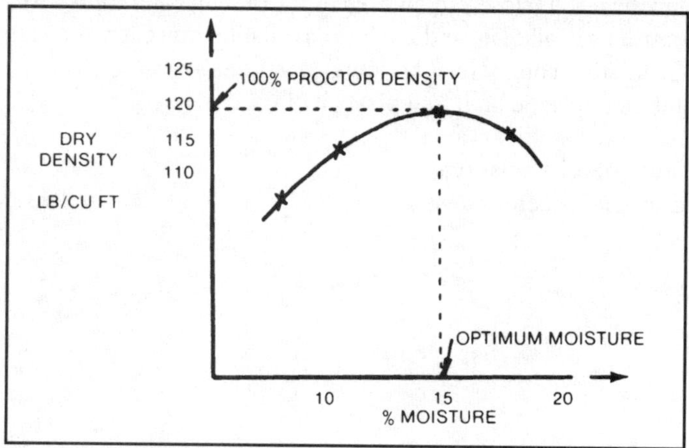


**Figure 7-7. Moisture-density curve.** *(Courtesy of Wacker Corporation)*

You will notice throughout all of the discussions regarding the various types of soil tests that moisture content and density is related to the dry weight of the soil. This is because the dry weight of a soil is constant, while the wet weight is variable, depending on the amount of moisture in the soil.

There is a simple test that you can perform to determine if soil contains the right amount of water for good compaction (Uniform gravel or mostly sandy soil does not react well to this test.). Take a handful of the soil and squeeze it into the size and shape of a tennis ball, then drop the soil ball to the ground through a distance of one foot. At optimum moisture content, the ball will break apart into a small number of fairly uniform fragments. If the soil is too dry, the soil will not form into a ball, and water must be added to the soil. If the soil is too wet, the soil will not break apart when dropped.

## COMPACTION EFFORT

*Compaction effort* refers to the method used by a compactor to impart energy into the soil to achieve compaction. Compactors use one, or a combination of the following types of compaction efforts:

1) Static weight (pressure)
2) Kneading (manipulation)
3) Impact (sharp blow)*
4) Vibration (shaking)

A greater compaction effort will result in greater soil density, and as the compaction effort is increased, the optimum moisture content will be reduced (Figure 7-14).

## SOIL TESTS MOST OFTEN SPECIFIED

Soil testing is a quantitative method of quality control for the compacted fill material, and the actual number and types of tests made will depend on the requirements specified by the designer. Soil testing should always be specified, conducted, and the results closely monitored by the designer and contractor. There is a wide spectrum of soil tests and analyses available; however, the tests most frequently specified in commercial building construction are:

1) Liquid limit, plastic limit and plasticity index
2) Unified soil classification
3) Moisture-density relations
4) In-place (field) density tests

* *Soil is compacted in a soil testing laboratory by impact.*

## LIQUID LIMIT

The *liquid limit* of a soil is the moisture content, expressed as a percentage of the dry sample weight, at which two halves of a soil cake will flow together through a distance of 1/2 inch along a groove separating the two halves in a liquid-limit test cup, as the cup is dropped a distance of one centimeter (0.3937 inches) 25 times at the rate of two blows per second (Photo 7-2). The liquid limit test is performed as follows:

1) The portion of the soil sample passing a No. 40 sieve (40 net openings per square inch) is mixed with water (Photo 7-3).
2) The soil is placed in a brass cup and separated into two pieces with a grooving tool (Photo 7-2).
3) The cup is lifted and dropped on a cam-crank apparatus until the two halves of the soil sample slide together, and the number of strokes required to close the gap is recorded.
4) Depending on the outcome of the first trial, moisture is added to, or evaporated from the sample and another test is performed.
5) This procedure is repeated until the groove closes at approximately 25 blows.
6) At this point, the sample is immediately weighed, dried, and re-weighed.

The liquid limit is defined as:

$$\text{Liquid Limit (\% of Dry Sample Weight)} = \frac{\text{Moisture Weight at Liquid Limit}}{\text{Dry Sample Weight}} \times 100\%$$

**(Equation 7.1)**

where:

Moisture Weight at Liquid Limit = Wet Sample Weight – Dry Sample Weight

**(Equation 7.2)**

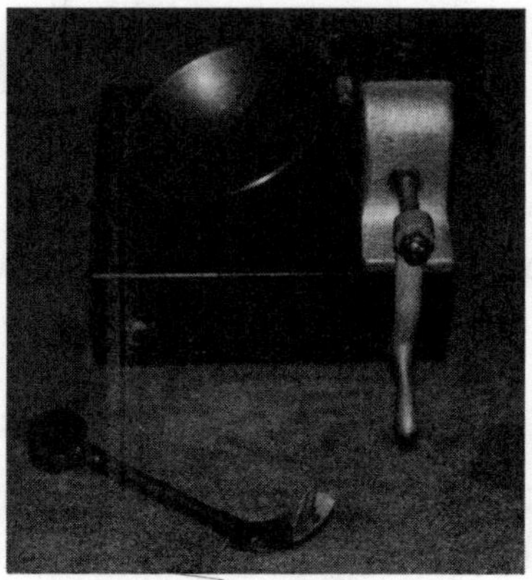

**Photo 7-2. Liquid limit test cup and grooving tool.** *(Photo by Dan Atcheson)*

**Photo 7-3. Various sizes of sieves.** *(Photo by Dan Atcheson)*

## PLASTIC LIMIT

The *plastic limit* is the moisture content, expressed as a percentage of the dry sample weight, at which the sample begins to crumble when rolled into a thread 1/8-inch in diameter (Photo 7-4). The plastic limit test is performed as follows:

1) The portion of the soil sample passing a No. 40 sieve is mixed with water.
2) The soil is hand-squeezed into an ellipsoidal-shaped mass. The mass is then rolled between the fingers and a flat glass plate into an 1/8-inch diameter thread.
3) The thread is broken into six or eight pieces, squeezed, and re-rolled into another thread. This process is repeated until the thread crumbles while being rolled.
4) At this point, the sample is immediately weighed, dried, and re-weighed.

The plastic limit is defined as:

$$\text{Plastic Limit (\% of Dry Sample Weight)} = \frac{\text{Moisture Weight at Plastic Limit}}{\text{Dry Sample Weight}} \times 100\%$$

**(Equation 7.3)**

where:

Moisture Weight at Plastic Limit = Wet Sample Weight – Dry Sample Weight (Eq. 7.2)

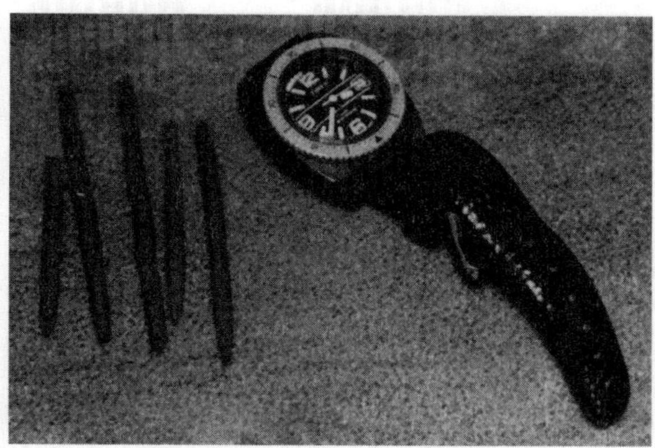

**Photo 7-4.**
**Plastic limit test.**
*(Photo by Dan Atcheson)*

## PLASTICITY INDEX

The *plasticity index* (P/I) of a given soil is the difference between the soil's plastic limit and liquid limit; i.e.,

Plasticity Index = Liquid Limit – Plastic Limit **(Equation 7.4)**

The plasticity index of a soil is an indication of the clay content and the soil's moisture-retaining capabilities. A large plasticity index indicates a high clay content and an increased ability to retain water, resulting in the tendency for the soil to swell when wet, and shrink when dry. As a general rule, soils with a plasticity index greater than 15 are "trouble-makers."

## THE UNIFIED SOIL CLASSIFICATION SYSTEM

There are several soil classification systems, but the Unified Soil Classification System is the most-widely used. In this system, each soil is given a two-letter designation, where the first letter describes the major soil constituents, and the second letter describes the soil's gradation, or plasticity (compressibility) (Table 7-1). The five basic soil constituents are abbreviated as follows:

Gravel (G)
Sand (S)
Silt (M)
Clay (C)
Organic Matter (O)

*Gravel* is rock-like material whose particle sizes range from 3 inches (76.2 mm) to 0.08 inches (2.03 mm). *Sand* ranges from 0.08 inches to 0.0029 inches (0.076 mm) and *silt* ranges from 0.0029 inches to 0.00024 inches (0.006 mm). *Clay* particles are 0.00024 inches and smaller (Figure 7-8).

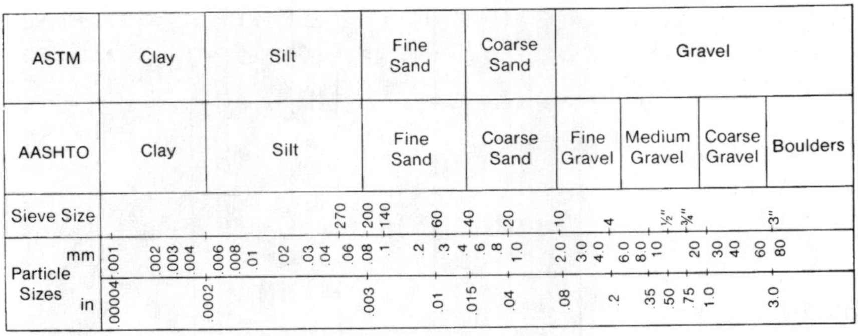

| ASTM | Clay | Silt | Fine Sand | Coarse Sand | Gravel | | | |
|---|---|---|---|---|---|---|---|---|
| AASHTO | Clay | Silt | Fine Sand | Coarse Sand | Fine Gravel | Medium Gravel | Coarse Gravel | Boulders |

| Scale | Values |
|---|---|
| Sieve Size | 270, 200, 140, 60, 40, 20, 10, 4, ½", ¾", 3" |
| Particle Sizes mm | .001, .002, .003, .004, .006, .008, .01, .02, .03, .04, .06, .08, .1, .2, .3, .4, .6, .8, 1.0, 2.0, 3.0, 4.0, 6.0, 8.0, 10, 20, 30, 40, 60, 80 |
| Particle Sizes in | .00004, .0002, .003, .01, .015, .04, .08, .2, .35, .50, .75, 1.0, 3.0 |

**Figure 7-8. Soil classification according to particle size.**

*(Courtesy of Wacker Corporation)*

Clay is a term that has been used to define both mineral type and particle size. In terms of mineralogy, a clay is a hydrated aluminum silicate mineral, and there are approximately a dozen distinct clay minerals.

If we had a microscope powerful enough to observe clay molecules, we would see that most clay minerals are composed of flat layers of molecules arranged like sheets in a book. Water can get between these molecular layers, increasing the distance between the layers, and this is why clay expands when it gets wet. If a small quantity of water gets between the layers, the water molecules will chemically bond themselves to the atoms in the clay mineral, causing soils containing clay to become cohesive.* However, if a large quantity of water is introduced, more than one layer of water molecules will form between each clay layer, causing the layers to slide past one another. This is why clay and soil containing clay is "slimy" when fully saturated with water.

In terms of particle size, any particle less than 0.006 mm is referred to as a clay. Calcite, quartz, garnet, dolomite, feldspar, pyrite, mica, wood chips, spores, and volcanic ash (all non-clay materials) have been found to be within the clay-size dimension. Since these relatively inert clay-sized particles are not actually expansive clay minerals, knowing the percentage of clay-sized particles alone is of little value when determining the expansiveness of the soil. Also, since the smallest sieve available in a soil testing lab is a No. 200 (0.0029 inches or 0.075 mm net opening), the clays cannot be separated from silts in a soil analysis. This is why it is necessary to know the soil's plasticity index when determining a soil's expansive behavior.

There is a simple test that you can perform to determine if a soil has a high clay content. Roll a moist sample of the soil into a ball approximately one inch in diameter and throw it against a wall. If the soil has a high clay content, the soil ball will stick to the wall like a paper "spit wad."

* *Some cohesion is also due to the surface tension of the water molecules in voids between soil particles.*

*Organic matter* is partially decomposed vegetable matter which will continue to decompose with time and leave voids; therefore, it should be removed from soil that is to be compacted.

Abbreviations used to describe gradation and plasticity are:
Well-graded (W)
Poorly-graded (P)
Low plasticity (L)
High plasticity (H)

From the previous discussion, we are now able to understand most of the soil classification symbols given in Table 7-1.

**Table 7-1: Symbols and Names (Unified Soil Classification System)**

| Symbol | Description | Stability as Construction Material |
|---|---|---|
| | Coarse-Grained Soils (Less than 50% pass No. 200 sieve) | |
| GW | Well-graded gravel | Excellent |
| SW | Well-graded sand | Excellent |
| GP | Poorly-graded gravel | Excellent to good |
| SP | Poorly-graded sand | Good |
| GM | Silty gravel | Good |
| SM | Silty sand | Fair |
| GC | Clayey gravel | Good |
| SC | Clayey sand | Good |
| | Fine-grained Soils (50% or more pass No. 200 sieve) | |
| ML | Low-plasticity silt | Fair |
| CL | Low-plasticity clay | Good to fair |
| OL | Low-plasticity organic | Fair |
| MH | High-plasticity silt | Poor |
| CH | High-plasticity clay | Poor |
| OH | High-plasticity organic | Poor |
| PT | Peat or organic | Unsuitable |

## GRADING SOIL PARTICLES

Soil *grading* is used to determine the distribution of grain sizes in the soil. To determine how well-graded a soil sample is, all cobbles (particles greater than 3 inches in diameter) are removed and the sample is washed to remove organic matter (Photo 7-5). Then the sample is dried and passed through a series of sieves, starting with a 3/4-inch sieve and ending with a No. 200. The sieves are linked together (racked) with the 3/4-inch sieve on top, the No. 200 at the bottom and intermediate sizes in between (Photo 7-6). A catch pan is placed below the No. 200 sieve to collect particles small enough to pass that sieve, and a lid is placed over the top of the rack to prevent spilling the sample. After the sample is poured into the rack, the unit is placed into a shaker for about 12 to 15 minutes (Photo 7-7). After shaking, the sieves are removed, one at a time, and starting with the 3/4-inch sieve, the amount of particles retained by each sieve is weighed and the results are entered onto a form as shown in Figure 7-9.

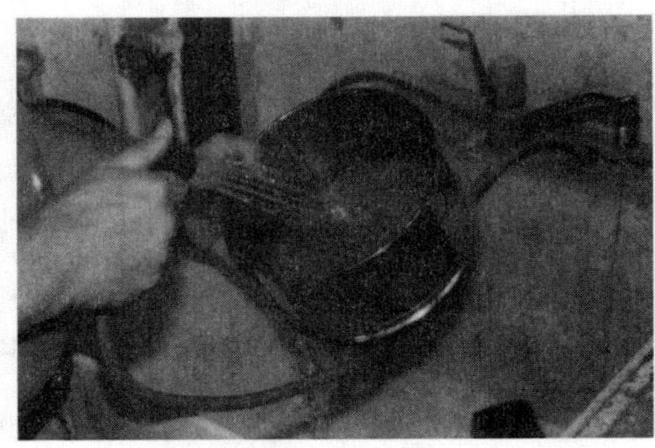

**Photo 7-5.**
**Washing the sample.**
*(Photo by Dan Atcheson)*

**Photo 7-6. Racked sieves.**
*(Photo by Dan Atcheson)*

**Photo 7-7. The shaker.**
*(Photo by Dan Atcheson)*

| TOTAL DRY SAMPLE ______ (gms.) | | | |
|---|---|---|---|
| SIEVE SIZE | GRAMS | % RETAINED | % PASSED |
| 3/4 | | | |
| 1/2 | | | |
| 3/8 | | | |
| 4 | | | |
| 10 | | | |
| 40 | | | |
| 60 | | | |
| 100 | | | |
| 200 | | | |

**Figure 7-9. Sieve analysis form.** *(Courtesy of Ardaman & Associates)*

The percentage of particles retained by a given sieve can be determined by:

$$\% \text{ Retained} = \frac{\text{Weight of Soil Retained}}{\text{Total Dry Sample Weight}} \qquad \textbf{(Equation 7.5)}$$

The percentage of particles passing a given sieve can be determined by:

$$\% \text{ Passing} = 100\% - \% \text{ Retained} \qquad \textbf{(Equation 7.6)}$$

How well a given soil sample is graded can be determined by analyzing numerical data, or by studying the data plotted as a *grain-distribution curve* (gradation curve) (Figure 7-10).

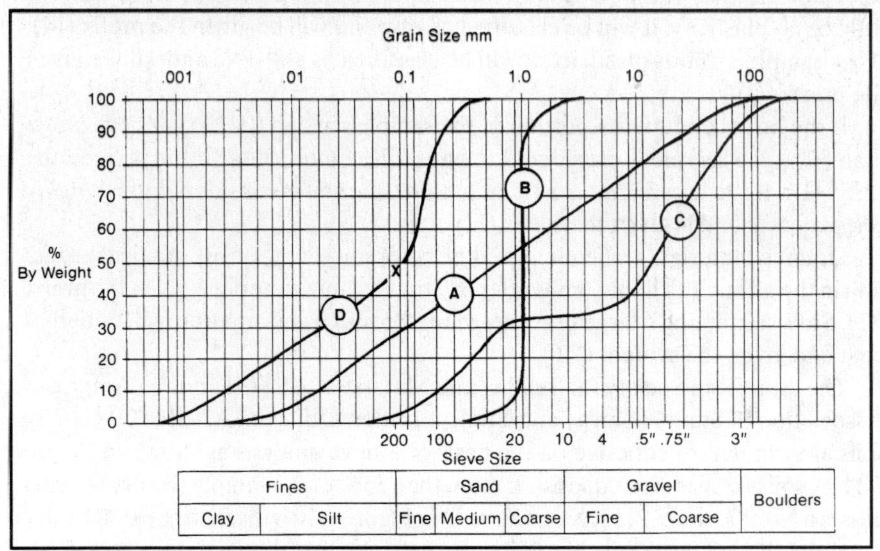


**Figure 7-10. Grain-distribution curve.** *(Courtesy of Wacker Corporation)*

**Coarse-Grained Soils**

If less than 50% of the total sample passes the No. 200 sieve, the soil is said to be *coarse-grained.* If 50% or more of the coarse-grained sample passes the No. 4 sieve (4 net openings per square inch), it is classified as sand and will be given the prefix (S). If less than 50% passes the No. 4 sieve, the soil is classified as gravel, and will be given the prefix (G).

If less than 5% of the total sample passes the No. 200 sieve, coarse-grained soil can be classified as *well-graded* (suffix W), or *poorly-graded* (suffix P). The proper suffix can be determined by analyzing sieve data (Figure 7-9), or a grain-distribution curve (Figure 7-10). A well-graded soil contains a broad and uniform range of grain sizes, and is distinguished by a curve with a uniform slope (Curve A in Figure 7-10). A poorly-graded soil contains a limited range of particle sizes and is distinguished by a steep curve (Curve B in Figure 7-10). A poorly-graded soil will often have a deficiency in one or more particle sizes and is referred to as *"gap graded"* (Curve C in Figure 7-10).

If more than 12% of the total sample passes the No. 200 sieve, the coarse-grained soil can be classified as silty, or clay-like. If the portion of the sample passing the No. 40 sieve exhibits little or no plasticity (or compressibility), it will be classified as *silty* and will be given the suffix (M). If it exhibits plasticity, it will be classified as *clay-like* and will be given the suffix (C).

If the portion of the coarse-grained sample passing the No. 200 sieve falls between 5 and 12%, a dual system of classification is used, such as GW-GM, SP-SC, etc.

**Fine-Grained Soils**

If more than 50% of the total sample passes the No. 200 sieve, the soil is said to be *fine-grained.* If the portion of the sample passing the No. 40 sieve exhibits little or no plasticity, it will be classified as silty and will be given the prefix (M). If the sample exhibits plasticity, it will be classified as clay-like and will be given the prefix (C).

If the liquid limit of the portion of the sample passing the No. 40 sieve is less than 50%, the sample is considered to have *low compressibility* and is given the suffix (L). If the liquid limit is 50% or greater, the soil is considered to be *highly compressible* and is given the suffix (H).

Organic material can be identified by color and odor, or by a radical reduction in the plastic and liquid limits after drying. Organic material is given the prefix (O). Soils containing a large percentage of fibrous organic matter is classified as peat and given the symbol (PT).

The American Society for Testing and Materials (ASTM) and the American Association of State Highway and Transportation Officials (AASHTO) classify soils as granular, or cohesive on the basis of a sieve analysis as shown in Figure 7-11. A soil is considered cohesive if more than 35% of the sample (by dry weight) passes a No. 200 sieve. Point X on Curve D in Figure 7-10 indicates that 48% by dry weight of the soil passes the No. 200 sieve, meaning that the soil is very cohesive.

| General Classification | Granular Material | | | | | | | Cohesive Material More than 35% of Total Sample Passing No. 200 | | | |
|---|---|---|---|---|---|---|---|---|---|---|---|
| Group Classification | A-1 | | A-3 | A-2 | | | | | | | A-7 |
| | A-1-a | A-1-b | | A-2-4 | A-2-5 | A-2-6 | A-2-7 | A-4 | A-5 | A-6 | A-7-5, A-7-6 |
| Sieve Analysis % Passing | | | | | | | | | | | |
| No. 40 ______ | 30 max | 50 max | 51 min | | | | | | | | |
| No. 200 ______ | 15 max | 25 max | 10 max | 35 max | 35 max | 35 max | 35 max | 36 min | 36 min | 36 min | 36 min |

**Figure 7-11. AASHTO group classification.** *(Courtesy of Wacker Corporation)*

*Granular soils* consist primarily of sands and gravels that are held in position by friction between the contact surfaces of the particles. When wet, granular soils can be formed into different shapes, but when dry, they crumble into separate, easily-identified particles.

*Cohesive soils* consist primarily of silts and clays held together by molecular attraction (An example of molecular attraction is water adhering to your skin.). Cohesive forces are strong, even when the soil is dry.

## MOISTURE-DENSITY RELATIONS (LABORATORY COMPACTION TESTS)

The compaction test is used to determine the maximum dry soil density obtainable under standard laboratory conditions. The maximum dry density is then used as a criterion with which all in-place (field) soil density tests are compared. The standard Proctor test or the modified Proctor test is usually used to determine the maximum laboratory dry soil density.

### Standard Proctor Test

The *standard Proctor test* was developed in the early 1930's by R.R. Proctor, a field engineer for the City of Los Angeles. It is now universally accepted throughout the construction industry. The standard Proctor test is also known as AASHTO test designation: T 99-70, or ASTM test designation: D 698. Soil labs normally refer to this test as a "T-99." The standard Proctor test is usually specified for fill material designated for use under building slabs and sidewalks, and in utility trenches under grassy areas.

To perform the standard Proctor test, a sample of the proposed fill material is crushed so it can pass a 3/4-inch sieve (Photos 7-8 and 7-9). Then a steel cylinder mold approximately 4 inches in diameter and 4.584 inches deep is filled with the sample in three layers, each layer being struck 25 blows from a 5.5-pound, 2-inch-diameter hammer dropped 12 inches. The hammer can be hand-operated (Photo 7-10), or mechanical (Photo 7-11). The volume of the mold is approximately 1/30 cubic feet (Figure 7-12).

**Photo 7-8. Crushing the sample.**
*(Photo by Dan Atcheson)*

**Photo 7-9. Passing the sample through a 3/4" sieve.**
*(Photo by Dan Atcheson)*

**Photo 7-10. Hand-operated standard Proctor test.**
*(Photo by Dan Atcheson)*

**Photo 7-11. Mechanical standard Proctor test.**
*(Photo by Dan Atcheson)*

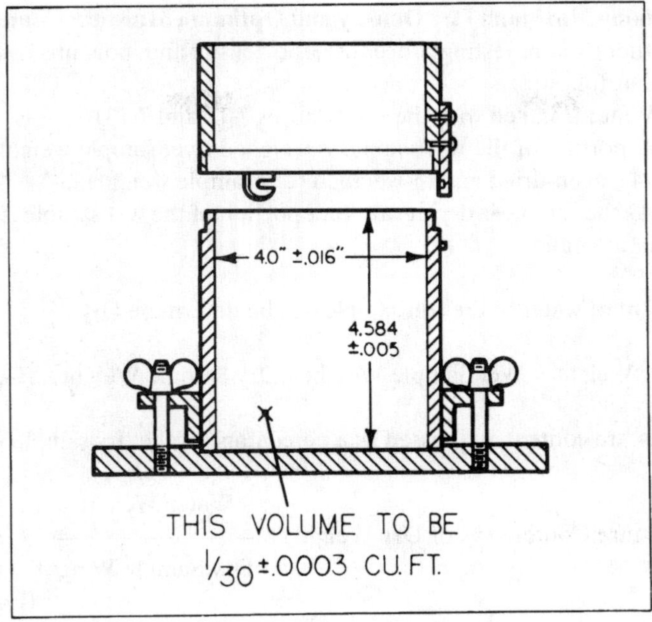


**Figure 7-12. Cylindrical mold for performing the standard Proctor soil test.**
*(Courtesy of American Society for Testing and Materials; Copyright ASTM. Reprinted with permission.)*

The amount of energy imparted to each cubic foot of soil during the standard Proctor test is:

$$\text{Energy} = \frac{\text{1 Ft. x 5.5 lb./Drop x 25 Drops/Layer x 3 Layers}}{\text{1/30 Cu. Ft.}}$$

$$= \text{12,375 foot-pounds/cubic foot}$$

**Modified Proctor Test**

The *modified Proctor test* is also known as Modified AASHTO T 180-70, or ASTM D 1557. Soil labs normally refer to this test as a "T-180." The modified Proctor test is usually specified for fill material designated for use under nuclear power plants, paved areas, concrete drives, airport runways, and any other area where high design loads are anticipated. The testing procedure is similar to that used for the standard Proctor test, except that the sample is placed in the steel cylinder mold in five layers, each layer being struck 25 times with a 10-pound hammer, mechanically dropped 18 inches. The energy imparted to each cubic foot of soil during the modified Proctor test is 56,250 foot-pounds

**Determining Maximum Dry Density and Optimum Moisture Content**

In either Proctor testing procedure, the density and moisture-content data is obtained as follows:

1) Water is mixed with the soil (Photos 7-12 and 7-13).
2) A portion of the wet sample is weighed (wet sample weight) (Photo 7-14), oven-dried and re-weighed (dry sample weight).
3) As the sample is drying, another portion of the wet sample is compacted in the mold.

The weight of water in the soil sample can be determined by:

$$\text{Water Weight} = \text{Wet Sample Weight} - \text{Dry Sample Weight} \qquad \textbf{(Equation 7.7)}$$

The moisture content, expressed as a percentage of the dry sample weight is:

$$\text{Moisture Content (\% of Dry Weight)} = \frac{\text{Water Weight}}{\text{Dry Sample Weight}} \times 100\%$$

**(Equation 7.8)**

The compacted sample weight can be determined by deducting the mold weight from the total sample-plus-mold weight (Photo 7-15); i.e.,

$$\text{Compacted Sample Wet Weight} = (\text{Mold} + \text{Compacted Sample Wet Weight}) - \text{Mold Weight}$$

**(Equation 7.9)**

*Note: The mold is weighed prior to running the test.*

The dry weight of the compacted sample is:

$$\text{Compacted Sample Dry Weight} = \frac{\text{Compacted Sample Wet Weight}}{1 + \text{Moisture Content (\% of Dry Weight)}}$$

**(Equation 7.10)**

The dry density of the soil expressed in pounds per cubic foot is:

$$\text{Dry Density (pcf)} = \frac{\text{Compacted Sample Dry Weight (lb.)}}{\text{Compacted Soil Volume (Cu. Ft.)}} \qquad \textbf{(Equation 7.11)}$$

*Note: The volume of the mold is constant. For example, the volume of the standard Proctor mold is approximately 1/30 (0.033) cubic feet (Figure 7-12).*

**Photo 7-12. Adding water to the sample.**
*(Photo by Dan Atcheson)*

**Photo 7-13. Mixing water into the sample.**
*(Photo by Dan Atcheson)*

**Photo 7-14. Weighing the wet sample.**
*(Photo by Dan Atcheson)*

**Photo 7-15. Weighing the compacted sample.**
*(Photo by Dan Atcheson)*

**Example 7-1:** Determine the dry density of a soil under the following conditions:

| | |
|---|---|
| Wet Sample Weight | = 300 grams |
| Dry Sample Weight | = 280 grams |
| Mold + Compacted Sample Weight | = 11.1 pounds |
| Mold Weight | = 6.8 pounds |
| Mold Volume | = 0.033 cubic feet |

**Solution:** The water weight is:

$$\text{Water Weight (g.)} = 300\text{ g.} - 280\text{ g.} \qquad \text{(Eq. 7.7)}$$
$$= 20\text{ grams}$$

The moisture content is:

$$\text{Moisture Content (\% of Dry Weight)} = \frac{20\text{ g.}}{280\text{ g.}} \times 100\% \qquad \text{(Eq. 7.8)}$$
$$= 7.1\%$$

The weight of the compacted wet sample is:

$$\text{Compacted Sample Wet Weight (g.)} = 11.1\text{ lb.} - 6.8\text{ lb.} \qquad \text{(Eq. 7.9)}$$
$$= 4.3\text{ pounds}$$

The dry weight of the compacted sample is:

$$\text{Compacted Sample Dry Weight (g.)} = \frac{4.3\text{ lb.}}{1.071} \qquad \text{(Eq. 7.10)}$$
$$= 4\text{ pounds}$$

The dry density of the soil is:

$$\text{Dry Density (pcf)} = \frac{4\text{ lb.}}{0.033\text{ Cu. Ft.}} \qquad \text{(Eq. 7.11)}$$
$$= 121\text{ pounds/cubic foot}$$

This procedure for determining dry density is repeated over a range of at least three moisture contents and the results plotted as dry density (in pounds per cubic foot) versus moisture content (as percentage of dry sample weight) (Figure 7-13). The optimum moisture content is the water content expressed as a percentage of dry sample weight that produces the greatest density under standard laboratory conditions. In Figure 7-13, the optimum moisture content for this particular soil is 13.7% and the maximum dry density is 115 pounds per cubic foot.

For any given soil type, the optimum moisture content decreases and the maximum dry density increases with an increase in compaction effort (Figure 7-14).

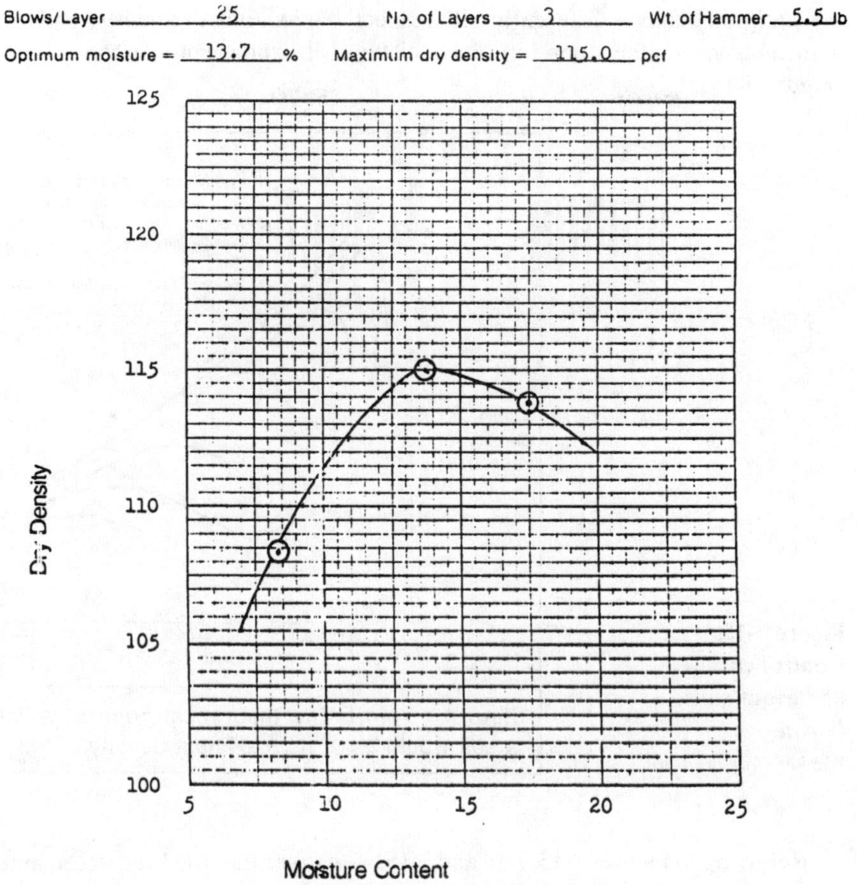

Blows/Layer 25
No. of Layers 3
Wt. of Hammer 5.5 lb
Optimum moisture = 13.7 %
Maximum dry density = 115.0 pcf
125
120
115
110
105
100
5
10
15
20
25
Dry Density
Moisture Content

**Figure 7-13. Moisture-density curve.** *(Illustrated by James Edward Atcheson)*

Modified AASHO
Standard AASHO
Dry density (lb/cu. ft)
Water content (% dry wt.)
120
115
110
105
100
95
5
10
15
20
25
30
35

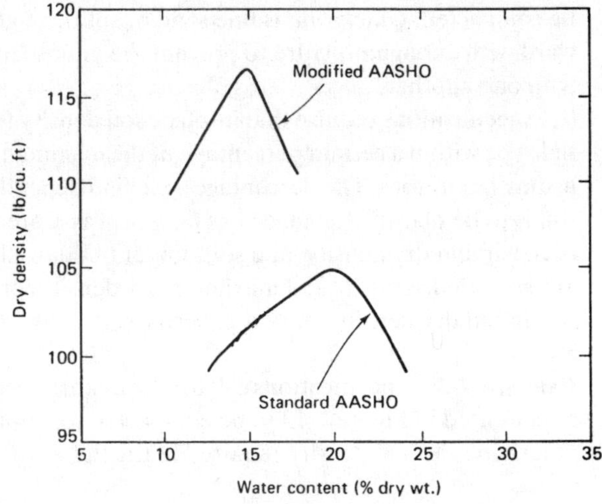

**Figure 7-14. Moisture content versus compaction effort.**

Each soil behaves differently with respect to maximum dry density and optimum moisture content; therefore, each soil type will exhibit a unique control curve (Figure 7-15).

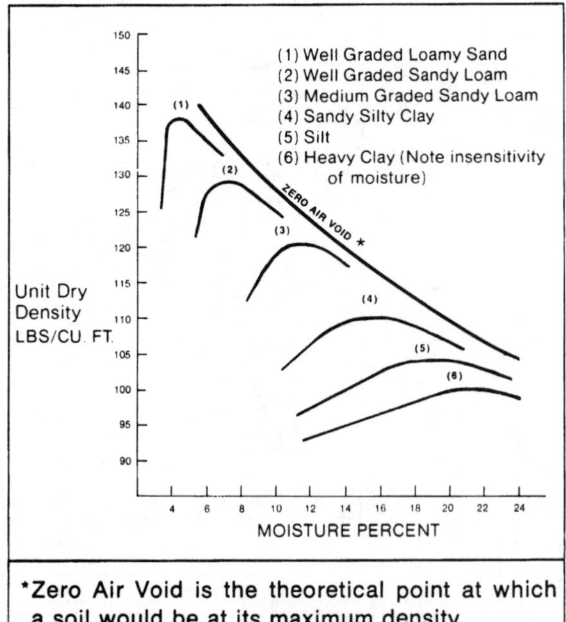


**Figure 7-15. Control curves of various soils.** *(Courtesy of Wacker Corporation)*

Referring to Figures 7-13, 7-14 and 7-15, we can see that too little or too much water reduces soil density. Too little water creates friction between soil particles and prevents them from sliding into a more compacted condition, while too much water causes soil to float out of the voids. This is why mud and quicksand cannot be compacted. *Quicksand* is fine sand or silt through which water is moving upward with enough pressure to prevent the grains from settling into firm contact with one another.

Specifications require that in-place soil density (density of fill material in the field) be within a certain percentage of the maximum dry density obtained in the testing laboratory. The percentage specified depends on soil type and where the soil is to be placed. Soil placed as fill in a grassy area is usually specified at 85% of maximum dry density, and soils under building slabs, sidewalks, etc. are usually specified as: a) 90% of maximum dry density for cohesive soils, or b) 95% of maximum dry density for non-cohesive soils.

**Example 7-2:** If specifications call for the material represented by laboratory test data plotted in Figure 7-13 to be compacted to within 95% of standard Proctor, determine the soil density required in the field.

**Solution:** One hundred percent of maximum dry density is 115 pounds per cubic foot (Figure 7-13); therefore, the required in-place soil density is:

$$\begin{aligned} 95\% \text{ of Maximum Dry Density} &= 0.95 \times 115 \text{ pcf} \\ &= 109.25 \text{ lb./cubic foot} \end{aligned}$$

**Example 7-3:** Determine the degree of compaction for the soil given in Figure 7-13 if the in-place density is 104 pounds per cubic foot.
**Solution:** One hundred percent of maximum dry density is 115 pounds per cubic foot (Figure 7-13); therefore, the extent of compaction is:

$$\begin{aligned} \text{Compaction (\% of Proctor)} &= \frac{104 \text{ lb./Cu. Ft.}}{115 \text{ lb./Cu. Ft.}} \\ &= 90\% \text{ of Proctor} \end{aligned}$$

By using special, or heavy compaction equipment in the field, the maximum in-place soil density is often greater than 100% of the maximum dry density obtained in the testing laboratory (Figure 7-16). However, avoid over-compacting the soil, since this wastes time and results in unnecessary equipment wear and fuel consumption. Also, the soil will begin to shift sideways. Excessive compaction often results from "method"-type specifications, where the designer dictates the type of compactor, lift thickness* and the number of passes required in order to fulfill a contract. With this type of specification, the designer should be legally obligated to accept the degree of compaction resulting from the specified method. A better type of specification is the "end-result"-type, where the contractor is given the freedom to achieve the desired compaction, regardless of the method used.

The optimum moisture content predicted from laboratory compaction tests is not necessarily the same as that actually required in the field. The optimum moisture content for a given soil tested in the field depends on the type and weight of the compactor used (Figure 7-16). When cohesive soils are compacted with pneumatic-tired rollers, the optimum moisture content required in the field is often slightly less than that predicted by laboratory tests, and when the soil is compacted with a sheepsfoot roller or tamping-foot roller, the optimum moisture content is often much lower than that predicted in lab tests. The optimum moisture content required in the field for non-cohesive soils is often only 80% of that predicted by lab tests, regardless of the type of roller used.

**A lift is a layer of loose soil to be compacted.*

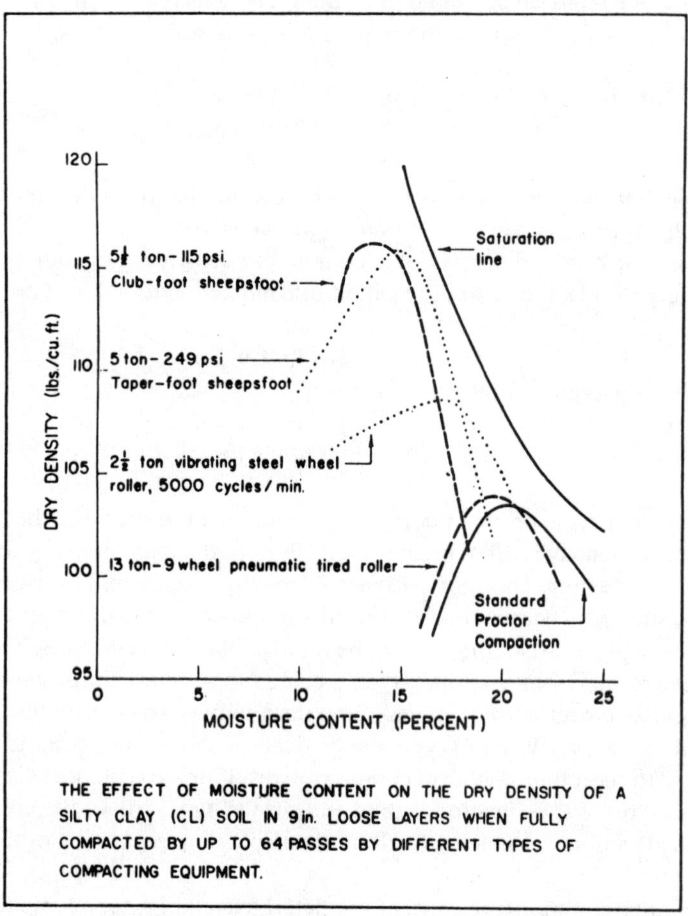


**Figure 7-16. Variation of optimum moisture content by roller type.**
*(D.L. Townsend)*

## FIELD (IN-PLACE) SOIL DENSITY TESTS

There are numerous tests available for determining in-place (field) soil density, some of which are described below. The dry density of any soil is:

$$\text{Dry Density} = \frac{\text{Compacted Soil Dry Weight}}{\text{Compacted Soil Volume}} \qquad \text{(Eq. 7.11)}$$

There is no difficulty in determining the wet and dry compacted soil weights, and thus, the moisture content of the soil. A primitive, but workable method is to extract a soil sample from a hole dug approximately 6 inches square and 6 inches deep. The soil sample is placed in sealed plastic bags, taken to the testing lab, weighed while wet and then weighed after drying. Soil sample weight can also be determined by weighing and drying the sample at the site. The weight of water in the soil can be determined by:

$$\text{Water Weight} = \text{Wet Sample Weight} - \text{Dry Sample Weight} \qquad \text{(Eq. 7.7)}$$

The moisture content expressed as a percentage of dry sample weight can be determined by:

$$\text{Moisture Content (\% of Dry Weight)} = \frac{\text{Water Weight}}{\text{Dry Sample Weight}} \text{ x } 100\% \qquad \text{(Eq. 7.8)}$$

The dry weight of the compacted soil can be determined by:

$$\text{Compacted Soil Dry Weight} = \frac{\text{Compacted Soil Wet Weight}}{1 + \text{Moisture Content (\% of Dry Weight)}} \qquad \text{(Eq. 7.10)}$$

There are several methods that can be used to determine the volume of the extracted soil. If the soil is impermeable, water can be poured into the hole from a container of known volume. If the soil is porous, thick oil or fine sand can be used instead of water. The test involving the use of sand is referred to as the *sand cone test* (Figure 7-17) (Photo 7-16). The volume of soil extracted is:

$$\text{Compacted Soil Volume} = \text{Initial Volume} - \text{Final Volume} \qquad \textbf{(Equation 7.12)}$$

where:

Initial Volume = Initial volume of water, oil or sand in the container

and,

Final Volume = Final volume of water, oil or sand in the container

The most convenient and accurate method of determining in-place soil density and moisture content is to use a device known as a nuclear density gauge *(nuclear meter)* (ASTM 2922). There are two types of nuclear testers. With one type, you simply place the device on the ground surface and read the instrument (Figure 7-18). The older device consists of a probe that emits radiation to two separate counters. To perform this test, a steel stake approximately 5/8-inches in diameter is driven into the compacted soil with a hammer, and removed. The probe is then placed into the hole and gamma radiation emitted (Photo 7-17). The emitted radiation is partially absorbed by the soil and partially transmitted to the counters. The percentage of radiation transmitted to the counters depends on the density of the soil, since dense soil absorbs more radiation than loose soil. One counter determines moisture content, and the other, soil density. Testing with the nuclear gauge is convenient and accurate, the soil is undisturbed, and the results are obtained within 3 minutes in the field. Nuclear testers can also be used to determine the density of asphalt paving.

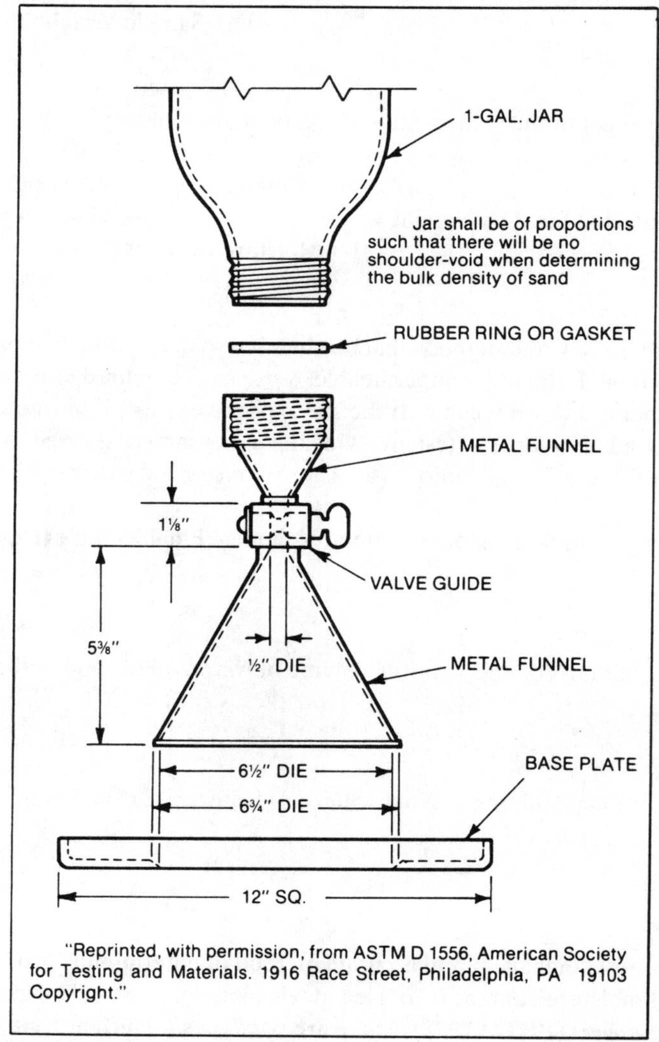


**Figure 7-17. Sand cone testing device.** *(Courtesy of American Society for Testing and Materials; Copyright ASTM. Reprinted with permission.)*

**Photo 7-16.**
**Sand cone testing apparatus.**
*(Photo by Dan Atcheson)*

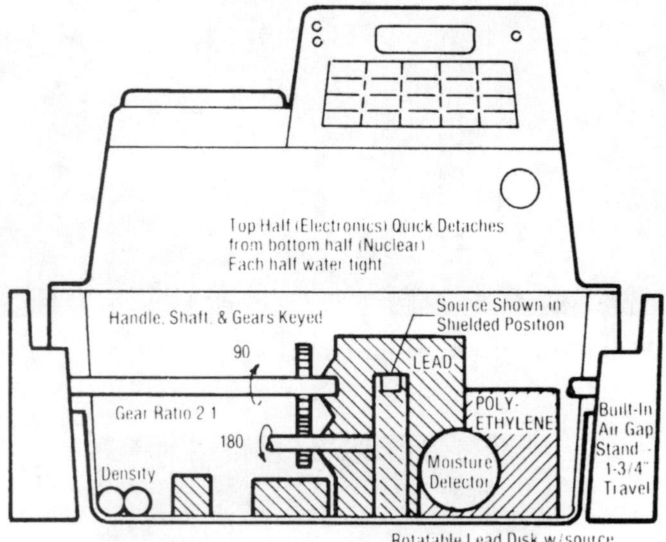


**Figure 7-18. Cross section of a nuclear meter.** *(Courtesy of Wacker Corporation)*

**Photo 7-17.**
**Nuclear meter.**
*(Photo by Dan Atcheson)*

The most recent development in soil- and asphalt-density testing is called *"density on the run."* When using this method, the nuclear density meter is attached to the roller and a display screen is positioned near the operator for a continuous, visual read-out of the soil or asphalt density (Photos 7-18 and 7-19).

**Photo 7-18. Density on the run.** *(Courtesy of Wacker Corporation)*

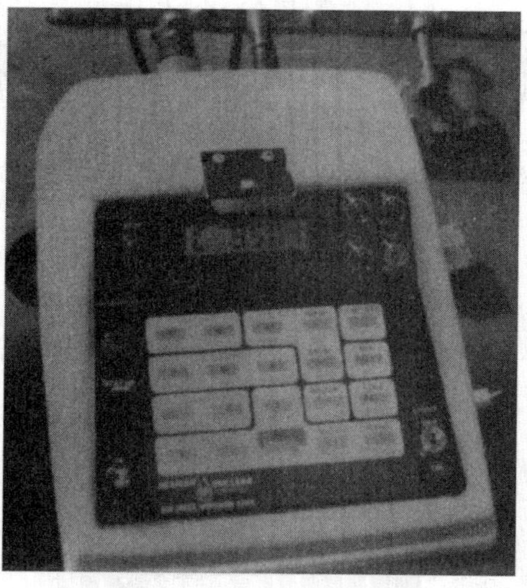

**Photo 7-19. Visual read-out.** *(Courtesy of Wacker Corporation)*

## DETERMINING TESTING COSTS

Soil testing costs can be determined if testing requirements and laboratory pricing policies are known. The costs for soil testing will vary form project to project, depending on the following major considerations:

1) The number of tests required per lift: Testing compacted fill is usually specified on a per-square-footage-of-fill basis, with a minimum number of tests required per lift. Therefore, testing costs will depend on the surface area of the fill, and the number of lifts placed.
2) Types of tests required: Different types of tests require different methodology, equipment and time to perform the test; therefore, costs vary from test to test.
3) The distance from the testing lab to the site: Most testing labs charge for mileage to and from the site. Also, if the trip requires overnight travel, overnight expenses will be charged.
4) The testing lab's minimum number of tests-per-trip policy: Most testing labs will charge for a minimum number of tests per trip, regardless of whether or not the additional tests are required, or performed.
5) The number of different types of fill material specified, or the number of separate sources of borrow: A soil analysis and laboratory compaction test is normally required for each type of soil placed.
6) The extent of overtime: Most testing labs charge time-and-a-half for labor required beyond normal working hours, on holidays and on weekends.
7) Special report preparation: Any test report or engineering analysis preparation requiring the expertise and/or stamp of a registered professional engineer is normally charged on an hourly basis.

**Example 7-4:** Determine the probable testing costs for the site diagrammed in Figure 7-19 under the following conditions:

- Initial soil samples at the borrow pits are obtained and delivered to the testing lab by the contractor.
- All fill material is to be imported.
- Fill material under the slab and in utility trenches is specified to be different from that used under the concrete paving.
- Round-trip distance from the testing lab to the site is 10 miles.
- Standard Proctor is to be used for testing fill under the slab and in the utility trenches.
- Modified Proctor is to be used for testing fill under the concrete paving.
- All utility trenches are to be excavated 3 feet deep.

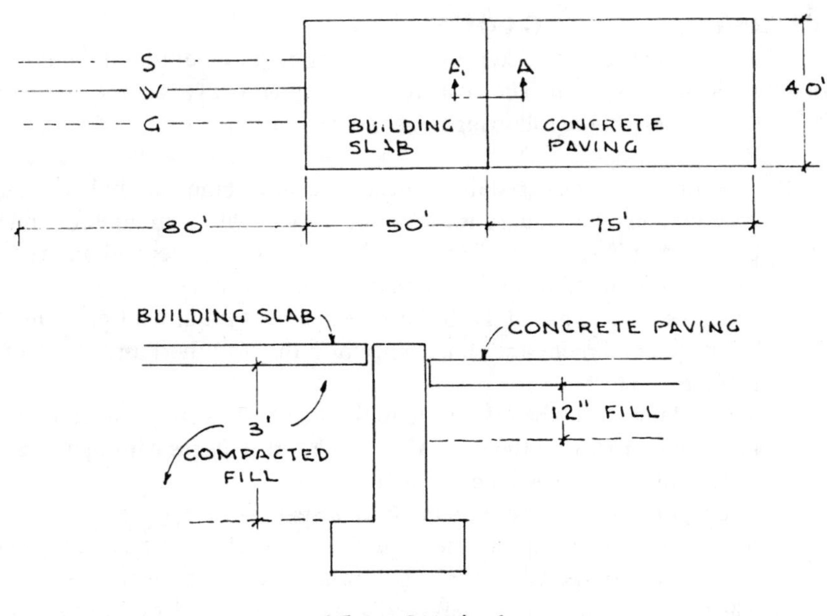


**Figure 7-19. Site plan for Example 7-4.** *(Illustrated by James Edward Atcheson)*

Testing requirements are specified as follows:

Under slabs and concrete paving:

A) One nuclear density test (ASTM-2922) per each 1000 square feet of slab, or fraction thereof, with a minimum of one test per 6-inch compacted lift.

B) One sand cone test (ASTM 1556) for each 3 nuclear tests performed, with a minimum of one test per each 6-inch compacted lift.

In utility trenches:

A) One nuclear density test per each 60 lineal feet of trench, or fraction thereof, with a minimum of one test per 6-inch compacted lift, per trench.

For each soil type:

A) Moisture-density relations (standard Proctor, or modified Proctor)

B) Unified Soil Classification with liquid limit, plastic limit and plasticity index

Laboratory unit testing costs are as follows:

Mileage charge ........................ $0.30/mile (each way)
Nuclear density tests .............. $15.00/each (Minimum of 3 tests per trip)
Sand cone tests ........................ $15.00/each (Minimum of 2 tests per trip)
Unified soil classification ....... $80.00/each
and plasticity index
Moisture-density relations ..... $60.00/each
(standard Proctor)
Moisture-density relations ..... $60.00/each
(modified Proctor)

**Solution:** Each trip will cost:

$$\text{Cost/Trip} = 10 \text{ Mi.} \times \$0.30/\text{Mi.}$$
$$= \$3.00$$

The number of 6-inch compacted lifts required at the slab is:

$$\text{Lifts at Slab} = \frac{36 \text{ In.}}{6 \text{ In./Lift}} \qquad \text{(Figure 7-19)}$$
$$= 6 \text{ lifts}$$

The slab area is:

$$\text{Area (Slab)} = 50' \times 40' \qquad \text{(Figure 7-19)}$$
$$= 2000 \text{ square feet}$$

The number of nuclear tests required per lift at the building slab is:

$$\text{Nuclear Tests Required at Slab} = \frac{2000 \text{ Sq. Ft.}}{1000 \text{ Sq. Ft./Test}}$$
$$= 2 \text{ tests}$$

A minimum of two nuclear tests is required per lift, but there is a minimum testing charge for three nuclear tests per trip; therefore, three nuclear tests must be paid for, per lift; i.e.,

$$\text{Nuclear Tests Performed at Slab} = 6 \text{ Lifts} \times 3 \text{ Tests/Lift}$$
$$= 18 \text{ tests}$$

A minimum of one sand cone test is required per lift, but there is a minimum testing charge for two sand cone tests per trip; therefore, two sand cone tests must be paid for, per lift; i.e.,

$$\text{Sand Cone Tests Performed at Slab} = 6 \text{ Lifts} \times 2 \text{ Tests/Lift}$$
$$= 12 \text{ tests}$$

The number of 6-inch compacted lifts required at the concrete paving is:

$$\text{Lifts at Paving} = \frac{12 \text{ In.}}{6 \text{ In./Lift}} \qquad \text{(Figure 7-19)}$$
$$= 2 \text{ lifts}$$

The concrete paving area is:

$$\text{Area (Paving)} = 75' \times 40' = 3000 \text{ square feet} \qquad \text{(Figure 7-19)}$$

The number of tests required per lift at the concrete paving is:

$$\text{Nuclear Tests Required at Paving} = \frac{3000 \text{ Sq. Ft.}}{1000 \text{ Sq. Ft./Test}} = 3 \text{ tests}$$

Since the minimum testing charge can be disregarded at the paved area, the number of nuclear tests paid for at the paving area is:

$$\text{Nuclear Tests Performed at Paving} = 2 \text{ Lifts} \times 3 \text{ Tests/Lift} = 6 \text{ tests}$$

A minimum of one sand cone test is required per lift, but there is a minimum testing charge for two sand cone tests per trip; therefore, two sand cone tests must be paid for, per lift; i.e.,

$$\text{Sand Cone Tests Performed at Paving} = 2 \text{ Lifts} \times 2 \text{ Tests/Lift} = 4 \text{ tests}$$

The number of compacted lifts required in the utility trenches is:

$$\text{Lifts at Trenches} = 3 \text{ Ea.} \times \frac{36 \text{ In.}}{6 \text{ In./Lift}} = 18 \text{ lifts} \qquad \text{(Figure 7-19)}$$

The number of nuclear tests required per lift at each utility trench is:

$$\text{Tests at Trenches} = \frac{80 \text{ Ft.}}{60 \text{ Ft./Test/Trench}} = 1.33 = 2 \text{ tests/trench}$$

Assuming that the project manager has the foresight to arrange for all three utility trenches to be backfilled simultaneously, the number of nuclear tests performed per trip is:

Tests/Trip = 3 Trenches x 2 Tests/Trench
= 6 tests

Since the minimum testing charge can be disregarded at the utility trenches, the number of nuclear tests paid for at the utility trenches is:

Nuclear Tests Performed at Trenches = 18 Lifts x 2 Tests/Lift
= 36 tests

The probable testing costs for the project are itemized as follows:

| Item | No. Trips | x $3.00 Total Trip Charge | Tests per Trip | Total No. Tests | Cost per Test | Total Item Costs |
|---|---|---|---|---|---|---|
| AT SLAB: | | | | | | |
| Standard Proctor | | (At lab) | | 1 | $60.00 | $60.00 |
| Unif. soil class & plasticity index | | (At lab) | | 1 | $80.00 | $80.00 |
| Nuclear density tests (6 lifts) | 6 | $18.00 | 3 | 18 | $15.00 | $288.00 |
| Sand cone tests (6 lifts) | 6 | (In above) | 2 | 12 | $15.00 | $180.00 |
| AT PAVING: | | | | | | |
| Modified Proctor | | (At lab) | | 1 | $60.00 | $60.00 |
| Unif. soil class & plasticity index | | (At lab) | | 1 | $80.00 | $80.00 |
| Nuclear density tests (2 lifts) | 2 | $6.00 | 3 | 6 | $15.00 | $96.00 |
| Sand cone tests (2 lifts) | 2 | (In above) | 2 | 4 | $15.00 | $60.00 |
| AT TRENCHES: | | | | | | |
| Nuclear density tests (6 lifts) | 6 | $18.00 | 6 | 36 | $15.00 | $558.00 |
| Total Testing Costs | | | | | | $1462.00 |

## THE FUNDAMENTAL STATES OF SOIL

Soil exists physically in three fundamental states: a) loose (in a stockpile), b) in the bank (natural, undisturbed condition) and c) compacted (building pads, paving base material, etc.) (Figure 7-20).

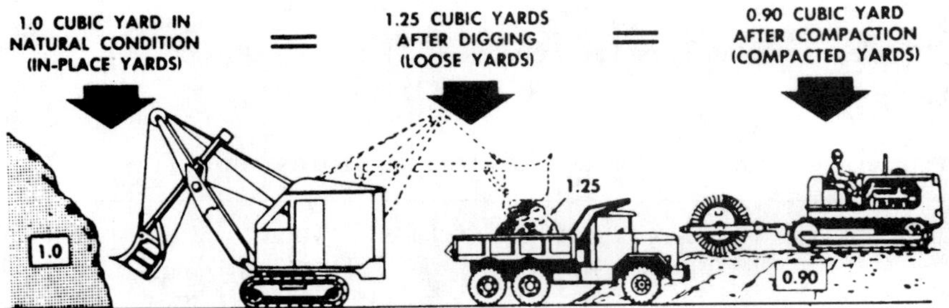


**Figure 7-20. Three fundamental states of soil.**
*(Courtesy of the U.S. Department of the Army)*

Soil swells (bulks) if disturbed, and shrinks under pressure, and volume corrections (swell and shrinkage factors) must often be used when estimating quantities, or production rates. Table 7-2 has been provided to show various swell and shrinkage factors for use in converting one soil volume condition to another soil volume condition. Table 7-2 indicates averages, only. For instance, soil shrinkage is dependent upon the type of equipment used in the compaction process, the compaction effort, moisture content, particle size and gradation, and the soil's initial density and cohesiveness. Due to unique local soil conditions, blanks have been provided in Table 7-2 for you to fill in values pertinent to your locale.

**Example 7-5:** How many compacted cubic yards will 1000 cubic yards of moist earth purchased from a soil bank yield? Assume that 100% of standard Proctor is required.
**Solution:** There are 1.07 bank cubic yards (BCY) of moist earth per compacted cubic yard (CCY) of moist earth (Table 7-2); therefore, the equivalent number of compacted cubic yards is:

$$\text{Volume (CCY)} = \frac{1000\ \text{BCY}}{1.07\ \text{BCY/CCY}} \qquad \text{(Table 7-2)}$$

$$= 935\ \text{CCY}$$

**Table 7-2: Soil Volume Conversion Factors**

| There are: LCY | BCY | per | | of |
|---|---|---|---|---|
| 1.13 ..... | 1.00 | 1.00 | BCY | Dry Sand |
| 1.32 ..... | 1.17 | ..... | CCY (95% Std. Proc.) | |
| 1.39 ..... | 1.23 | ..... | CCY (100% Std. Proc.) | |
| 1.38 ..... | 1.22 | ..... | CCY (95% Mod. Proc.) | |
| 1.45 ..... | 1.28 | ..... | CCY (100% Mod. Proc.) | |
| 1.13 ..... | 1.00 | 1.00 | BCY | Damp Sand |
| 1.16 ..... | 1.02 | ..... | CCY (95% Std. Proc.) | |
| 1.22 ..... | 1.07 | ..... | CCY (100% Std. Proc.) | |
| 1.21 ..... | 1.06 | ..... | CCY (95% Mod. Proc.) | |
| 1.27 ..... | 1.12 | ..... | CCY (100% Mod. Proc.) | |
| 1.14 ..... | 1.00 | 1.00 | BCY | Damp Gravel |
| 1.23 ..... | 1.07 | ..... | CCY (95% Std. Proc.) | |
| 1.29 ..... | 1.13 | ..... | CCY (100% Std. Proc.) | |
| 1.32 ..... | 1.16 | ..... | CCY (95% Mod. Proc.) | |
| 1.39 ..... | 1.22 | ..... | CCY (100% Mod. Proc.) | |
| 1.31 ..... | 1.00 | 1.00 | BCY | Dry Clay** |
| 1.18 ..... | 0.90* | .... | CCY (85% Std. Proc.) | |
| 1.25 ..... | 0.95* | .... | CCY (90% Std. Proc.) | |
| 1.39 ..... | 1.06 | ..... | CCY (100% Std. Proc.) | |
| 1.39 ..... | 1.06 | ..... | CCY (90% Mod. Proc.) | |
| 1.54 ..... | 1.18 | ..... | CCY (100% Mod. Proc.) | |
| 1.32 ..... | 1.00 | 1.00 | BCY | Dry Common Earth |
| 1.31 ..... | 1.00 | ..... | CCY (85% Std. Proc.) | |
| 1.39 ..... | 1.05 | ..... | CCY (90% Std. Proc.) | |
| 1.54 ..... | 1.17 | ..... | CCY (100% Std. Proc.) | |
| 1.45 ..... | 1.10 | ..... | CCY (90% Mod. Proc.) | |
| 1.61 ..... | 1.22 | ..... | CCY (100% Mod. Proc.) | |
| 1.28 ..... | 1.00 | 1.00 | BCY | Moist Common Earth |
| 1.17 ..... | 0.91* | .... | CCY (85% Std. Proc.) | |
| 1.23 ..... | 0.96* | .... | CCY (90% Std. Proc.) | |
| 1.37 ..... | 1.07 | ..... | CCY (100% Std. Proc.) | |
| 1.29 ..... | 1.00 | ..... | CCY (90% Mod. Proc.) | |
| 1.43 ..... | 1.11 | ..... | CCY (100% Mod. Proc.) | |

*Indicates that the material in the bank has a greater density than that required for the compacted material.

**Due to clay's ability to retain such a wide range of moisture contents, it is difficult to define "wet" clay; therefore, dry clay only, is considered here.

*(Excerpted from ESTIMATING EARTHWORK QUANTITIES by Dan Atcheson)*

**Example 7-6:** Determine the total number of haul trips necessary for a project requiring 20,000 cubic yards of compacted sand fill, assuming the following conditions:

- Fill material is damp sand purchased from stockpiles
- Compaction is specified at 95% of modified Proctor
- Haul-unit capacity is 24 loose cubic yards (LCY)

**Solution:** There are 1.21 loose cubic yards of damp sand per each compacted cubic yard of damp sand (Table 7-2); therefore, the number of loose cubic yards of sand required is:

$$\text{Volume (LCY)} = 20{,}000 \text{ CCY} \times 1.21 \text{ LCY/CCY} \qquad \text{(Table 7-2)}$$
$$= 24{,}200 \text{ LCY}$$

The total number of haul trips required is:

$$\text{No. Haul Trips} = \frac{24{,}200 \text{ LCY}}{24 \text{ LCY/Trip}}$$
$$= 1009 \text{ trips}$$

The factors in Table 7-2 are based on the soil weights given in Table 7-3.

**Table 7-3: Average Soil Weights**

| | Weight (Pounds per cubic yard) | | | |
|---|---|---|---|---|
| Soil Type | LCY | BCY | CCY 100% Std. Proctor | CCY 100% Mod. Proctor |
| Dry sand | 2420 | 2740 | 3362 | 3510 |
| Damp sand | 2760 | 3130 | 3362 | 3510 |
| Damp gravel | 2623 | 2980 | 3375 | 3645 |
| Dry clay | 2050 | 2675 | 2835 | 3159 |
| Dry common earth | 2185 | 2883 | 3375 | 3510 |
| Moist common earth | 2463 | 3160 | 3375 | 3510 |

*(Excerpted from ESTIMATING EARTHWORK QUANTITIES by Dan Atcheson)*

**Example 7-7:** Moist common earth has been selected from stockpiles for use as compacted fill and is to be compacted to within 95% of maximum density obtained in a soil testing laboratory. Assuming that maximum density is 3100 pounds per cubic yard, determine the number of loose cubic yards that must be purchased in order to place 11,000 compacted cubic yards.

**Solution:** Density at 95% of optimum is:

$$\text{Density} = 0.95 \text{ x } 3100 \text{ lb./CCY}$$
$$= 2945 \text{ lb./CCY}$$

The soil shrinkage is:

$$\text{Shrinkage (LCY to CCY)} = \frac{2463 \text{ lb./LCY}}{2945 \text{ lb./CCY}} \qquad \text{(Table 7-3)}$$
$$= 0.84 \text{ CCY/LCY}$$

Therefore, the required quantity of loose cubic yards to be purchased is:

$$\text{LCY (Purchased)} = \frac{11{,}000 \text{ CCY}}{0.84 \text{ CCY/LCY}}$$
$$= 13{,}095 \text{ LCY}$$

**Example 7-8:** The 13,095 LCY obtained in the previous example equates to how many tons of moist common earth?
**Solution:** Moist common earth weighs:

$$\text{Weight (Tons/LCY)} = \frac{2463 \text{ lb./LCY}}{2000 \text{ lb./Ton}} \qquad \text{(Table 7-3)}$$
$$= 1.23 \text{ tons/LCY}$$

Therefore, the equivalent tonnage is:

$$\text{Weight (Tons)} = 13{,}095 \text{ LCY x } 1.23 \text{ Tons/LCY}$$
$$= 16{,}107 \text{ tons}$$

**Example 7-9:** How many square yards of 6-inch-deep compacted damp gravel will the payload of a 24-LCY dump truck yield? Assume that 100% of modified Proctor is specified.
**Solution:** Each payload contains:

$$\text{Payload (Cu. Ft.)} = \frac{24 \text{ LCY}}{1.39 \text{ LCY/CCY}} \qquad \text{(Table 7-2)}$$
$$= 17.3 \text{ CCY x } 27 \text{ Cu. Ft./CCY}$$
$$= 467 \text{ cubic feet}$$

The volume of the required solid is:

$$V = l \times w \times h$$
$$= \text{Area} \times h$$

or,

$$\text{Area} = \frac{V}{h}$$

$$= \frac{467 \text{ Cu. Ft.}}{0.5 \text{ Ft.}}$$
$$= 934 \text{ Sq. Ft.} \div 9 \text{ Sq. Ft./S.Y.}$$
$$= 104 \text{ square yards}$$

## COMPACTION FORCES AND COMPACTION EQUIPMENT

Compaction is attained by static weight, kneading, impact, and/or vibration. The weight of a compactor *(static weight)* squeezes the soil particles together. All compactors use static weight (pressure) in combination with one or more of these forces to achieve compaction. *Kneading* (manipulation) is especially effective in compacting plastic soils. *Impact* involves applying sharp blows to the soil. *Vibration* (shaking) is similar to impact, except that impact is delivered at a low frequency (usually less than 10 cycles per second), while vibration is delivered at a high frequency (up to and beyond 80 cycles per second). Vibration is especially effective for compacting non-cohesive soils such as sand or gravel. When vibrated, soil particles become momentarily separated from one another, allowing them to twist and turn until they can assume a position that limits their movement (Figure 7-21).

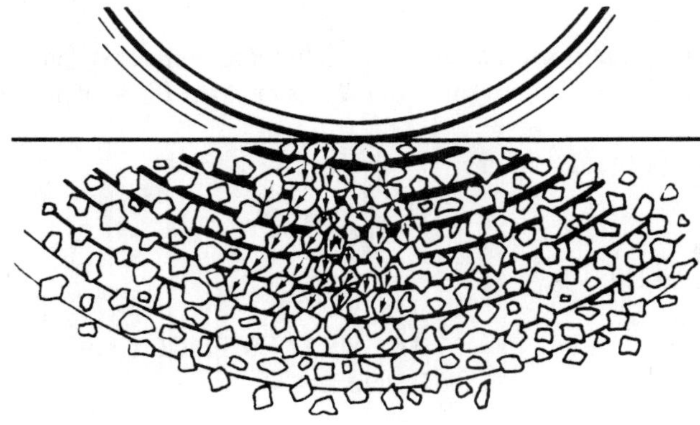

**Figure 7-21. Vibration of non-cohesive soil.** *(Courtesy of Barber-Greene)*

Cohesive soils do not settle under vibration because of the molecular attraction of the tiny soil particles. Also, the pancake shape of clay particles prevents them from dropping into voids under vibration (Figure 7-22). Instead, the shearing force of impact must be used to squeeze clay particles together.

**Figure 7-22. Clay particles.**
*(Courtesy of Wacker Corporation)*

For ease of comparison, eight types of compactors and their zones of application are shown in Figure 7-23. Each type of equipment has been positioned in what is considered to be its most effective and economical zone of application. However, it is not uncommon to find equipment working beyond the given application zones, and the exact positioning of the zones can vary, depending on material conditions.

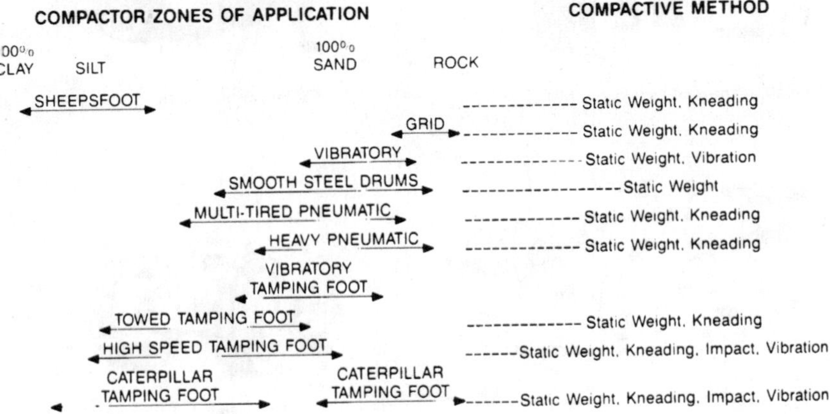


**Figure 7-23. Compactor zones of application.** *(Courtesy of Caterpillar Inc.)*

Regardless of the type of compaction effort used, the soil must be placed in layers (lifts) thin enough so that water and air can be expelled during the compaction process. The lift thickness will vary, depending on the type of compactor used and the soil type. Coarse-grained soils allow air and water to escape more easily than will fine-grained, clay-type soils that tend to retain air and water.

## PADFOOT ROLLERS

*Padfoot rollers* include sheepsfoot, tamping-foot segmented-pad and grid rollers (Figure 7-24). These rollers are often used for the initial compaction pass since they impart a kneading action that mixes the soil and eliminates excessively wet or dry pockets. These rollers are good for drying saturated soil because they create multiple indentations in the soil, increasing the exposed surface area to speed up drying. The forces of static weight and impact are also involved to help crush rock chips and hard soil lumps. Some padfoot rollers are equipped with a vibrator to help compact non-cohesive soils.

**Figure 7-24. Typical compaction equipment.** *(Courtesy of Caterpillar Inc.)*

Padfoot rollers can be used as a single drum (Photo 7-21), or multiple drums connected in tandem and towed by a tractor (Photo 7-24), but self-propelled units are also available. Some units are available with "drum drive" for added traction to ascend grades up to 62%. Some self-propelled units are available with an articulated frame and can be equipped with an optional dozer blade (Photo 7-20).

The foot pattern on the front and rear drums are usually offset to prevent the rear feet from overprinting the front feet (Photo 7-22), and cleaner bars (scraper bars) can be mounted over the drums to prevent soil from clogging between the feet (Photos 7-20 and 7-22). Some padfoot drums are interchangeable with smooth drums.

**Photo 7-20. Articulated padfoot roller with dozer blade.** *(Courtesy of Dynapac)*

Some rollers consist of a closed hollow drum whose weight can be increased by pouring water or damp sand ballast into the drum. Other rollers are mounted on an open-ended cylinder, and concrete blocks must be placed on the frame to increase the roller weight (Photo 7-24).

The speed at which a padfoot roller can operate depends on the size and type of equipment used, the soil type and the depth of foot penetration. Speed has little effect on soil density as long as soil shearing and displacement are minimized.

**Sheepsfoot Rollers**

*Sheepsfoot rollers* consist of a hollow steel drum, 4 to 5 feet long and 36 to 60 inches in diameter, equipped with protruding tapered shanks (7 to 9-1/2 inches long) with round, square or angular feet (lugs) welded to the outside surface of the drum (Figure 7-24) (Photo 7-21). Foot pressures can vary from 150 to 750 psi, depending on the size of the roller and the amount of ballast added. Tests indicate that ground contact pressure has little effect on the resulting soil density; however, density can be increased by increasing the size of the feet mounted on the roller.

Some sheepsfoot rollers are towed behind tractors; however, self-propelled units are becoming very popular and widely used. The speed of a towed roller is dependent upon tractor speed, but a self-propelled unit can travel in excess of 20 mph in both forward and reverse.

**Photo 7-21. Sheepsfoot roller.** *(Photo by Dan Atcheson)*

As the sheepsfoot roller makes its initial pass, the feet completely penetrate full-depth into the soil. With repeated passes, the depth of foot penetration decreases until the feet rise to the surface, or "walk out" of the soil as the soil becomes more compacted. The roller should be within one inch of the surface after three or four passes, and failure to walk out indicates soil shearing due to excessive ground contact pressure, or excessive soil moisture content.

A sheepsfoot roller is very effective for compacting plastic soils; however, it is virtually useless for compacting non-cohesive soils such as sand or gravel, unless the roller is equipped with a vibrator. Since the depth of soil that can be effectively compacted is limited to the length of the shanks, the maximum loose lift thickness should not exceed 2 inches beyond the combined length of the shanks and feet.

### Tamping-Foot Roller

A *tamping-foot roller* is self-propelled with rollers similar to a sheepsfoot roller, except that the feet are more closely spaced and consist of a solid, tapered foot (no shank) with a chevron-, or rectangular-shaped end (Figure 7-24) (Photo 7-22). Tamping-foot rollers are more versatile than sheepsfoot rollers, since they can compact a wider variety of soils, especially when equipped with a vibrator (Figure 7-23). They are also very effective at crushing lumpy soils and relatively soft rock.

### Segmented-Pad Rollers

A roller equipped with rectangular pads (soles) is called a *segmented-pad roller* (Figure 7-24) (Photo 7-23). Segmented-pad rollers create less surface disturbance than sheepsfoot rollers, and with a given contact pressure and equal number of passes, the segmented-pad roller will produce greater compaction. Sole pressures range from 250 to 750 psi.

**Photo 7-22. Tamping-foot roller.** *(Photo by Dan Atcheson)*

**Photo 7-23. Segemented-pad roller.** *(Photo by Dan Atcheson)*

**Grid (Mesh) Rollers**

A *grid roller* consists of an open-ended cylinder whose surface has a weaved grid or mesh pattern (Figure 7-24) (Photo 7-24). When ballasted with concrete blocks, grid rollers produce high soil pressures, and are especially useful for breaking down lumps of cohesive soil and crushing relatively soft rocks. The grid roller can be operated at high speeds since it does not scatter the soil.

**Photo 7-24. Grid rollers towed in tandem.** *(Hyster Company)*

## SMOOTH-WHEEL ROLLERS

*Smooth-wheel rollers* are used primarily for smoothing and finishing the surface being compacted; therefore, they are used for finishing base courses, gravel roads and bituminous pavements. They are also effective for compacting non-cohesive soils and small rock; however, smooth-wheel rollers have very little traction and are not suitable for working over a rough travel surface.

A smooth-wheel roller tends to form a crust over the surface of cohesive soil, which hinders compaction at the lower portion of the lift; therefore, the lift should first be compacted by a machine such as a tamping-foot roller and the loose lift thickness should not exceed 6 inches. The sealed surface also allows very little water penetration; therefore, if the soil requires additional moisture, add the moisture and mix it with the soil prior to rolling the surface with a smooth-wheel roller.

Static weight is the primary force involved when compacting soil with a smooth-wheel roller. Smooth-wheel rollers are often classified according to their weight, and a weight range will often be given for a particular roller, such as 15-20 tons. This means that the minimum weight is 15 tons, and that ballast can be added to increase the weight to a maximum of 20 tons. The weight can be increased by adding wheel weights, or by filling the drums with water or damp sand ballast. Due to the heavy weight required to compact even moderate lift thicknesses (increasing component costs and problems with handling and transportation), static rollers are being replaced with more efficient vibrating rollers (Photo 7-27). However, since static rollers produce a smooth surface, they will probably always be used for asphalt rolling (Photo 7-25).

**Photo 7-25. Smooth-wheel static roller.** *(Courtesy of Dynapac)*

Smooth-wheel rollers are available in a variety of models. A three-wheel two-axle roller is shown in Figure 7-24 and Photo 7-26. The front (bull) wheel is used for steering while the rear wheels drive the unit. The weights of three-wheel rollers normally range from 5 to 15 tons. Two- and three-axle tandem rollers are also available. The weights of two-wheel tandem rollers normally range from 1-1/2 to 20 tons. A self-propelled vibrating roller is shown in Figure 7-24 and Photo 7-27. Smooth-wheel rollers normally travel at speeds ranging from 2 to 10 miles per hour.

**Photo 7-26. Three-wheel static roller.** *(Photo by Dan Atcheson)*

**Photo 7-27. Vibrating smooth-wheel roller.** *(Courtesy of Dynapac)*

A *"combi-roller"* is manufactured with pneumatic tires mounted on the rear axle and a smooth steel drum mounted on the front (Photos 7-28 and 7-29).

**Photo 7-28. Combi-roller with small tires.** *(Courtesy of Dynapac)*

**Photo 7-29. Combi-roller with large tires.** *(Courtesy of Dynapac)*

## PHEUMATIC-TIRED ROLLERS

*Pneumatic-tired rollers* are ballast boxes supported by wheels equipped with smooth-tread tires. There are two basic types of pneumatic roller, including multi-tired rollers and heavy pneumatic rollers. *Multi-tired rollers* can be towed (Photo 7-30); however, they are usually self-propelled, and are manufactured with two tandem axles with up to five tires mounted on one axle, and one less tire mounted on the other (Figure 7-24) (Photo 7-31). The rear wheels are offset to travel over the surface between the front wheels so that there are no spaces in surface coverage. Some multi-tired rollers are manufactured with an oscillating wobble-wheel design to impart a kneading action to the soil, and some are equipped with a vibrator to help compact non-cohesive soils. The speed of a multi-tired roller varies from 2.5 to 16 mph, depending on the size and power of the machine, gear range, total resistance, traction and traffic at the fill.

Multi-tired rollers are used primarily for smoothing and finishing the surface they are compacting; therefore, they are used for finishing base courses and bituminous pavements. However, they are also effective for compacting granular soils. As compared to a smooth-wheel roller, the multi-tired roller is better for finishing sticky soils or wet rock, since the tires have less tendency to pull up the soil.

A multi-tired roller tends to form a crust over the surface of cohesive soil, which hinders compaction at the lower portion of the lift; therefore, the lift should first be compacted by a machine such as a tamping foot roller, and the compacted lift thickness should not exceed 6 inches. The sealed surface also allows very little water penetration; therefore, if the soil requires additional moisture, add the moisture and mix it with the soil prior to rolling the surface with a pneumatic-tired roller.

**Photo 7-30. Towed pneumatic roller.** *(Photo by Dan Atcheson)*

**Photo 7-31. Tandem-axle pneumatic roller.** *(Courtesy of Dynapac)*

*Heavy pneumatic* (large-tired) *rollers* are normally towed behind a tractor and usually have a single row of two, or four tires beneath the center of the weight box (Figure 7-24) (Photo 7-32). They are used to compact thick layers of soil to great density. Loose lifts can be more than 24 inches thick, and compacted lifts can be more than 12 inches thick. The speed of a heavy pneumatic roller varies from 2 to 5 mph, depending on the size and power of the machine, total resistance, traction and traffic at the fill.

**Photo 7-32. Heavy pneumatic-tired roller ballasted with water.**
*(Photo by Dan Atcheson)*

Static weight is the primary force involved in compacting the soil with a pneumatic-tired roller. Heavy pneumatic rollers are available with weights ranging from 15 to 200 tons, and the weight of a given machine can be varied by adding or removing water, soil, sand, gravel, stone, concrete blocks and/or steel ballast.

With a given machine weight, ground contact pressure can be varied by adjusting the tire inflation pressure. Depending on the type of tire, the inflation pressures can be varied from 35 to 150 psi. The first passes over the fill should be made with lower tire pressures to prevent excessive rutting and to increase ground coverage. As the soil begins to compact, the tire inflation pressure should be increased until it is at maximum during the final pass. Some pneumatic rollers are designed to allow the operator to vary the tire inflation pressure without stopping the machine ("air-on-the-go" system). Without this option, the contractor must to stop the machine and change the tire inflation pressure, vary the weight of the ballast, or keep a variety of rollers on hand with different weights and/or tire pressures.

The pressure imposed on a travel surface is measured in pounds per square inch (psi), pounds per square foot (psf), or tons per square foot (tsf). The pressure imposed on a travel surface can be determined by:

$$\text{Ground Contact Pressure (psi)} = \frac{\text{Load Over Wheel (lb.)}}{\text{Ground Contact Area (Sq. In.)}}$$

**(Equation 7.13)**

Ground contact pressure given in terms of psi can be converted into terms of psf as follows:

$$\text{Ground Contact Pressure (psf)} = \text{Ground Contact Pressure (psi)} \times 144 \text{ Sq. In./Sq. Ft.}$$

**(Equation 7.14)**

Ground contact pressure can also be converted into terms of tons per square foot as follows:

$$\text{Ground Contact Pressure (tsf)} = \frac{\text{Ground Contact Pressure (psf)}}{\text{2000 lb./Ton}}$$

**(Equation 7.15)**

**Example 7-10:** Determine the ground contact pressure exerted by a 15,000-pound multi-tired roller equipped with 7 wheels if each wheel has a ground contact area of 38.5 square inches.
**Solution:** The load imposed over each wheel is:

$$\text{Wheel Load (lb.)} = \frac{\text{15,000 lb.}}{\text{7 Wheels}}$$
$$= \text{2142.9 lb./wheel}$$

The ground contact pressure is:

$$\text{Ground Contact Pressure (psi)} = \frac{\text{2142.9 lb.}}{\text{38.5 Sq. In.}} \qquad \text{(Eq. 7.13)}$$
$$= \text{55.7 psi}$$

When determining the tire load, we can assume that 10% of the wheel load is carried by the sidewalls of the tires and 90% is carried by the portion of the tire in contact with the travel surface; therefore, the ground contact area for a tire is:

$$\text{Ground Contact Area (Sq. In.)} = \frac{0.90 \times \text{Wheel Load (lb.)}}{\text{Inflation Pressure (psi)}}$$

**(Equation 7.16)**

**Example 7-11:** Determine the ground contact pressure of the pneumatic roller given in the previous example if the tire inflation pressure is 55 psi.
**Solution:** The ground contact area is:

$$\text{Ground Contact Area (Sq. In.)} = \frac{0.90 \times \text{2142.9 lb.}}{\text{55 psi}} \qquad \text{(Eq. 7.16)}$$
$$= \text{35.1 square inches}$$

The ground contact pressure is:

$$\text{Ground Contact Pressure (psi)} = \frac{2142.9 \text{ lb.}}{35.1 \text{ Sq. In.}} \qquad \text{(Eq. 7.13)}$$

$$= 61 \text{ psi}$$

Thus, the ground contact pressure can be varied by varying the weight of the machine, or by varying the contact area between the tire and the ground surface. With any given machine weight, increasing the tire inflation pressure reduces the ground contact area and increases the ground contact pressure (Figure 7-25).

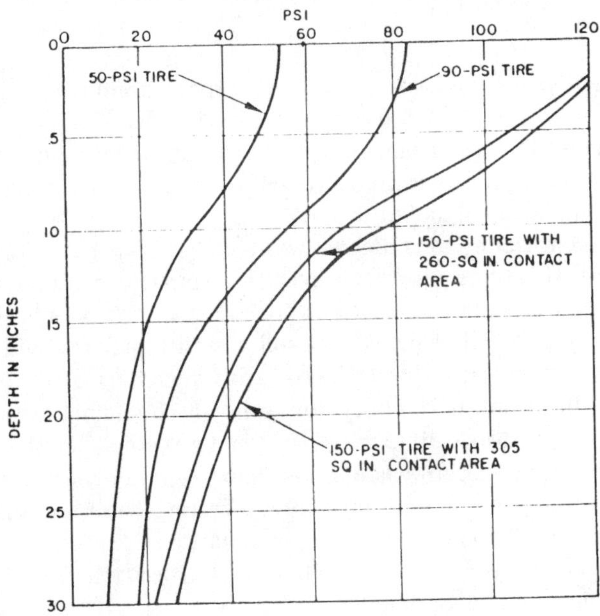


**Figure 7-25. Vertical stresses versus depth; pneumatic roller.**
*(Courtesy of the U.S. Department of the Army)*

## VIBRATING COMPACTORS

Several types of compactors can be equipped with a vibrator, including padfoot rollers, smooth-wheel rollers and pneumatic-tired rollers. The vibration is usually activated by a separate motor mounted in the roller drum (Figure 7-26), or on the frame. The motor is mechanically attached to an unbalanced eccentric weight (exciter unit) which spins, causing the device to vibrate forcefully (Figure 7-27). Vibration combined with static weight is especially useful for compacting non-cohesive soils such as sand, gravel and stone; however, it has little effect on silt or clay.

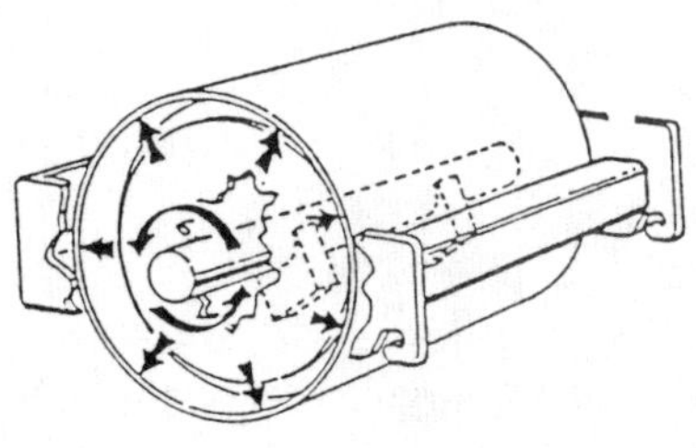

Figure 7-26.
**Drum-mounted vibrating unit.**
*(Courtesy of Dynapac)*

Direction of Travel
rpm
Direction of Rotation
W
CENTER OF GRAVITY
r
Direction of Centrifugal Force

**Figure 7-27.**
**Exciter unit.**
*(Courtesy of Wacker Corporation)*

For a given compaction task, a vibrating compactor requires only about one-half as much weight as that required for a non-vibrating compactor. Vibration penetrates deep into the soil, meaning that thick lifts can be compacted. Also, in addition to static machine weight, vibration produces centrifugal force that increases the compaction ability.

Vibrating compactors can compact non-cohesive soils to the required density after only two or three passes; however, the compactor must be run at a relatively slow speed. Generally, the speed varies from 1.5 to 2.5 mph, although the actual speed required will depend upon soil type, lift thickness and the type of machine used. The maximum lift thickness varies, depending on the soil type and machine used; however, the thickness for aggregate should not exceed 6 inches.

Vibrating compactors vibrate the soil into a more compacted state, and some machines such as the smooth-wheel roller help to shatter rock particles at the ground surface to form a smooth surface. A vibrating smooth-wheel roller will yield the fastest production when rolling aggregate base material.

Depending on the type of machine, vibrations can be varied from 500 to 14,000 cycles per minute, and the optimum number of vibrations depends on the natural resonant frequency of the soil being compacted. The smaller the soil particle, the higher the natural frequency; the larger the particle, the lower the natural frequency (Figure 7-28). Cohesive soils can sometimes be effectively compacted through vibration by running the machine at a low frequency and high amplitude.

## ESTIMATING COMPACTOR PRODUCTION

Compactor production rates can be determined by:

$$\text{Production (CCY/Hr.)} = \frac{W \times S \times L \times 16.3 \times E}{P} \qquad \textbf{(Equation 7.17)}$$

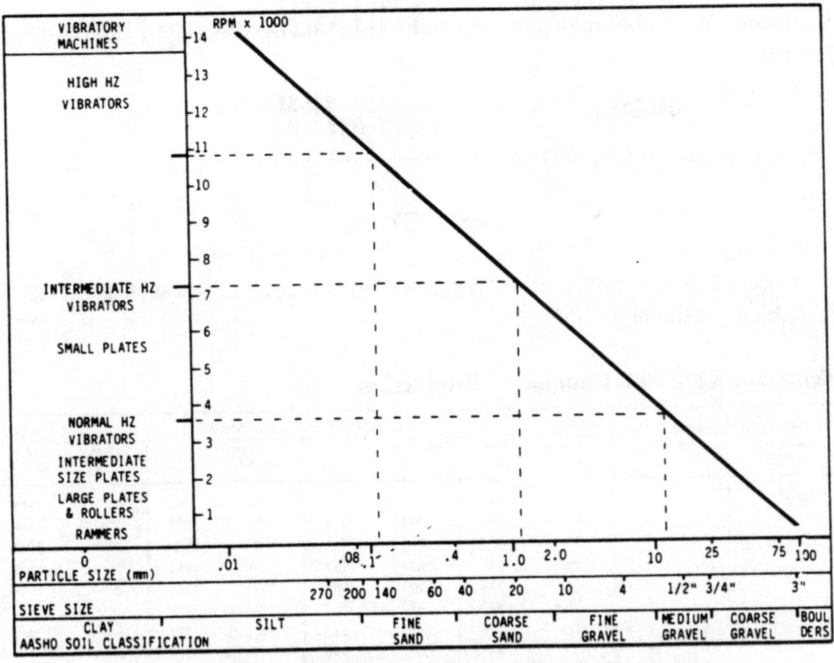


VIBRATION FREQUENCY — PARTICLE SIZE

**Figure 7-28. Vibration frequency — particle size.**
*(Courtesy of Wacker Corporation)*

where: W = Compacted width per pass (ft.)
S = Average speed (mph)
L = Lift thickness (in.)
P = Number of passes required
E = Job efficiency

and,

$$16.3 = \frac{5280 \text{ Ft.}}{12 \text{ In. x } 27 \text{ Cu. Ft.}}$$

**Example 7-12:** Determine the production rate of a tamping-foot roller working under the following conditions:

W = 7 feet
S = 6 mph
L = 6 inches
P = 5 each
E = 0.83

**Solution:** The probable production rate expressed in compacted cubic yards per hour is:

$$\text{Production (CCY/Hr.)} = \frac{7 \times 6 \times 6 \times 16.3 \times 0.83}{5} \qquad \text{(Eq. 7.17)}$$

$$= 682 \text{ CCY/hour}$$

Caterpillar Inc. publishes soil compactor production tables such as that shown in Tables 7-4 through 7-6.

## Table 7-4: 825C Soil Compactor Production

| MODEL AND MACHINE PASSES* | AVERAGE SPEED km/h | mph | COMPACTED LIFT THICKNESS 100 mm m³/h | 4 in yd³/hr | 150 mm m³/h | 6 in yd³/hr | 200 mm m³/h | 8 in yd³/hr | 250 mm m³/h | 10 in yd³/hr |
|---|---|---|---|---|---|---|---|---|---|---|
| **825C** 3 | 6.5 | **4** | 488 | **642** | 731 | **962** | 975 | **1283** | 1219 | **1604** |
| | 9.5 | **6** | 713 | **962** | 1069 | **1444** | 1425 | **1925** | 1781 | **2406** |
| | 13.0 | **8** | 975 | **1283** | 1463 | **1925** | 1950 | **2566** | 2438 | **3208** |
| 4 | 6.5 | **4** | 366 | **481** | 534 | **722** | 731 | **962** | 914 | **1203** |
| | 9.5 | **6** | 534 | **722** | 802 | **1083** | 1069 | **1444** | 1336 | **1804** |
| | 13.0 | **8** | 731 | **962** | 1097 | **1444** | 1463 | **1925** | 1828 | **2406** |
| 5 | 6.5 | **4** | 293 | **385** | 439 | **577** | 585 | **770** | 731 | **962** |
| | 9.5 | **6** | 428 | **577** | 641 | **866** | 855 | **1155** | 1069 | **1444** |
| | 13.0 | **8** | 585 | **770** | 878 | **1155** | 1170 | **1540** | 1463 | **1925** |
| 6 | 6.5 | **4** | 244 | **321** | 366 | **481** | 488 | **642** | 609 | **802** |
| | 9.5 | **6** | 356 | **481** | 534 | **722** | 713 | **962** | 891 | **1203** |
| | 13.0 | **8** | 488 | **642** | 731 | **962** | 975 | **1283** | 1219 | **1604** |

*The number of machine passes required is dependent on soil type, moisture content, desired compaction and machine weight

*(Courtesy of Caterpillar Inc.)*

## Table 7-5: CS-323 Smooth Drum Vibratory Soil Compactor Production

| MODEL AND MACHINE PASSES | | AVERAGE SPEED MPH | COMPACTED LIFT THICKNESS 150 mm m³/hr | 6 in yd³/hr | 200 mm m³/hr | 8 in yd³/hr | 250 mm m³/hr | 10 in yd³/hr | 300 mm m³/hr | 12 in yd³/hr |
|---|---|---|---|---|---|---|---|---|---|---|
| CS-323 | 3 | 2.5 | 249.5 | 326.0 | 332.7 | 434.7 | 415.9 | 543.3 | 499.0 | 652.0 |
| | 3 | 3.0 | 299.4 | 391.2 | 399.2 | 521.6 | 499.0 | 652.0 | 598.8 | 782.4 |
| | 3 | 3.5 | 349.3 | 456.4 | 465.8 | 608.5 | 582.2 | 760.7 | 698.7 | 912.8 |
| | 3 | 4.0 | 399.2 | 521.6 | 532.3 | 695.5 | 665.4 | 869.3 | 798.5 | 1043.2 |
| | 4 | 2.5 | 187.1 | 244.5 | 249.5 | 326.0 | 311.9 | 407.5 | 374.3 | 489.0 |
| | 4 | 3.0 | 224.6 | 293.4 | 299.4 | 391.2 | 374.3 | 489.0 | 449.1 | 586.8 |
| | 4 | 3.5 | 262.0 | 342.3 | 349.3 | 456.4 | 436.7 | 570.5 | 524.0 | 684.6 |
| | 4 | 4.0 | 299.4 | 391.2 | 399.2 | 521.6 | 499.0 | 652.0 | 598.8 | 782.4 |
| | 5 | 2.5 | 149.7 | 195.6 | 199.6 | 260.8 | 249.5 | 326.0 | 299.4 | 391.2 |
| | 5 | 3.0 | 179.7 | 234.7 | 239.5 | 313.0 | 299.4 | 391.2 | 359.3 | 469.4 |
| | 5 | 3.5 | 209.6 | 273.8 | 279.5 | 365.1 | 349.3 | 456.4 | 419.2 | 547.7 |
| | 5 | 4.0 | 239.5 | 313.0 | 319.4 | 417.3 | 399.2 | 521.6 | 479.1 | 625.9 |

*(Courtesy of Caterpillar Inc.)*

**Table 7-6: CP-323 Padded Drum Vibratory Soil Compactor Production**

| MODEL AND MACHINE PASSES | | AVERAGE SPEED MPH | COMPACTED LIFT THICKNESS | | | | | | | |
|---|---|---|---|---|---|---|---|---|---|---|
| | | | 150 mm $m^3$/hr | 6 in $yd^3$/hr | 200 mm $m^3$/hr | 8 in $yd^3$/hr | 250 mm $m^3$/hr | 10 in $yd^3$/hr | 300 mm $m^3$/hr | 12 in $yd^3$/hr |
| CP-323 | 3 | 2.5 | 249.5 | 326.0 | 332.7 | 434.7 | 415.9 | 543.3 | 499.0 | 652.0 |
| | 3 | 3.0 | 299.4 | 391.2 | 399.2 | 521.6 | 499.0 | 652.0 | 598.8 | 782.4 |
| | 3 | 3.5 | 349.3 | 456.4 | 465.8 | 608.5 | 582.2 | 760.7 | 698.7 | 912.8 |
| | 3 | 4.0 | 399.2 | 521.6 | 532.3 | 695.5 | 665.4 | 869.3 | 798.5 | 1043.2 |
| | 4 | 2.5 | 187.1 | 244.5 | 249.5 | 326.0 | 311.9 | 407.5 | 374.3 | 489.0 |
| | 4 | 3.0 | 224.6 | 293.4 | 299.4 | 391.2 | 374.3 | 489.0 | 449.1 | 586.8 |
| | 4 | 3.5 | 262.0 | 342.3 | 349.3 | 456.4 | 436.7 | 570.5 | 524.0 | 684.6 |
| | 4 | 4.0 | 299.4 | 391.2 | 399.2 | 521.6 | 499.0 | 652.0 | 598.8 | 782.4 |
| | 5 | 2.5 | 149.7 | 195.6 | 199.6 | 260.8 | 249.5 | 326.0 | 299.4 | 391.2 |
| | 5 | 3.0 | 179.7 | 234.7 | 239.5 | 313.0 | 299.4 | 391.2 | 359.3 | 469.4 |
| | 5 | 3.5 | 209.6 | 273.8 | 279.5 | 365.1 | 349.3 | 456.4 | 419.2 | 547.7 |
| | 5 | 4.0 | 239.5 | 313.0 | 319.4 | 417.3 | 399.2 | 521.6 | 479.1 | 625.9 |

*(Courtesy of Caterpillar Inc.)*

**Example 7-13:** Use Table 7-4 to determine the production rate of a Caterpillar 825C soil compactor under the following conditions:

S = 4 mph
L = 8 inches
P = 6 each

**Solution:** Referring to Table 7-4, the probable production rate is:

Production (CCY/Hr.) = 642 CCY/hour (Table 7-4)

## PNEUMATIC-TIRED ROLLER PRODUCTION

The production rate of a pneumatic-tired roller can be approximated by using Figure 7-29. To use Figure 7-29, enter the chart from the left side and find the appropriate lift thickness. Project a horizontal line to the right until it intersects the appropriate diagonal line representing the roller speed. From this intersection, drop vertically to obtain the cubic yards of soil or paving material compacted per hour, per pass. The actual production rate can then be determined by:

$$\text{Production (C.Y./Hr.)} = \frac{\text{C.Y./Pass}}{\text{No. Passes Required}} \qquad \textbf{(Equation 7.18)}$$

**Example 7-14:** Assuming a 6-inch lift thickness, determine the production rate of a Caterpillar PS-130 pneumatic-tired roller traveling at 2 mph if four passes are required per lift.

**Solution:** Referring to Figure 7-29, the production rate per hour per pass is:

Production (C.Y./Hr./Pass) = 1000 C.Y./Hr./Pass (Figure 7-29)

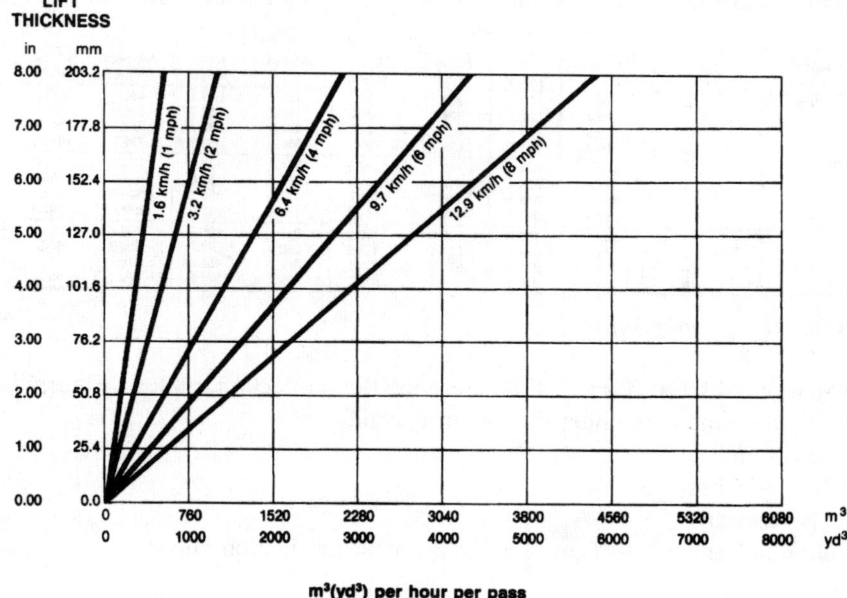


**Figure 7-29. PS-130 or PS-180 pneumatic roller production (C.Y./Hr./Pass)**
*(Courtesy of Caterpillar Inc.)*

The actual production rate using four passes is:

$$\text{Production (C.Y./Hr.)} = \frac{1000 \text{ C.Y./Hr./Pass}}{4 \text{ Passes}} \qquad \text{(Eq. 7.18)}$$

$$= 250 \text{ C.Y./hour}$$

## NUMBER OF PASSES REQUIRED

Soil continues to compact until internal resistance within the soil reaches equilibrium with the compaction pressure. The number of machine passes required to compact soil is dependent upon soil type, moisture content, machine weight and type, and the degree of compaction required; therefore, P in Equation 7.17 is often difficult to determine, beforehand. Figure 7-30 compares the performance of various types of compactors while compacting sandy clay, and Figure 7-31 shows the performance of a tamping foot roller while compacting a variety of soil types.

THE EFFECT OF NUMBER OF PASSES OF COMPACTING EQUIPMENT ON THE DRY DENSITY OF A SANDY CLAY SOIL COMPACTED AT THE OPTIMUM MOISTURE CONTENT FOR THE PLANT IN 9in. LOOSE LAYERS.

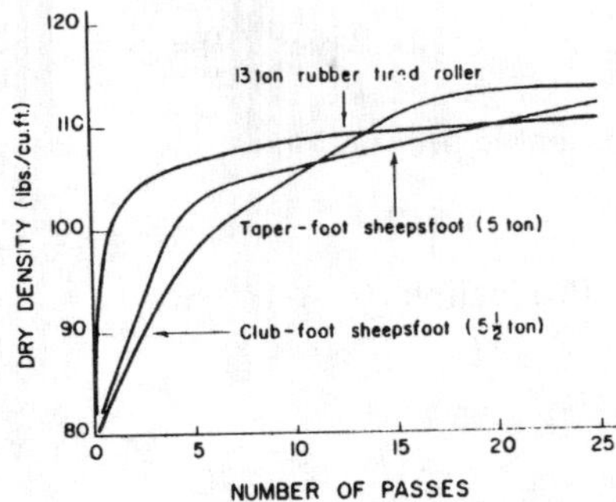


Standard Proctor Compaction for the Sandy Clay Soil is 115 lbs./cu Ft at 14 % optimum moisture content.

| Compacting unit | Percent Compaction | Number of Passes | Cubic yds compacted per hour at 2 mph. |
|---|---|---|---|
| 13 T wobble wheel. | 95 | 14 | 140 |
| | 100 | – | – |
| 5½ T club - foot | 95 | 14 | 110 |
| | 100 | 38 | 56 |
| 5 T taper - foot | 95 | 19 | 90 |
| | 100 | 44 | 50 |

**Figure 7-30. Effect of number of passes.**
*(D.L. Townsend)*

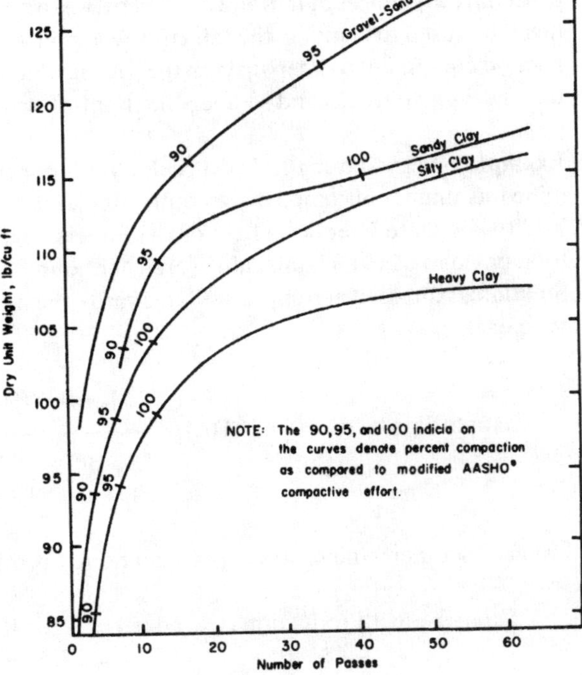


**Figure 7-31. Number of passes versus soil type.**
*(Courtesy of the U.S. Department of the Army)*

**Example 7-15:** Use data from Figure 7-31 to determine the production rate of the compactor given in Example 7-12 if the compactor is required to compact sandy clay to a dry density of 110 pounds per cubic foot.
**Solution:** Referring to Figure 7-31, the number of passes required is:

No. Passes (P) = 14 passes (Figure 7-31)

The probable production rate is:

$$\text{Production (CCY/Hr.)} = \frac{7 \times 6 \times 6 \times 16.3 \times 0.83}{14}$$

(Eq. 7.17; Example 7-12)

$$= 244 \text{ CCY/hour}$$

Compactor production is most efficient when several areas are worked simultaneously. Begin by working the subgrade. For good compaction, the subgrade should be ripped to a depth of 6 inches, watered and mixed, and then compacted. After enough subgrade area has been compacted, haul in and spread the first lift of fill material over this area. At the same time, start preparing a second subgrade area. As the first lift of fill material is being spread over the second area, the first lift of fill in the first area can be watered and compacted. Then, the compactor can compact the first lift in the second area as the second lift is spread over in the first area. Proceed in this manner until all fill is brought up to final grade. This procedure minimizes delays and conflict between machines hauling in fill and the machines used to compact the fill. However, you will have to have enough compaction equipment to keep up with the rate at which fill material is being hauled in. Always give trucks and scrapers the right-of-way at the fill.

**Example 7-16:** Assuming 1.15 bank cubic yards per compacted cubic yard, determine the number of compactors required under the conditions given in Example 7-12 to adequately service a fleet of 12 scrapers if each scraper hauls an average total payload of 164.3 bank cubic yards per hour.
**Solution:** Expressed in compacted cubic yards, the hourly production rate of each scraper is:

$$\text{Scraper Production (CCY/Hr.)} = \frac{164.3 \text{ BCY/Hr.}}{1.15 \text{ BCY/CCY}} \quad \text{(Given)}$$

$$= 142.9 \text{ CCY/hour}$$

Using 12 scrapers, the total scraper fleet hourly production rate is:

$$\text{Scraper Fleet Production (CCY/Hr.)} = 12 \times 142.9 \text{ CCY/Hr.}$$

$$= 1715 \text{ CCY/hour}$$

The compactor production rate is:

Compactor Production (CCY/Hr.) = 682 CCY/Hr. (Example 7-12)

Therefore, the number of compactors required is:

$$\text{No. Compactors} = \frac{1715 \text{ CCY/Hr.}}{682 \text{ CCY/Hr.}}$$
$$= 2.5$$
$$= 3 \text{ compactors}$$

**Example 7-17:** Assuming each compactor given in the previous example costs \$80.00 per hour, including operator's wages, determine the unit production cost.
**Solution:** The unit production cost for compaction is:

$$\text{Unit Production Cost} = \frac{3 \text{ x } \$80/\text{Hr.}}{1715 \text{ CCY/Hr.}} \quad \text{(Eq. 1.18; Example 7-16)}$$
$$= \$0.14/\text{CCY}$$

## ADJUSTING MOISTURE CONTENT

If the soil is too wet, it can be loosened and aerated with rippers or a disk harrow (Photo 7-38), and then re-rolled. If the soil fails a compaction test because it is too wet, it can sometimes be left in place to dry, re-rolled and tested again. However, if it fails the compaction test because it is too dry, you must rip it, add water from a water truck (Photo 7-33) and re-roll it with a sheepsfoot or tamping-type roller that has the ability to work the water into the soil.

**Photo 7-33. Water truck.** *(Photo by Dan Atcheson)*

Endeavor to compact a lift as soon as possible after water has been added. If you delay, the upper portion of the lift will evaporate more than the lower portion, and rolling a dry top layer over a wetter bottom layer will cause the top layer to crumble, leaving a scaly surface.

When adding water, keep in mind that different soils require varying amounts of water to change the moisture content. A given volume of sandy soil or poorly-graded soil will require more water than clay-like or well-graded soil, because sandy or poorly-graded soils contain more voids that can accommodate the added water. Therefore, it is difficult to add the proper amount of water to a soil consisting of clay and sand, since each constituent requires a differing amount of water in order to achieve maximum density.

**Example 7-18:** If earth containing 4% moisture is placed as fill at the rate of 200 compacted cubic yards per hour and the dry weight of the soil is 2900 pounds per compacted cubic yard, determine the amount of water that must be supplied each hour to increase the moisture content to 10%.

**Solution:** The moisture content of the soil can be determined by:

$$\text{Moisture Content (\%)} = \frac{\text{Wet Weight} - \text{Dry Weight}}{\text{Dry Weight}} \text{ x } 100\% \qquad \text{(Eq. 7.8)}$$

Solving for the wet soil weight, we have:

$$\text{Wet Weight} = \frac{\text{Moisture Content (\%) x Dry Weight}}{100\%} + \text{Dry Weight}$$

**(Equation 7.19)**

At 4% moisture content, the wet soil weighs:

$$\text{Wet Weight (4\%)} = \frac{4\% \text{ x } 2900 \text{ lb./CCY}}{100\%} + 2900 \text{ lb./CCY} \qquad \text{(Eq. 7.19)}$$

$$= 3016 \text{ lb./CCY}$$

At 4% moisture content, each compacted cubic yard of soil contains water weighing:

$$\text{Water Weight (4\%)} = 3016 \text{ lb./CCY} - 2900 \text{ lb./CCY} \qquad \text{(Eq. 7.7)}$$

$$= 116 \text{ lb./CCY}$$

At 10% moisture content, the wet soil weighs:

$$\text{Wet Weight (10\%)} = \frac{10\% \times 2900 \text{ lb./CCY}}{100\%} + 2900 \text{ lb./CCY} \qquad \text{(Eq. 7.19)}$$
$$= 3190 \text{ lb./CCY}$$

At 10% moisture content, each compacted cubic yard of soil contains water weighing:

$$\text{Water Weight (10\%)} = 3190 \text{ lb./CCY} - 2900 \text{ lb./CCY} \qquad \text{(Eq. 7.7)}$$
$$= 290 \text{ lb./CCY}$$

The amount of water that must be added to each compacted cubic yard to increase the moisture content from 4% to 10% is:

$$\text{Water Added (lb.)} = 290 \text{ lb./CCY} - 116 \text{ lb./CCY}$$
$$= 174 \text{ lb./CCY}$$

In terms of gallons, the amount of water that must be added to each compacted cubic yard is:

$$\text{Water Added (Gal.)} = \frac{174 \text{ lb./CCY}}{8.345 \text{ lb./Gal.}} \qquad \text{(Appendix B)}$$
$$= 21 \text{ gallons/CCY}$$

Since the soil is being placed at the rate of 200 compacted cubic yards per hour, we will have to supply water at the rate of:

$$\text{Water Added (Gal./Hr.)} = 200 \text{ CCY/Hr.} \times 21 \text{ Gal./CCY}$$
$$= 4200 \text{ gallons/hour}$$

The number of water trucks required to adequately service the project is:

$$\text{N (Number of Trucks Required)} = \frac{\text{Haul Unit Cycle Time}}{\text{Fill Time}} \qquad \text{(Eq. 4.3)}$$

where:

$$\text{Fill Time} = \frac{\text{Haul Unit Capacity (Gal.)}}{\text{Fill Rate (Gal./Unit Time)}} \qquad \textbf{(Equation 7.20)}$$

However, the extent to which a project can be furnished with water is limited by the filling rate at the water source.

**Example 7-19:** Assuming a haul unit cycle time of 1.5 hours (excluding fill time), determine the number of 2000-gallon water trucks required to service the project conditions given in the previous example if it takes 0.5 hours to fill each truck.
**Solution:** The number of water trucks required is:

$$N = \frac{1.5\text{ Hr.} + 0.5\text{ Hr.}}{0.5\text{ Hr.}} \quad \text{(Eq. 4.3)}$$
$$= 4\text{ trucks}$$

The filling rate at the water source is:

$$\text{Fill Rate (Gal./Hr.)} = \frac{2000\text{ Gal.}}{0.5\text{ Hr.}}$$
$$= 4000\text{ gallons/hour}$$

The project requires 4200 gallons per hour (Example 7-18); however, no more than 4000 gallons can be supplied per hour, due to the filling rate at the water source. Therefore, the rate at which the fill is supplied must be reduced in order to keep the equipment balanced.

**Example 7-20:** Work Example 7-18 assuming that only 4000 gallons of water can be supplied per hour.
**Solution:** The production must be reduced to:

$$\text{Production (CCY/Hr.)} = \frac{4000\text{ Gal./Hr.}}{21\text{ Gal./CCY}} \quad \text{(Example 7-18)}$$
$$= 190\text{ CCY/hour}$$

## COMPACTION IN TIGHT AREAS

There are instances when it is impractical or impossible to use large compaction equipment, such as in narrow trenches, curb and gutters, bridge supports, graves, against walls, near manholes, or over a sand cushion installed under a concrete slab. In such areas, manually-operated hand compactors can often be used effectively.

Examples of manually-operated compactors include vibrating plate compactors (Photo 7-34), vibrating tamping-foot compactor (Photo 7-35), vibrating roller compactors (Photo 7-36) and impact rammer compactors (Photo 7-37). These units can be powered by gasoline, diesel, air or electricity. Electric motors are normally used where noise or fumes must be minimized. Gasoline engines used are normally two-cycle engines ( up to 10 horsepower). The advantages of this engine include the fact that oil mixed with the fuel helps lubricate moving parts, and there is no engine damage caused when the machine is tilted. Also, the engine is lightweight, since it has no valves. These machines weigh from 75 to more than 300 pounds, and deliver 50 to 9000 cycles per minute.

**Photo 7-34. Vibrating plate compactor.** *(Courtesy of Wacker Corporation)*

**Photo 7-35. Vibrating tamping-foot compactor.** *(RayGo, Inc.)*

**Photo 7-36. Vibrating roller compactor.** *(Courtesy of Wacker Corporation)*

**Photo 7-37. Impact rammer compactor.** *(Courtesy of Wacker Corporation)*

Soil used for backfill in trenches and other difficult locations is sometimes compacted by injecting water into the soil with a hose equipped with a solid pipe screwed onto the end of the hose. This process is referred to as *jetting,* and is especially useful in sand or gravel which allows the water to escape after lubricating and shifting the soil particles into a more compacted position.

Backfill near the top of a trench can be effectively compacted by rolling over the backfill with the tires of a heavy machine. Wheel loaders and backhoe loaders are the most effective since their tires fit over most trenches, and their weight can be increased by adding soil into the bucket.

## DISK HARROWS

In addition to loosening soil for aeration, *disk harrows* can be used for mixing and pulverizing soil, and for loosening soil for faster scraper loading (Photo 7-38). Many disk harrow frames are equipped with rubber tires whose positions can be adjusted to control the depth of disk penetration.

**Photo 7-38. Disk harrow.** *(Courtesy of Rome Plow Company)*

## SOIL STABILIZATION

Soils are stabilized to reduce expansion and contraction, prevent physical displacement, and to increase bearing strength and resistance to moisture. Soils can be stabilized with regard to physical movement by mixing different soil types, or by mixing soil particles to produce an even distribution of soil particles. This helps to fill voids, which increases internal friction and impedes movement. Also, angular-shaped particles will increase internal friction and provide a more interlocking structure. Soil can also be stabilized by blending asphalt (*asphalt soil,* or "black base"), or portland cement with the soil (*soil cement,* cement-treated base, or cement-stabilized base). Asphalt soil is produced by mixing an MC-3 or RC-3 asphalt into granular soil at the rate of 5 to 7% of the soil volume. Soil cement is produced by adding portland cement at the rate of 3 to 16% of the soil weight. As a general rule, as the clay content increases, the amount of cement required also increases.

Soils can be stabilized with regard to expansion and contraction due to a high clay content by blending hydrated lime with the soil at the rate of 3 to 5% of the soil volume *(lime-treated soil)*. In most soils, 18% water must be added to start the reaction of the lime.

With any soil stabilization process, the soil should first be ripped or scarified prior to adding the stabilizing agent. The mixing can be accomplished with a motor grader, disk harrow or a stabilizer/reclaimer (Photo 7-39). The reclaimer shown in Photo 7-39 can be used to remove old asphalt paving, as well as for mixing soils, and it can add water to the soil as it mixes.

**Photo 7-39. Stabilizer/reclaimer.** *(Courtesy of Caterpillar Inc.)*

# Chapter 8

# MOTOR GRADERS

Motor (road) graders are used for surface finishing, shaping, trimming, bank sloping, road maintenance, material blending and spreading, ditching, backfilling, scarifying (loosening) and ripping. Some graders are towed behind a tractor; however, most graders are self-propelled.

Machine shipping dimensions are normally published with manufacturer's specifications for any given grader (Figure 8-1).

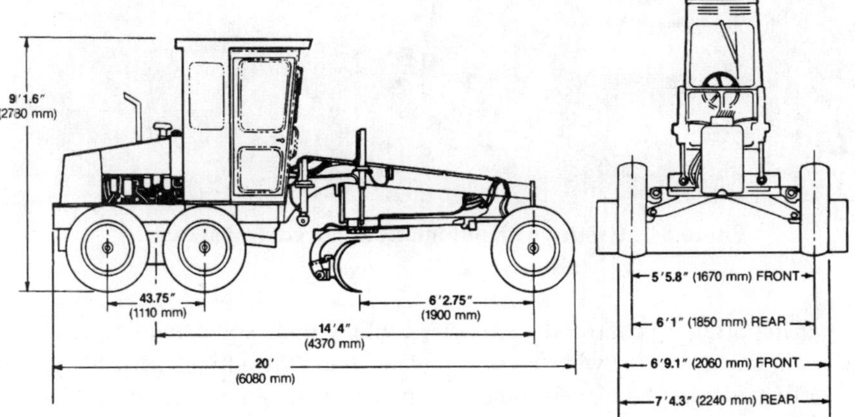


**Figure 8-1. Grader dimensions.** *(Courtesy of Fiat-Allis North America, Inc.)*

## FINISHING

Finishing operations include balancing and trimming. *Balancing* consists of cutting down high spots while filling in low areas, and *trimming* consists of shaping the road surface to a specified elevation (Photo 8-1). Final grade tolerances are normally specified as a given amount of deviation in elevation over a given horizontal distance, and the grade tolerance specified for roads depends on the road component being graded; e.g., subgrade ("basement soil," or "foundation soil"), subbase, or base (Figure 8-2). Final grade tolerances depend on the type of construction, and for most road projects, the grade tolerances are:

Subgrade: 1/2" to 1" in 10 feet
Subbase: 1/4" to 1/2" in 10 feet
Base: 1/8" in 10 feet

Airport runways and federal highway projects have even stricter grade tolerances.

Some motor graders are equipped with an automatic blade control sensing system that follows an established surface line by automatically raising and lowering the blade, resulting in precise grade control (Photo 8-1).

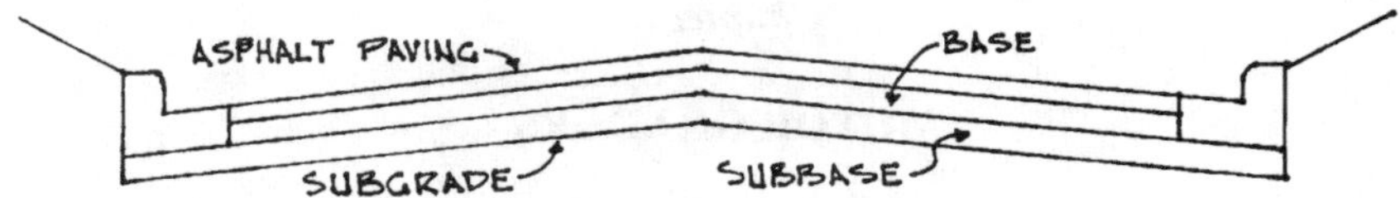


**Figure 8-2. Road components.**

**Photo 8-1. Grader with automatic blade control system.**
*(Courtesy of CMI Corporation)*

It is important to control the moisture content in the soil being finished because soil that is too wet will stick to the blade and leave a rough surface. Also, the cutting edge of the blade should be replaced when it becomes badly worn, since a worn cutting edge is inefficient and will not cut properly, or cleanly.

Finish grading is seldom designated as a *pay item* in a unit-price contract, so be careful not to omit the finishing cost from the estimate. An easy way to include the finishing cost is to incorporate it within the cost for installing a lift (layer) of fill material. Since the surface area to be finish-graded remains constant, regardless of the lift thickness, the finish-grading costs can be determined on the basis of cubic yards of lift material installed. I have provided Table 8-1 to help you determine the square feet and square yards of surface area created per cubic yard of soil installed.

**Example 8-1:** Assuming it costs \$0.10 per square yard for finish grading, express this cost on a per-cubic-yard basis if the compacted lift thickness being graded is: a) 6 inches, or b) 10 inches.
**Solution:** The per-cubic-yard costs are:

a) Cost/CCY = 6 S.Y./CCY x \$0.10/S.Y. (Table 8-1)
= \$0.60/CCY

b) Cost/CCY = 3.56 S.Y./CCY x \$0.10/S.Y.
= \$0.36/CCY

**Table 8-1: Surface Area Per Cubic Yard**

| Lift Thickness (Inches) | Sq. Ft. of Surface Area/C.Y. | S.Y. of Surface Area/C.Y. |
|---|---|---|
| 1 | 324 | 36.00 |
| 2 | 162 | 18.00 |
| 3 | 108 | 12.00 |
| 4 | 81 | 9.00 |
| 5 | 65 | 7.22 |
| 6 | 54 | 6.00 |
| 7 | 46 | 5.11 |
| 8 | 41 | 4.56 |
| 9 | 36 | 4.00 |
| 10 | 32 | 3.56 |
| 12 | 27 | 3.00 |
| 14 | 23 | 2.56 |
| 16 | 20 | 2.22 |
| 18 | 18 | 2.00 |
| 20 | 16 | 1.78 |
| 22 | 15 | 1.67 |
| 24 | 14 | 1.56 |

## HAUL-ROAD MAINTENANCE

Depending on the nature of the haul road surface and the amount of traffic, it is often economically advantageous to periodically or constantly maintain the haul road with motor graders, and occasionally, with water trucks.

**Example 8-2:** Assuming that haul-road maintenance can reduce the rolling resistance of a haul road from 80 pounds per ton to 60 pounds per ton of gross vehicle weight, determine the production of the scraper given in Example 6-5. Also, assume that the turn & dump time and spot & delay times can be reduced to 0.4 and 0.2 minutes, respectively.

**Solution:** The load time is:

Load Time (Min.) = 0.7 minutes (Example 6-5)

The total fixed time is:

Fixed Time (Min.) = 0.7 Min. + 0.4 Min. + 0.2 Min. (Given)
= 1.3 minutes

We have no grade resistance since the haul road is level; therefore, the total resistance (effective grade) is reduced to:

$$\text{Effective Grade (\%)} = \frac{60 \text{ lb./Ton}}{20} \qquad \text{(Eq. 2.16)}$$
$$= 3\%$$

Referring to Figures 6-5 and 6-6 while using a one-way travel distance of 3000 feet, the travel times are:

Travel Time (Loaded) = 1.60 minutes (Figure 6-5)

and,

Travel Time (Empty) = 1.27 minutes (Figure 6-6)

The total travel time is:

$$\text{Total Travel Time (Min.)} = 1.60 \text{ Min.} + 1.27 \text{ Min.}$$
$$= 2.87 \text{ minutes}$$

The total scraper cycle time is:

$$\text{Total Scraper Cycle Time (Min.)} = 1.30 \text{ Min.} + 2.87 \text{ Min.} \qquad \text{(Eq. 6.3)}$$
$$= 4.17 \text{ minutes}$$

Working at 100% efficiency, the number of cycles per hour is:

$$\text{Cycles/Hr.} = \frac{60 \text{ Min./Hr.}}{4.17 \text{ Min./Cycle}} \qquad \text{(Eq. 1.6)}$$
$$= 14.4 \text{ cycles/hour}$$

Working at a job efficiency of only 50 minutes per hour, the probable production rate is:

$$\text{Production (BCY/Hr.)} = 15.4 \text{ BCY/Cycle} \times 14.4 \text{ Cycles/Hr.} \times 50/60 \qquad \text{(Example 6-5)}$$
$$= 184.8 \text{ BCY/hour}$$

Therefore, haul-road maintenance increases the hourly production rate by:

Production Increase (BCY/Hr.) = 184.8 BCY/Hr. - 166.8 BCY/Hr. (Example 6-5)
= 18 BCY/hour

Since each scraper payload is 15.4 bank cubic yards (Example 6-5), the production increase equates to more than one scraper payload per hour.

**Example 8-3:** Assuming the scraper given in the previous example costs $200.00 per hour and a motor grader costs $75.00 per hour, including operators' wages, determine the unit costs and production rates when running various numbers of scrapers over a maintained haul road, versus a rough haul road. Assume that the motor grader is used only 20 minutes per hour maintaining the haul road, while the remainder of each hour is dedicated to spreading fill, or other miscellaneous tasks.

**Solution:** Since the motor grader is used for road maintenance only 20 minutes per hour (1/3-hour), the hourly cost attributable to haul-road maintenance is:

Grader Cost/Hour (Road Maintenance) = 1/3 x $75.00/Hr.
= $25.00/hour

The unit production cost while using one scraper traveling over the rough haul road is:

$$\text{Unit Production Cost} = \frac{\$200.00/\text{Hr.}}{166.8 \text{ BCY/Hr.}} \quad \text{(Eq. 1.18; Example 8-2)}$$

$$= \$1.20/\text{BCY}$$

The unit production cost while using one scraper traveling over the maintained haul road is:

$$\text{Unit Production Cost} = \frac{\$200.00/\text{Hr.} + \$25.00/\text{Hr.}}{184.8 \text{ BCY/Hr.}}$$

(Given; Example 8-2)

$$= \$1.22/\text{BCY}$$

Continuing in this fashion, we can construct the following cost-production table to analyze unit production costs:

| Scrapers | Haul Road Not Maintained | | | Haul Road Maintained | | |
|---|---|---|---|---|---|---|
| | Fleet Cost/Hr. | BCY/Hr. | Cost/BCY | Fleet Cost/Hr. | BCY/Hr. | Cost/BCY |
| 1 | $200.00 | 166.8 | $1.20 | $225.00 | 184.8 | $1.22 |
| 2 | $400.00 | 333.6 | $1.20 | $425.00 | 369.6 | $1.15 |
| 4 | $800.00 | 667.2 | $1.20 | $825.00 | 739.2 | $1.12 |
| 6 | $1200.00 | 1000.8 | $1.20 | $1225.00 | 1108.8 | $1.10 |
| 8 | $1600.00 | 1334.4 | $1.20 | $1625.00 | 1478.4 | $1.10 |
| 10 | $2000.00 | 1668.0 | $1.20 | $2025.00 | 1848.0 | $1.10 |
| 12 | $2400.00 | 2001.6 | $1.20 | $2425.00 | 2217.6 | $1.09 |

It is obvious that it is economically advantageous to maintain the haul road, provided that sufficient numbers of scrapers are using the haul road. Also, economy is increased by increasing the number of scrapers.

If we look at the overall picture with regard to haul-road maintenance, we will realize benefits beyond increased production and reduced unit costs, as demonstrated in the following example.

**Example 8-4:** Assuming you must excavate 500,000 bank cubic yards and 10 scrapers are available (as given in Example 8-3), determine the cost savings for the project if the field overhead such as supervision, job trailer, insurance, etc. costs $400.00 per day.

**Solution:** Assuming an 8-hour day, the project overhead costs are:

$$\text{Overhead Cost} = \frac{\$400.00/\text{Day}}{8\ \text{Hr./Day}} = \$50.00/\text{hour}$$

*Without* haul-road maintenance, the project duration will be:

$$\text{Project Duration (Hr.)} = \frac{500{,}000\ \text{BCY}}{10 \times 166.8\ \text{BCY/Hr.}} = 300\ \text{hours} \qquad \text{(Eq. 1.17; Example 8-3)}$$

The scraper portion of the project cost will be:

$$\text{Total Project Cost} = (500{,}000\ \text{BCY} \times \$1.20/\text{BCY}) + (300\ \text{Hr.} \times \$50.00/\text{Hr.}) \quad \text{(Example 8-3)}$$

$$= \$615{,}000.00$$

*With* haul-road maintenance, the project duration will be:

$$\text{Project Duration (Hr.)} = \frac{500{,}000 \text{ BCY}}{10 \times 184.8 \text{ BCY/Hr.}} \qquad \text{(Eq. 1.17; Example 8-3)}$$

$$= 271 \text{ hours}$$

The scraper portion of the project cost will be:

Total Project Cost = (500,000 BCY x \$1.10/BCY) + (271 Hr. x \$50.00/Hr.) (Example 8-3)

= \$563,550.00

The cost savings resulting from haul-road maintenance is:

Cost Savings = \$615,000.00 – \$563,550.00
= \$51,450.00!

Another cost-saving feature often overlooked in a haul-road maintenance analysis is the reduction in tire wear when vehicles travel over a smooth surface.

**Example 8-5:** Assuming the following conditions, determine the cost savings resulting from reduced tire wear for the equipment given in the previous example.

Scraper Tire Replacement Cost = \$15,000.00
Scraper Tire Life (Rough Haul Road) = 2500 hours
Scraper Tire Life (Maintained Haul Road) = 4000 hours
Grader Tire Replacement Cost = \$12,000.00
Grader Tire Life = 5000 hours
Tire Repair Cost = 15% of tire replacement costs

**Solution: a)** The hourly tire replacement cost for each scraper traveling over the *rough* haul road is:

$$\text{Hourly Tire Cost} = \frac{\$15{,}000.00}{2500 \text{ Hr.}} \qquad \text{(Given)}$$

$$= \$6.00\text{/hour}$$

The tire replacement cost for all 10 scrapers is:

Total Tire Replacement Cost = 10 x \$6.00/Hr. x 300 Hr. (Example 8-4)
= \$18,000.00

The tire repair cost for all 10 scrapers is:

Tire Repair Cost = 0.15 x \$18,000.00 (Given)
= \$2700.00

The total tire cost is:

Total Tire Cost = \$18,000.00 + \$2700.00
= \$20,700.00

**b)** The hourly tire replacement cost for each scraper traveling over the *maintained* haul road is:

$$\text{Hourly Tire Cost} = \frac{\$15{,}000.00}{4000 \text{ Hr.}} \quad \text{(Given)}$$
$$= \$3.75/\text{hour}$$

The tire replacement cost for all 10 scrapers is:

Total Tire Replacement Cost = 10 x \$3.75/Hr. x 271 Hr. (Example 8-4)
= \$10,162.50

The tire repair cost for all 10 scrapers is:

Tire Repair Cost = 0.15 x \$10,162.50
= \$1524.38

The total tire cost is:

Total Tire Cost = \$10,162.50 + \$1524.38
= \$11,686.88

The hourly tire replacement cost for the motor grader is:

$$\text{Hourly Tire Cost} = \frac{\$12{,}000.00}{5000 \text{ Hr.}} \quad \text{(Given)}$$
$$= \$2.40/\text{hour}$$

Since the motor grader maintains the haul road only 20 minutes per hour (1/3-hour), the total grader operation time attributable to haul-road maintenance is:

Grader Operation Time = 1/3 x 271 Hr. (Example 8-4)
= 90 hours

The tire replacement cost for the motor grader is:

Tire Replacement Cost = \$2.40/Hr. x 90 Hr.
= \$216.00

The tire repair cost for the motor grader is:

Tire Repair Cost = 0.15 x \$216.00
= \$32.40

The total tire cost is:

Total Tire Cost = \$216.00 + \$32.40
= \$248.40

The total fleet tire cost is:

Total Fleet Tire Cost = \$11,686.88 + \$248.40
= \$11,935.28

The tire cost savings resulting from haul-road maintenance is:

Tire Cost Savings = \$20,700.00 – \$11,935.28
= \$8764.72

Combining the results of this example and those of Example 8-4, the total cost savings resulting from haul-road maintenance is:

Total Cost Savings = \$51,450.00 + \$8764.72 (Example 8-4)
= \$60,214.72!

## MATERIAL BLENDING AND SPREADING

The ability to vary the position of the blade *(moldboard)* makes the motor grader one of the most versatile earthmoving machines available (Figure 8-3). Blade adjustments can be made manually, or with hydraulic controls. The blade can be tilted to a forward pitch position which imparts a dragging action for blending soils (Figure 8-3), or it can be adjusted to a backward pitch position to increase cutting action for fine grading. Some graders are equipped with a second blade mounted on the front (Photo 8-2).

The concave shape of the moldboard imparts a rolling action to the soil for spreading and mixing material. When spreading material, the blade is normally set perpendicular to, or at a slight angle relative to the direction of travel. When spreading aggregate, it is important not to overwork the material, since too much blading will separate the fine material from the larger aggregate, resulting in aggregate at the surface that cannot be held in place, due to the absence of finer material to fill the voids. Large aggregate that has been overworked is often referred to as *"bones."*

There are many instances where soil must be blended and mixed. Some soil mixing takes place as the grader spreads the soil during leveling and grading operations. Wet, unstable areas can be bladed and mixed with dry soil, then bladed back in place as suitable material.

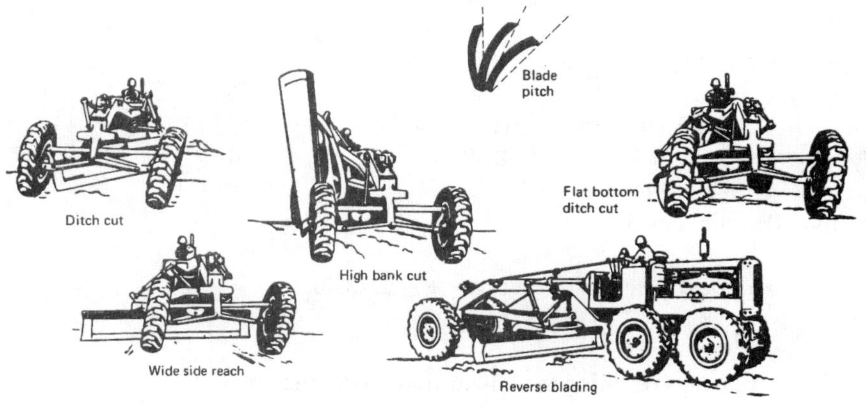


**Figure 8-3. Motor grader blade positions.**
*(Courtesy of the U.S. Department of the Army)*

**Photo 8-2. Blade mounted on the front of a grader.** *(Photo by Dan Atcheson)*

## SIDECASTING AND BACKFILLING

Soil can be moved laterally *(sidecasting)* by setting the blade at an angle relative to the direction of travel (Figure 8-4) (Photo 8-3). When sidecasting, set the blade so that the resulting *windrow* (longitudinal soil pile) of sidecast soil does not form in the path of the rear wheels of the motor grader, since this will increase rolling resistance and decrease traction (Photo 8-4). Sidecasting can be used for backfilling ditches and trenches, or for creating windrows of ripped asphalt to be crushed by a compactor. When lime must be added to clay-type soil for stabilization, the lime is often added to windrows of sidecast soil, and mixed by rolling the soil several times with a grader blade.

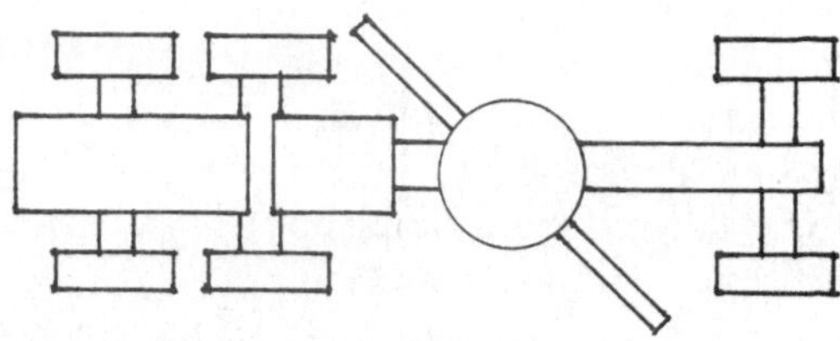

**Figure 8-4.**
**Sidecasting blade position.**

**Photo 8-3. Sidecasting.** *(Courtesy of Caterpillar Inc.)*

**Photo 8-4. Sidecasting beyond the rear wheels.** *(Photo by Dan Atcheson)*

Traction is dependent upon the condition of the road surface, and the weight imposed over the drive wheels of the motor grader. Since the drive wheels of a motor grader are located at the rear of the machine, it is often advantageous to move the blade laterally *(wide side reach)* as shown in Figure 8-3, or to offset the front and rear wheels *(crab mode)* as shown in Figure 8-5, to keep the drive wheels on firm ground. Keeping one set of rear wheels on firm ground will also prevent the motor grader from sliding down slope as it cuts a ditch. The front wheels can also be leaned to help prevent sliding down slope (Photo 8-5).

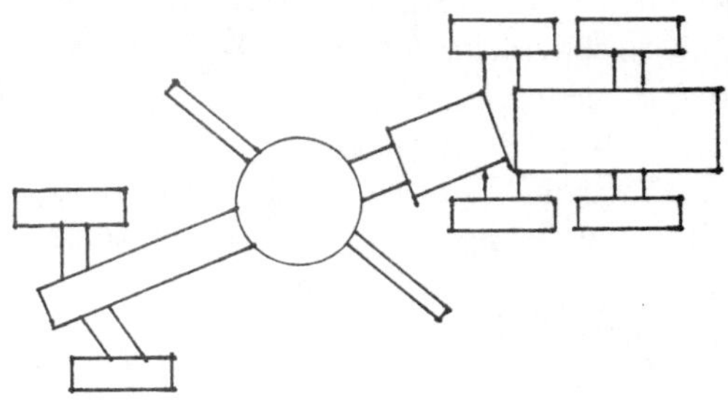

**Figure 8-5. Crab position.**

Even when the blade is set perpendicular to the direction of travel, material will spill beyond the ends of the blade, forming windrows on each side of the motor grader. Attaching wings to the ends of the blades will reduce the amount of spillage by keeping more material in front of the blade.

**Photo 8-5. Leaning the front wheels.**
*(Courtesy of Fiat-Allis North America, Inc.)*

## BANK SLOPE TRIMMING

The blade can be oriented to an almost vertical position to cut and trim bank slopes (Figure 8-3). The bank slope must be trimmed as the road cut proceeds downward. As a general rule, the subgrade can be taken down no more than 5 feet before the side slopes must be trimmed by a grader.

A *grader sloper* is manufactured specifically to be mounted to the side of a grader for dressing road shoulders, cleaning ditches and shaping banks (Photo 8-6). The 16-foot blade allows the grader to remain on a level, firm footing while sloping banks from 60 degrees above, to 60 degrees below the horizontal.

**Photo 8-6. Grader sloper.** *(Courtesy of Rome Plow Company)*

## DITCHING

A motor grader can be used to cut ditches up to 3 feet deep. Beyond this depth, it is normally more economical to use a more productive excavator such as a scraper. Graders can be equipped with an elevating conveyor belt to help remove graded material (Photo 8-7).

**Photo 8-7. Grader with elevating conveyor belt.** *(Courtesy of Caterpillar Inc.)*

## SCARIFYING AND RIPPING

Scarifier teeth can be mounted behind and between the front wheels to aid the motor grader in loosening tough or frozen soil, ice or gravel, and for breaking up asphalt paving (Photo 8-4). Rippers can also be mounted at the rear of a motor grader (Photo 8-8), but the ripping ability is limited by the weight and power of the grader; therefore, the ripping is usually only light-duty. Subgrade material is normally trimmed to 0.10 feet above final subgrade elevation, ripped and then compacted (Figure 8-2). Since the grader is already working the area, it is usually used for ripping the subgrade material.

**Photo 8-8. Rippers mounted on motor grader.** *(Photo by Dan Atcheson)*

## MOTOR-GRADER PRODUCTION

Motor-grader production can be estimated in terms of area graded per hour as follows:

$$\text{Production (Sq. Ft./Hr.)} = \frac{\text{Blade Width (Ft.)} \times \text{Ave. Speed (mph)} \times 5280 \times \text{Job Efficiency}}{\text{Number of Passes}}$$

**(Equation 8.1)**

or,

$$\text{Production (Sq. Yd./Hr.)} = \frac{\text{Blade Width (Ft.)} \times \text{Ave. Speed (mph)} \times 5280 \times \text{Job Efficiency}}{9 \times \text{Number of Passes}}$$

**(Equation 8.2)**

where: Blade width is the width of the blade measured perpendicular to the direction of travel (Figure 8-6). The physical blade width is normally 10, 12, 14 or 16 feet, and 2-foot extensions are sometimes attached to provide increased width.

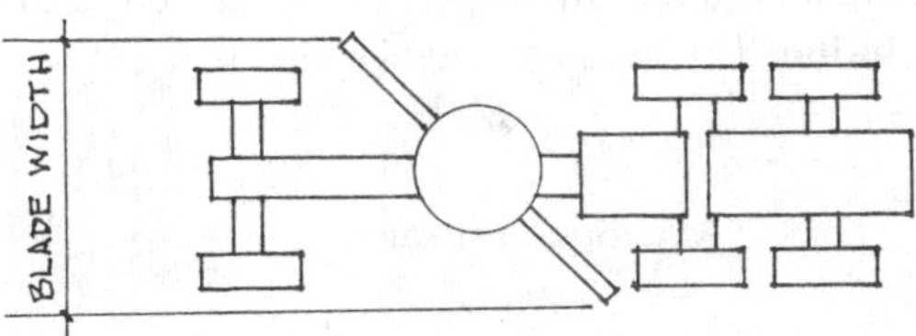


**Figure 8-6.**
**Blade width perpendicular to direction of travel.**

Production can also be expressed in terms of lineal feet graded per hour as follows:

$$\text{Production (L.F./Hr.)} = \frac{\text{Blade Width (Ft.)} \times \text{Ave. Speed (mph)} \times 5280 \times \text{Job Efficiency}}{\text{Road Width} \times \text{Number of Passes}}$$

**(Equation 8.3)**

or,

$$\text{Production (Mi./Hr.)} = \frac{\text{Blade Width (Ft.)} \times \text{Ave. Speed (mph)} \times \text{Job Efficiency}}{\text{Road Width} \times \text{Number of Passes}}$$

**(Equation 8.4)**

where: Blade width is the width of the blade measured perpendicular to the direction of travel (Figure 8-6).

Motor-grader travel speed depends primarily on the type of work, as shown in Table 8-2.

**Table 8-2: Typical Grader Speeds**

| Type of Work | Speed (mph) |
|---|---|
| Ditching | 2.0 - 4.0 |
| Bank sloping | 1.5 - 2.5 |
| Road cutting and maintenance | 3.5 - 6.0 |
| Finish grading | 3.5 - 9.0 |
| Spreading | 5.5 - 9.0 |
| Mixing | 8.0 - 20.0 |
| Snow removal | 10.0 - 20.0 |

**Example 8-6:** A 30,000-square-yard site is to be finish-graded, requiring five passes. Assuming two passes at 3 mph, two at 4 mph and one at 6 mph, estimate the production rate and the time required to grade the site if job efficiency is 0.75 and blade length measured perpendicular to the direction of travel is 12 feet.
**Solution:** The average travel speed is:

$$\text{Ave. Speed (mph)} = \frac{(2 \times 3 \text{ mph}) + (2 \times 4 \text{ mph}) + (1 \times 6 \text{ mph})}{5}$$

$$= 4 \text{ mph}$$

The probable production rate will be:

$$\text{Production (Sq. Yd./Hr.)} = \frac{12 \times 4 \times 5280 \times 0.75}{9 \times 5} \qquad \text{(Eq. 8.2)}$$

$$= 4224 \text{ S.Y./hour}$$

The duration of the grading portion of the project is:

$$\text{Project Duration (Hr.)} = \frac{30{,}000 \text{ S.Y.}}{4224 \text{ S.Y./Hr.}} \qquad \text{(Eq. 1.17)}$$

$$= 7.1 \text{ hours}$$

**Example 8-7:** Ten miles of a two-lane dirt road is to be graded, requiring four passes in each 10-foot-wide lane. Assuming a job efficiency of 0.83, an average speed of 5 mph and a blade width of 10 feet measured perpendicular to the direction of travel, determine the production rate and the time required to grade the road.
**Solution:** The probable production rate will be:

$$\text{Production (Mi./Hr.)} = \frac{10 \times 5 \times 0.83}{20 \times 4} \qquad \text{(Eq. 8.4)}$$

$$= 0.52 \text{ miles/hour}$$

The time required to grade the road is:

$$\text{Project Duration (Hr.)} = \frac{10 \text{ Mi.}}{0.52 \text{ mph}} \qquad \text{(Eq. 1.17)}$$

$$= 19.2 \text{ hours}$$

Maximum grader travel speeds for various types of Caterpillar motor graders are shown in Table 8-3. Since the reverse speed is normally equal to, or greater than the forward speed for any given gear, it is usually more efficient for a grader to back up, rather than to turn around for passes less than 1000 feet. If turning room is restricted, it is often advantageous for a grader to back up for passes longer than 1000 feet.

**Table 8-3: Motor Grader Ground Speeds**

| GEAR | 1ST | | 2ND | | 3RD | | 4TH | | 5TH | | 6TH | | 7TH | | 8TH | |
|---|---|---|---|---|---|---|---|---|---|---|---|---|---|---|---|---|
| | km/h | mph | km/h | mph | km/h | mph | km/h | mph | km/h | mph | km/h | mph | km/h | mph | km/h | mph |
| 120G — Forward | 3.9 | **2.4** | 6.2 | **3.9** | 9.8 | **6.1** | 16.2 | **10.1** | 25.9 | **16.1** | 40.9 | **25.4** | — | | — | |
| Reverse | 3.7 | **2.3** | 5.8 | **3.6** | 9.2 | **5.7** | 15.1 | **9.4** | 24.3 | **15.1** | 38.3 | **23.8** | — | | — | |
| 130G — Forward | 3.8 | **2.3** | 6.0 | **3.7** | 9.5 | **5.9** | 15.6 | **9.7** | 25.0 | **15.5** | 39.4 | **24.5** | — | | — | |
| Reverse | 3.5 | **2.2** | 5.6 | **3.5** | 8.9 | **5.5** | 14.7 | **9.1** | 23.3 | **14.5** | 36.9 | **22.9** | — | | — | |
| 12G | 3.8 | **2.3** | 6.0 | **3.7** | 9.5 | **5.9** | 15.6 | **9.7** | 25.0 | **15.5** | 39.4 | **24.5** | — | | — | |
| 140G, 140G VHP, 140G AWD** | 3.9 | **2.4** | 6.2 | **3.9** | 9.8 | **6.1** | 16.2 | **10.1** | 26.0 | **16.1** | 41.0 | **25.5** | — | | — | |
| 14G — Forward | 3.8 | **2.3** | 5.3 | **3.3** | 7.2 | **4.4** | 10.4 | **6.5** | 15.6 | **9.7** | 22.0 | **13.6** | 29.8 | **18.5** | 43.0 | **26.8** |
| Reverse | 4.4 | **2.7** | 6.1 | **3.8** | 8.3 | **5.2** | 12.0 | **7.5** | 18.2 | **11.3** | 25.5 | **15.9** | 34.6 | **21.5** | 50.1 | **31.1** |
| 16G | 3.8 | **2.4** | 5.4 | **3.3** | 7.3 | **4.3** | 10.5 | **6.5** | 15.9 | **9.8** | 22.3 | **13.8** | 30.1 | **18.7** | 43.6 | **27.1** |

Unless otherwise specified, speeds are the same in forward and reverse.

*(Courtesy of Caterpillar Inc.)*

## TOWED GRADERS

York Modern Corporation manufactures a multi-purpose *towed grader* that can be used for scarifying, grading and raking in one low-cost operation (Photo 8-9). The blade can also be used for ditching, and since it is towed, the tractor or truck towing the unit can remain of firm footing while the grading unit works through water or mud.

**Photo 8-9. Towed grader.** *(Courtesy of York Modern Corporation)*

## TOWED RAKES

*Towed rakes* are also available for road and grounds maintenance (Photo 8-10). This machine is also multi-purpose since it can be equipped with a scarifier and flip-down blade. The multi-purpose unit can be used for spreading topsoil and other materials, ditching and terracing, removing roots, twigs, stones and debris, breaking up hard-packed soil, and grading and leveling. This unit can be hitched to a truck, tractor, or any other motive power unit.

**Photo 8-10. Towed rake.** *(Courtesy of York Modern Corporation)*

## GRADE EXCAVATORS AND TRIMMERS

*Grade trimmers* have the ability to rip soil or asphalt, excavate, reclaim, grade and compact base material (Photo 8-11). Grade trimmers can trim and finish surfaces faster and more accurately than can motor graders. They can also lay asphalt, or concrete paving. The trimmer shown can also be equipped with a belt conveyor for moving excess soil beyond the excavation area.

**Photo 8-11. Finishing base material with a grade trimmer.**
*(Courtesy of CMI Corporation)*

# *Chapter 9*

# PAVING EQUIPMENT

Bituminous paving is composed of a bituminous material (binder) and aggregate. Since asphalt is normally used as the binder, bituminous paving is also referred to as asphalt paving, or flexible paving. Compared with concrete paving, bituminous paving is more flexible and less affected by temperature changes. At the present time, bituminous pavements are less costly to construct than concrete pavements, and additional pavement layers and surface treatments can be installed as traffic loads increase and/or more building funds are available (*stage construction*). Stage construction allows normal wear, settlement, and design and construction deficiencies to be corrected as new layers of paving are added.

## AGGREGATE

Aggregates usually make up about 90 to 95% by weight, and 75 to 85% by volume of a paving mix, and they normally consist of crushed or natural stone, slag or gravel, sand and mineral filler. An experimental paving mix in which a portion of the aggregate is made up of crushed glass, is called *galsphalt*. Slag is a by product of blast furnace operations and consists essentially of silicates and alumino silicates of lime. Slag is very porous and absorbs more asphalt than does gravel; however, it is also lightweight, and the decreased transportation expense often offsets the cost of the additional asphalt required. The economy of bituminous paving often results from the use of local aggregate materials, provided they are acceptable.

Aggregate should be angular, providing high internal friction which helps to prevent movement (creep). Aggregates should also be properly graded, clean, dry, hydrophobic (or non-water absorbing), durable, weather-resistant and resistant to stripping. *Stripping* refers to the separation of asphalt film from the aggregate in the presence of water. When the moisture content of the aggregate exceeds 1-1/2 to 2%, most asphalt materials will strip away from the aggregate. Resistance to stripping can be increased by additives, or a mineral filler such as hydrated lime. The *immersion-compression test* is used to determine the resistance of aggregate to stripping.

*Coarse aggregate* is retained by a No. 10 sieve, while *fine aggregate* is retained by a No. 200 sieve. *Mineral filler* (fines, or granular dust) passes the No. 200 sieve and can consist of rock dust, portland cement, hydrated lime, or other inert non-plastic materials such as flyash, a by product of burning pulverized coal.

Aggregate must normally be blended to achieve the desired gradation. There are two basic types of gradation, including dense-graded and open-graded aggregate. *Dense-graded aggregate* yields a more stable and water-resistant paving mix, and requires less asphalt since there are fewer voids to fill. However, it also requires more sophisticated plant equipment to produce the mix, and stockpiled material tends to hold moisture, due to the lack of air circulation and drainage. Also, it is less flexible than open-graded aggregate paving.

*Open-graded aggregate* is used in lower-cost road construction. Open-graded aggregate lacks the fines which help hold the binder to the larger aggregates; therefore, care must be taken to ensure that the aggregate is not overheated prior to mixing with the binder, since excessive heat will create voids in the binder. This could allow water to enter the pavement and ultimately destroy it. Since a small percentage of aggregate is in physical contact with the base, a tack coat is usually required between open-graded asphalt paving and an existing pavement or rigid base, in order to maintain a strong bond. Some voids in an asphalt mix are desirable (3 to 4% by volume), since this allows for expansion in hot weather; however, excessive voids will cause the road to experience fatigue after repeated bending under traffic loads.

An open-graded asphalt concrete base course is often placed over cracked concrete paving to reduce crack reflection of the concrete into the overlying paving (*crack-relief pavement*). A second course is often installed to seal the open-graded pavement surface. Open-graded asphalt concrete is sometimes used without a sealing course to improve skid resistance, and to reduce tire splash, spray and hydroplaning, since water passes through the pavement and into the base, instead of remaining on the surface. This pavement is often referred to as a *porous friction course, plant-mix seal, or popcorn mix.*

Some open-graded aggregate mixes are referred to as *harsh*, meaning that they contain a high percentage of coarse aggregate, making the mix difficult to compact, with little or no lateral flow during the final stages of compaction. On the other hand, too much filler material can make the mix gummy, which is also difficult to compact. *Tender mixes* are too easily worked, not readily supporting the weight of a paver screed. Tender mixes are caused by the absence of mineral filler, too much medium-sized sand, smooth and rounded aggregate particles, and/or too much moisture in the mix.

Francis N. Hveem, a Materials and Research Engineer with the California Division of Highways, examined many existing roads, good and bad, and established recommended aggregate gradations for good asphalt paving mixes. The results of his study are shown in Figure 9-1. When the gradation falls within the solid lines of Figure 9-1, the roads hold up well; however, gradations falling outside the solid curves and within the areas bounded by a dotted curves tend to produce weak roads. The *Hveem method* is a mix design based on the cohesion and friction of a compacted asphalt concrete specimen.

Aggregates transmit traffic load to the base course (Figure 9-2), resist wear and provide a skid-resistant road surface. To test for strength, a sample is weighed and dumped into a rotating drum containing steel balls. After 500 revolutions, the portion of the sample passing a No. 12 sieve is weighed, and the percentage of the original sample is calculated. The percentage passing is referred to as the *Los Angeles abrasion value*, and aggregate used for hot mix must have an abrasion value of 50% or less, while that used for surface treatment must have an abrasion value ranging between 27% and 35%.

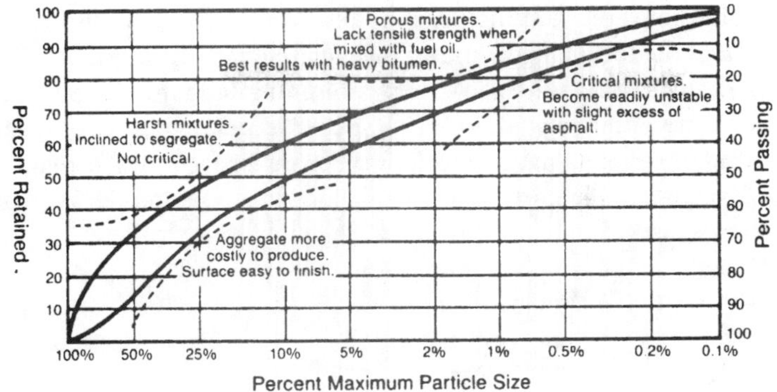


**Figure 9-1. Limits established for aggregate gradations which give good asphalt mixes.**
(*Courtesy of Barber-Greene*)

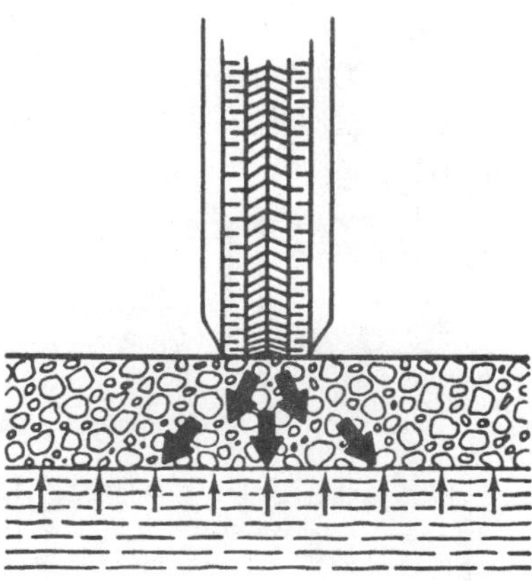

**Figure 9-2. Transmitting traffic loads.** (*Courtesy of Barber-Greene*)

Other tests used to determine the required strength and thickness of paving material include the *California bearing ratio, equivalent axle loading* (EAL), *plate bearing, resistance value*, and *resilient modulus tests*. These, and many other tests dealing with mix design mentioned throughout this chapter are not discussed in great detail, since they are usually under the jurisdiction of the engineer and/or the testing lab.

## BINDER

The *binder* holds the aggregate together, cushions and provides a waterproof surface which retards the flow of water to the subbase material. Water accumulating in the subbase layer does not easily evaporate, and it can undermine the support value of the road and cause subsidence and/or heaving. Asphalt does not contribute substantially to the mechanical strength of the surface, but it does provide a cushion to absorb the grinding and kneading action of traffic. Binder material also helps prevent the breakup of the road surface (disintegration, or spalling), and the loosening of the surface aggregate (raveling).

Binder can be composed of a variety of bituminous materials, but the primary constituents are either asphalt, or tar. *Asphalt* occurs in nature, or it can be distilled during the processing of petroleum (Figure 9-3). Natural asphalt was used as a building material as early as 3000 B.C. Liquid asphalt has an average weight of 8.7 pounds per gallon.

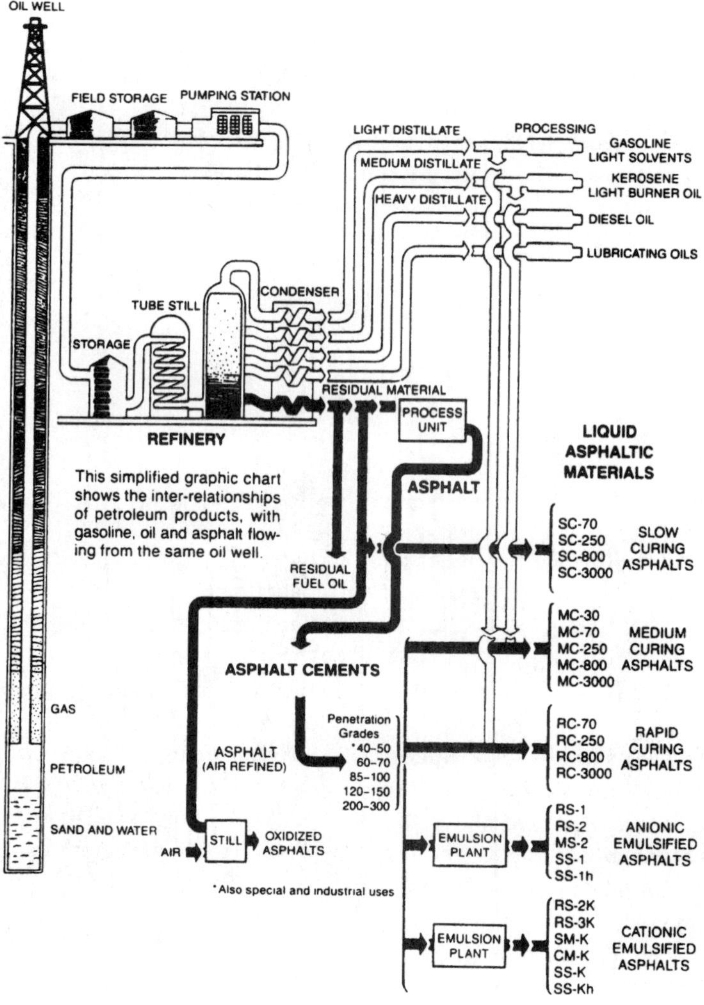


**Figure 9-3. Petroleum asphalt flow chart.** (*Courtesy of Barber-Greene*)

Numerous types of asphalts are manufactured with varying softening-point temperatures. The *ring and ball test* is used to determine the softening point of asphalt material.

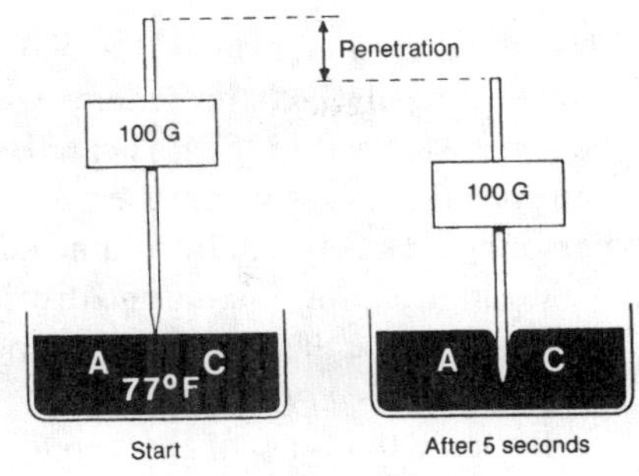


**Figure 9-4. Penetration test.**
(*Courtesy of The Asphalt Institute*)

*Asphalt cement* (AC) is often graded by *viscosity* (the resistance to flow), ranging from AC-2.5 (soft) to AC-40 (hard). It can also be graded by the results of a *penetration test*, where a 100-gram load is placed on asphalt at a temperature is 77°F. After a 5-second interval, the penetration by a standard needle under the load is measured in 100'ths of a centimeter (Figure 9-4). *Penetration grades* of asphalt cements range from 200 through 300 (soft) to 40 through 50 (hard). An AP-3 designation means that an asphalt cement has a penetration grade of 300. The 77° criterion was originally selected, since it could be maintained under standard laboratory conditions. However, 140°F. is considered to be the maximum temperature reached by an asphalt pavement in actual service. This is also the temperature at which asphalt will be displaced beyond a critical point (deformation); therefore, viscosity grading of asphalt products is measured when the bitumen is heated to 140°F.

To eliminate some confusion, keep in mind that when an asphalt is *penetration-graded*, the higher the penetration number, the softer the asphalt. But when the asphalt is *viscosity-graded*, the higher the viscosity number (viscosity is defined as the *resistance* to flowing), the harder the asphalt. The two units used for measuring viscosity are the *poise*, and the *stoke*. The test often used to measure viscosity is performed with a *Zeitfuchs cross-arm viscometer*.

The ambient temperature is a very important consideration when selecting the viscosity or grade of asphalt. In cold regions, softer asphalts should be used, and in hot regions, harder asphalts should be used.

*Coal-tar bitumen* (coal-tar pitch) is a product of the distillation of coal tar in the manufacture of coke, or gas. There are two types of tar, including coal tar and water-gas tar, which is usually combined with coal tar as a flux. Coal-tar pitch is softer than asphalt, and after cooling, has excellent self-healing properties. Also, it is virtually unaffected by water, due to a tight molecular structure; however, coal-tar pitch is more expensive than asphalt and is subject to extreme variations in consistency with changes in temperature. Coal-tar pitch is recommended in highly corrosive environments, and it is not soluble in petroleum products; therefore, asphalt and coal-tar pitch are generally incompatible. However, this property allows coal-tar pitch to be used in areas where fuel spillage is likely to occur. Road tars are often graded by viscosity at room temperature. Tars are available in 12 viscosities, ranging from RT-1 (very fluid) to RT-12 (solid). Liquid coal-tar pitch has an average weight of 10.6 pounds per gallon.

The *flash point* of bitumen is the lowest temperature at which it produces sufficient vapor to form an ignitable mixture which will instantaneously flash at any point on the surface, in the presence of an open flame. Most asphalts have a flash point ranging from 473 to 500°F. The flash point is usually well below the temperature at which the material will burn; therefore, safety precautions should be taken when working with heated bitumens. Keep spark-producing equipment and open flames away from the material, and keep a fire extinguisher handy. Train personnel to safely handle bitumen, and use only approved equipment designed for storing, heating, mixing and distributing bituminous materials. Tests used for determining the flash point of bitumen include the *Tagliabue open cup*, and the *Cleveland open cup tests.*

Bitumen tends to congeal rapidly in cold weather; however, do not overheat the bitumen to compensate for the conditions, because the waterproofing quality of bitumen is greatly reduced when it is heated excessively and/or heated for long periods of time. Store and protect bituminous materials from rain and snow, because moisture absorbed into the bitumen can cause heating and application problems.

*Emulsified asphalts* consist of fine droplets of water dispersed in asphalt with the aid of an emulsifier such as silicates, vegetable oil, soap, or bentonite clay to retard separation. Some emulsions also contain glass fibers or other materials. Since emulsions contain water, they are non-flammable; however, do not allow them to freeze, and do not heat them above 212°F. Liquid emulsions solidify as the water evaporates, leaving the bitumen as residue. One advantage to using an emulsified asphalt is that it can be applied over wet aggregate.

Emulsified asphalts are classified as *rapid-set* (RS), *medium-set* (MS), or *slow-set* (SS). There is also a *high-float* (HF) grade which contains chemical additives, allowing the application of a thicker film coating without danger of runoff. The rapid-set classification is followed by the numeral 1 (thin), or 2 (thick). The *quick-set* (QS) designation is also used for emulsions designed primarily for use in slurry seals. The test used to determine the viscosity of an emulsified asphalt is the *Saybolt Furol test.*

Emulsified asphalt is normally classified as an *anionic type* (electro-negatively-charged asphalt), or a *cationic type* (electro-positively-charged asphalt), depending on the type of emulsion used. The letter C is normally used to designate the cationic emulsion, such as CSS (cationic slow-setting), CMS and CRS. Traditional theory holds that the anionic emulsions perform best with aggregates having mostly positive surface charges, such as limestone and dolomite. Conversely, cationic emulsions perform best with aggregates having mostly negative surface charges, such as quartz and silica. A *nonionic emulsion* is an emulsified asphalt in which the asphalt globules are neither electro-negatively, nor electro-positively charged.

*Cutback bitumens* are asphalt cements thinned with organic solvents and light oils to make them flow freely. Since cutbacks contain flammable solvents, do not expose them to an open flame. Cutbacks solidify as the solvent evaporates, leaving the bitumen as residue. Relatively recent clean-air legislation has sought to reduce solvent evaporation into the air, and has reduced the use of cutback asphalts for surface treatments. Cutbacks are classified as *rapid-curing* (RC) (cut back with naphtha or gasoline), *medium-curing* (MC) (cut back with kerosene), or *slow-curing* (SC) (cut back with residual oil). They are also classified according to viscosity, ranging from 30 (similar to water) to 3000 (very thick); therefore, the classification MC-70 indicates a medium-curing cutback with a viscosity of 70. The higher the viscosity number, the greater the asphalt content.

*Powdered asphalts* are low-penetration-grade asphalts (less than 10) pulverized into a very fine powder (100% passing a No. 10 sieve, and at least 50% passing a No. 100 sieve). The powder is mixed with road oil (SC-250, or SC-800) and aggregate to produce a *cold mix*. Under heat and pressure in the road, the mixture amalgamates with the road oil to yield a consistency similar to asphalt cement. *Colprovia* is a patented asphalt mix formula utilizing powdered asphalt.

*Modified asphalts* consist of a mixture of asphalt and a polymer such as synthetic rubber (rubberized asphalt), or plastic. Modified asphalts produce resilient pavements, and help prevent asphalt from liquifying in extreme heat, and from becoming brittle in extreme cold. Modified asphalts do not add substantially to the cost of production; however, storage can be a problem, since some mixes require constant agitation, while other will solidify if over-mixed. Many experts predict that modified asphalts will be widely used in the future.

## BITUMINOUS SURFACE TREATMENTS

*Bituminous surface treatments* (overlays) are normally less than one inch thick. Surface treatments include dust palliatives, road oiling, preliminary treatments and seal coats. A *dust palliative* (dust abatement, or dust laying) is sprayed onto the surface of unpaved roads to reduce dust and provide some waterproofing of the road surface. Slow-, or medium-curing asphalt cutbacks, or slow-setting diluted emulsions are frequently used for dust palliatives. Cutbacks are applied at the rate of 0.1 to 0.5 gallons per square yard, depending on the texture and dryness of the existing road surface. Several applications are required when emulsions are used.

*Road oiling* is similar to applying a dust palliative, except that road oiling refers to a continuous build-up of a dust palliative over a period of at least three years. Road oil is usually a cutback asphalt (SC-70, SC-250, MC-70, or MC-250). The road surface should be compacted and dampened prior to oiling, because a dry, dusty surface will not absorb the oil. After oiling the surface, it should be compacted and allowed to cure for at least 24 hours before traffic is permitted over the surface. Road oiling is not economical except under the lightest traffic conditions, since the grinding and kneading action of traffic will easily disturb the soil beneath the coating, causing the surface to break up. However, after three or more years of oiling treatments, an earthen road will change into a stable road. The first year's treatment should be applied at the rate of one gallon per square yard (total), applied with three applications. The first application should be 1/2 the total (0.5 gallons per square yard) and the second and third applications should each be applied at the rate of 1/4 of the total (0.25 gallons per square yard, per application). The treatments should be spaced several weeks apart.

*Preliminary treatments* such as prime coats and tack coats (*binder courses*) provide a bond between an existing road surface and a new bituminous surface.

*Prime coats* are applied to an untreated base material to harden and seal the surface, fill voids and bond loose mineral particles. Materials most often used for prime coats include low-viscosity asphalts or tars, diluted asphalt emulsions, or cutbacks (MC-30 and heavier, or SC-70 and heavier). An MC-250 grade should be applied over open-graded aggregate lacking fine sand. Under average conditions, a prime coat is usually applied at the rate of 0.20 to 0.60 gallons per square yard. When applied over dense-graded aggregate, the application rate can normally be reduced to 0.25 to 0.35 gallons per square yard. The minimum air temperature recommended while applying a prime coat is 60°F.

When applied to an untreated granular base, a preliminary treatment normally consists of a prime coat followed by one or more seal coats (Type B surface treatment). If the prime coat is not completely absorbed by the aggregate within 24 hours, spread sand to blot the excess asphalt. The most common application is the double surface treatment, which consists of a prime coat to hold an aggregate coating, followed by a seal coat. The thickness of a double surface treatment is approximately equal to the thickness of the maximum size aggregate applied over the prime coat.

A *tack coat* is applied to an existing paved road surface to develop a bond between an existing pavement and the overlying course. A tack coat usually consists of rapid-curing asphalt cutbacks (RC-70 through RC-250), low-viscosity road tars (RT-7, or RT-8), asphalt cements, or diluted emulsions (SS-1, or RS-1). Of all the asphalt products, emulsified asphalts seem to work best. A tack coat is applied at the rate of 0.05 to 0.15 gallons per square yard. Never apply a tack coat in cold weather, or when the road surface temperature is below 80°F. Always allow the tack coat enough time to cure and become tacky, or it will behave more like a lubricant than an adhesive, causing the overlying course to creep or push. If an excessive tack coat is applied, blot it with sand or screenings before the subsequent pavement course is placed.

A *seal coat* is applied over an existing surface treatment or pavement to rejuvenate, seal and skid-proof the surface. There are several types of seal coats, including a fog seal, slurry seal, sand seal, hot-mix seal, and standard seal coat. A *fog seal* rejuvenates and seals existing asphalt surfaces, and consists of a diluted, slow-setting asphalt emulsion without aggregate cover (Type A surface treatment), applied at the rate of 0.1 to 0.15 gallons per square yard, depending on the texture and dryness of the existing pavement. A fog seal should be applied to a clean, dry surface (with holes patched) while the surface temperature is at least 60°F. No traffic should be permitted on the road surface until the emulsion has set, as indicate by a black color, and do not apply more emulsion than can be absorbed by the surface. If surface bleeding (*flushing*) develops, apply sand as a blotter course.

A *slurry seal* is a low-cost method used to rejuvenate and seal existing asphalt surfaces exhibiting cracks and spalled areas. This mix normally consists of a lightly-diluted asphalt emulsion (SS-1), sand and mineral filler. Prior to placing a slurry seal, a tack coat is sometimes required, consisting of a dilute emulsified asphalt of the same type and grade used in the slurry. A slurry seal is used in the preventive and corrective maintenance of asphalt surfaces, and is not intended to increase structural strength. All holes, bumps, weak areas, cracks and drainage problems should be corrected prior to placing a slurry seal. A slurry seal surface does not contain loose stones, thereby protecting windshields. Slurry is usually applied 1/8 to 1/4 inches thick and should be applied at a minimum air temperature of 50°F.

A *sand seal* (anti-skid coat, or sand asphalt) is used to provide a waterproof and skid-resistant road surface. A sand seal consists of a bituminous coating covered with fine angular aggregate. The bitumen used in a sand seal is usually an emulsified asphalt (RS-1, or MS-1) applied at the rate of 0.15 to 0.20 gallons per square yard. The sand seal coat is subsequently compacted using a pneumatic-tired roller. The minimum air temperature during application should be 50°F. Sand seal coats are generally used on city streets and on aprons for general aviation, where large stones can damage propellers and control surfaces.

A *standard seal coat* consists of an application of asphalt material, followed immediately by a single layer of aggregate of uniform size. The quantities of asphalt and aggregate required for a standard seal coat can be determined by a calculation known as the *Kearby method.*

A *hot-mix seal coat* consists of a one-inch layer of aggregate, hot-mixed with asphalt. This type of seal coat is also referred to as a *friction course*, carpet coat, plant-mixed seal, plant-mixed surface treatment, light asphalt resurfacing, or a thin overlay. Hot-mix seal coats are used in difficult situations that make standard seal coats impractical, such as high traffic volumes, wide temperature variations, short sections of sun and shade, and existing road surface conditions making adhesion difficult. The aggregate can be dense-, or open-graded, and the asphalt cement used in the mix is normally an AC-20. A tack coat should be applied to the existing surface prior to placing the hot mix, and the air temperature should be at least 60°F.

## AGGREGATE SURFACE TREATMENTS

*Aggregate surface treatments* consist of alternate applications of bitumen and aggregate. A single surface treatment (Type C) consists of bitumen covered by an aggregate layer approximately one stone thick. A double surface treatment (Type D) has an additional layer of bitumen and aggregate applied over the first layer, and a triple surface treatment has a third layer of bitumen and aggregate. The maximum aggregate size used in a given layer should be about one-half the size of that used in the previous layer.

To install a single surface treatment, the existing road is swept and a prime coat applied, if required. After a curing period of at least 24 hours, a binder is applied and aggregate is immediately rolled into the binder material. When multiple surface treatments are installed, more binder and aggregate are applied. Following a single or multiple surface treatment, the finished surface should be swept to prevent stones from being thrown by fast-moving vehicles. Road cleaning can be accomplished with self-propelled, or towed rotary brooms, blowers, flush trucks, or hand brooms.

Aggregate can be applied by various types of *spreaders*, including the whirl-type spreader, vane (gravity) spreader, hopper spreader (Photo 9-1), or self-propelled spreader (Figure 9-5). Single-pass surface-treatment machines that apply the binder and aggregate in one continuous operation are also available. It is essential that the asphalt be covered within about two minutes, since the bonding ability is reduced as the asphalt cools. Also, in a single application, aggregate will not stick more than one particle thick to the asphalt; therefore, it is useless and wasteful to apply aggregates at a rate greater than a single layer in thickness. Also, traffic moving over excess aggregate will help grind up the adhered particles.

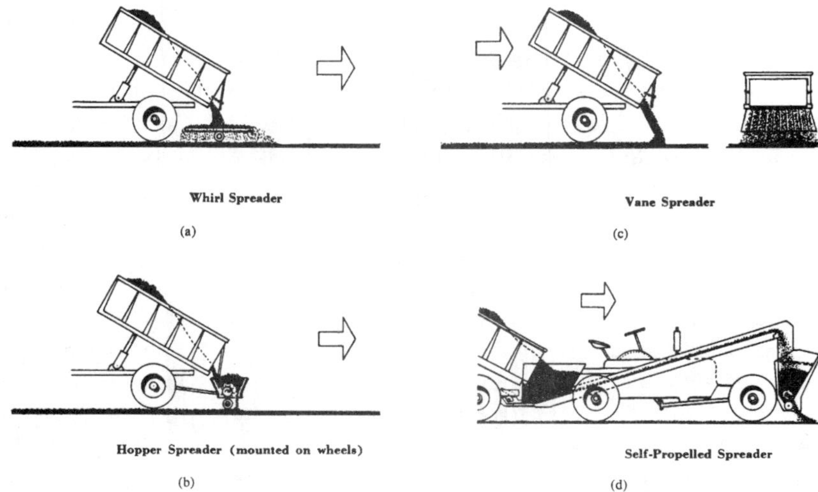


**Figure 9-5. Aggregate spreaders.** (*Courtesy of The Asphalt Institute*)

**Photo 9-1. Hopper spreader.** (*Photo by Archibald Berasso*)

## MACADAM

A *macadam* (penetration macadam) is a common method of paving using crushed stone, and it can be used as a base course, or as a pavement. This paving method is named after John L. McAdam (1756-1836), a Scottish engineer. A macadam is installed by compacting a single layer of coarse aggregate, followed by an application of bitumen, into which intermediate-sized aggregate is compacted. The coarse aggregate layer is normally 3 to 4 inches thick, while the finer aggregate course is usually 2-1/2 to 3 inches thick. The overlying layer of finer aggregate is often applied in small quantities and serves only to fill the surface voids of the coarser aggregates, which does not substantially increase the thickness of the coarse layer of aggregate. The smaller aggregate applied to fill surface voids is referred to as a *choker*. The size of the coarse aggregate should be about 1/2 inch less than the layer thickness, and the smaller aggregate is normally 3/8-inch crushed rock. The bitumen binder is usually a hot, viscosity-graded asphalt cement (AC-10), a cutback (RC-800, or RC-3000), or an emulsion (RS-1, or RS-2). The binder is usually applied at the rate of 1.2 to 2 gallons per square yard, depending on the amount of surface voids. A second macadam course can be installed over the first course in the same manner. Macadam pavements often consist of three aggregate applications, separated by two binder applications. If the macadam is to be used as a base, a tack coat is installed, followed by paving material. If the macadam is to function as a pavement, a single or multiple surface treatment is installed over the macadam. The minimum air temperature for macadam work is normally 60°F.

## BITUMINOUS PAVEMENTS

*Bituminous pavements* are more than one inch thick, and are composed of a pre-mixed aggregate and binder. This type of paving is also referred to as blacktop, asphalt concrete paving, or bituminous concrete paving. Asphalt paving is a high type of asphalt mix (dense-graded aggregate with 4 to 9% asphalt cement) used for constructing roadways for high traffic volumes and heavy axle loads. Asphalt concrete paving is often installed in three layers, consisting of a base, binder, and surface (wearing, or finish) course. Mixing can be accomplished along the road using portable mixing plants (road mixes), or at stationary mixing plants (plant mixes). *Road mixes* are the most economical; however, quality is reduced, due to a lack of uniform mixing of aggregate and binder, and weather can affect the moisture content of the aggregate. Most plant mixes are *hot mixes*, which means that both the aggregate and binder are heated and dried prior to mixing.

*Sheet asphalt* is a hot mixture of asphalt cement and clean, angular, well-graded sand and mineral filler. The maximum size aggregate (95 to 100%) will normally pass a No. 8 sieve, and the binder is usually a viscosity-graded asphalt cement (AC-10, AC-20, or AC-40) which normally comprises about 8 to 12% of the total mix. Sheet asphalt results in a dense, smooth and extremely quiet pavement, widely used on city streets. Due to high costs and the lack of skid resistance at high speeds, sheet asphalt is seldom used on highways. When used in new construction, the road usually consists of a 1-1/2-inch coarse binder course and a 1-1/2-inch sheet surface course. Sheet asphalt is difficult to handle and must be laid with a paving machine.

*Stone-filled sheet asphalt* is similar to sheet asphalt, except that coarser aggregates are mixed with the sand. The sand comprises about 75 to 80% of the total aggregate, and the binder comprises about 6 to 9% of the total mix.

*Plant mixes* are of the highest quality; however, the mix must be transported to the site, sometimes in tremendous quantities per hour. This can become critical when a hot-mix asphalt (HMA) is installed, since the mix cannot be applied if it cools beyond a specified temperature. In some instances, hot mix must be stored temporarily in insulated storage tanks located on the site. Specialized trucks are manufactured for hauling hot mix, built with double bottoms and sides to help provide insulation (Photo 9-2). Even on short haul trips, an unprotected surface will be cooled by wind, forming a hard crust over the top of the mix; therefore, hauling units should be equipped with a heavy tarp to cover the surface of the load.

**Photo 9-2. Hot-mix truck.** (*J.H. Holland Company*)

*Cold-mix asphalt* consists of aggregate and an emulsion, or cutback asphalt binder. Cold mixes can be transported, then stockpiled until needed. Cold mixes produce no smoke, and dust emission is low, and fumes and odors are eliminated when an emulsified asphalt binder is used. However, a cold mix does not produce as high quality a pavement as does a hot mix. Since a cold mix is placed at ambient temperature, cold weather inhibits proper mixing, which makes placing the mix difficult, and retards the curing of the cutback or emulsion. Also, it is difficult to properly compact a cold mix during cold weather, and it is not as resistant as a hot mix to damage caused by moisture and frost. The minimum air temperature normally allowed during the placement of a cold mix is 60°F. A cold mix requires curing time and has low initial stability, but a hot-mix pavement can support traffic as soon as it has cooled to ambient temperature.

## BITUMINOUS DISTRIBUTOR

Bituminous material is normally applied by a *bituminous distributor* (Figure 9-6). A bituminous distributor is also referred to as an asphalt tank truck, asphalt distributor, pitch truck, or boot truck (Photo 9-3).

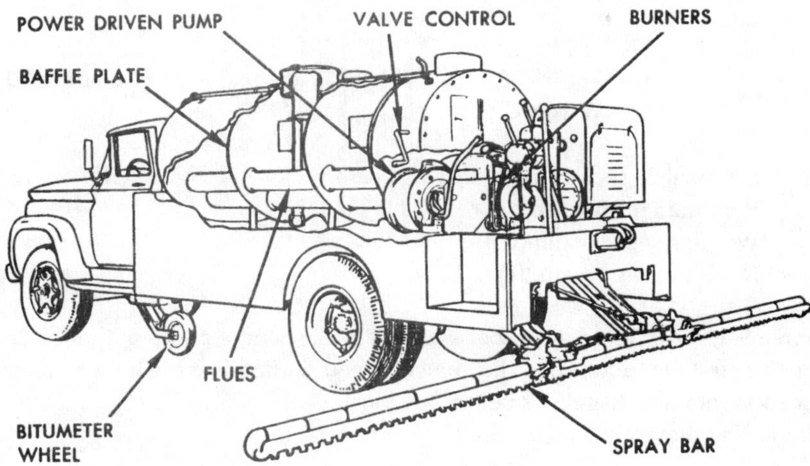


**Figure 9-6. Bituminous distributor.** (*Courtesy of The Asphalt Institute*)

**Photo 9-3. Bituminous distributor.** (*Photo by Dan Atcheson*)

A uniform rate of application of asphalt requires proper asphalt spraying temperature, correct pressure throughout the full length of the spraying bar, correct nozzle angle, correct nozzle height above the surface and a constant distributor speed.

The rate of application is usually expressed in terms of gallons per square yard, and is dependent upon the pump output, vehicle speed and the length of the spray bar. The application rate is usually indicated on a tachometer chart supplied by the distributor manufacturer; however, if no chart is available, the vehicle speed required to obtain a specified application rate can be determined by:

$$S = \frac{9 \times P}{W \times R}$$ **(Equation 9.1)**

where: S = Vehicle speed (fpm)
P = Pump output (gpm)
W = Spray bar width (ft.)
and, R = Application rate (gal./S.Y.).

**Example 9-1:** Assuming a spray bar width of 20 feet and a pump output of 220 gallons per minute, determine the vehicle speed required to obtain a bitumen application rate of 0.40 gallons per square yard.
**Solution:** The required vehicle speed is:

$$S = \frac{9 \times 220}{20 \times 0.40}$$ (Eq. 9.1)

$$= 248 \text{ feet/minute}$$

The length of road surface that can be covered by a bituminous distributor (equipped with a sump pump) can be determined by:

$$L = \frac{9 \times Q}{W \times R}$$ **(Equation 9.2)**

where: L = Length of spread (ft.)
Q = Quantity of bitumen (gal.)
W = Spray bar width (ft.)
and, R = Application rate (gal./S.Y.).

**Example 9-2:** Assuming a 2000-gallon tank capacity, determine the length of spread for the distributor given in the previous example, if the distributor is equipped with a sump pump.
**Solution:** The length of spread is:

$$L = \frac{9 \times 2000}{20 \times 0.40}$$ (Eq. 9.2)

$$= 2250 \text{ lineal feet}$$

If a bituminous distributor is not equipped with a sump pump, 50 gallons of bitumen should be left in the tank to prevent an uneven application at the end of the spread, due to air entering the pump suction line. Therefore, if the distributor is not equipped with a sump pump, the length of spread can be determined by:

$$L = \frac{9 \times (Q - 50)}{W \times R} \qquad \textbf{(Equation 9.3)}$$

**Example 9-3:** Assuming no sump pump, determine the length of spread for the distributor given in the previous example.
**Solution:** The length of spread is:

$$L = \frac{9 \times (2000 - 50)}{20 \times 0.40} \qquad \text{(Eq. 9.3)}$$
$$= 2194 \text{ lineal feet}$$

The volume of asphalt varies with temperature. Most specifications are based on the volume at 60°F; therefore, a temperature correction factor must be applied to determine the volume at other temperatures. The factor is actually the specific gravity of the asphalt at a given temperature, relative to the specific gravity of the asphalt at 60°F. I have included Table 9-1 with limited values in order to work the following example. A comprehensive set of tables are given in the *Asphalt Construction Handbook*, published by Barber-Greene.

**Table 9-1: Temperature-Volume-Weight Conversion Data for Bituminous Materials***

| Observed Temp. (Degrees, F.) | Relative Specific Gravity |
|---|---|
| 60 | 1.0000 |
| 100 | 0.9841 |
| 140 | 0.9686 |
| 180 | 0.9532 |
| 220 | 0.9382 |
| 260 | 0.9234 |
| 300 | 0.9088 |

**Limited to bitumens with specific gravities ranging from 0.850 to 0.966 at 60°F.* (*Courtesy of Barber-Greene*)

**Example 9-4:** Determine the quantity of asphalt required for a bituminous distributor at 60°F. in order to yield a capacity of 2000 gallons at a temperature of 260°F.
**Solution:** The quantity of asphalt required at 60°F. is:

$$\text{Quantity (At 60°F.)} = 2000 \text{ Gal.} \times 0.9234 \quad \text{(Table 9-1)}$$
$$= 1847 \text{ gallons}$$

## INSTALLING ASPHALT PAVING

Prior to the 1930's, asphalt mixes were placed primarily with hand labor. In 1930, Harry H. Barber, co-founder of Barber-Greene Company, invented the first piece of equipment designed specifically for laying these materials (Photo 9-4). Barber's paving machine was equipped with a floating screed with leveling arms attached to a towpoint on each side of a tractor, which allowed the screed to float on the asphalt mix (Figures 9-7 and 9-8). This design permitted gradual thickness changes and automatic leveling of the mix.

**Photo 9-4. H.H. Barber's first paving machine.** (*Courtesy of Barber-Greene*)

A self-propelled *paving machine* (paver, or finisher) simultaneously pushes the dump truck delivering the mix into a receiving hopper (Photos 9-6 and 9-7). Using twin drag-slat conveyors, the mix is carried to the rear of the machine through a set of adjustable metering gates and is deposited on the road surface directly in front of the paver screed. The material is then transported transversely across the width to be paved by a pair of screw conveyors (spreading screws), or variable-speed augers. An automatic feeder control in the screw chamber maintains a constant level (or head) of asphalt mix ahead of the screed.

The *screed* floats on the mix, determining both the mat thickness and texture. The floating screed permits the paver to lay a uniformly thick *mat* (paved strip), regardless of base irregularities. Paving thickness is controlled primarily by adjusting the thickness control screw (Figures 9-7 and 9-9). The screed compacts the mix to a certain extent using a vibrating and tamping motion (*combination screed*). Some pavers can be equipped with a *high-density screed*, designed to achieve greater compaction than can be obtained with standard screeds. Some screeds are equipped with a heater to prevent the mix from sticking to the screed plate.

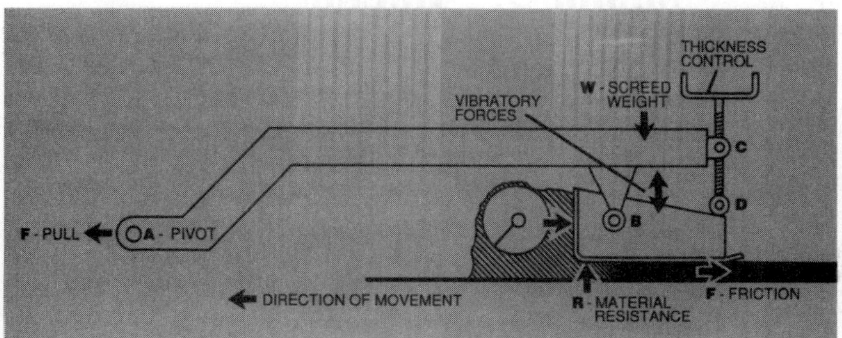


**Figure 9-7. Forces acting on the screed.** (*Courtesy of Barber-Greene*)

Most paving machines pave a mat 10 to 12 feet wide, and have hydraulically-controlled *extensions* to widen the paved area up to 20 feet. Hydraulic extensions are referred to as self-extending screeds, or variable-width strike-off screeds. Standard screed extensions ("hard extensions") are 6", 1', 2' and 5', and can be placed on one or both sides of the main screed. Some extensions provide heat and vibration, but very little compaction. Larger pavers are capable of paving areas up to 28 feet wide. *Cut-off shoes* can be installed to decrease the width of paving by cutting off the flow of paving material. The narrowest paving width available with cut-off shoes is normally about 3 feet, depending on the size of the paver. An *end gate* is attached to the end of a screed to prevent material from spilling beyond the paving width, and to provide a smooth edge to the mat..

Standard pavers are equipped with an automatic screed-height control (leveler) and screed-slope control (Figure 9-8). The *screed height control* senses the average surface configuration, and is available in 30- to 40-foot lengths. The automatic screed control can be adjusted to override the self-leveling action of the screed in order to pave to a predetermined grade or slope, using a mobile or rigid grade reference such as a stringline, curb, gutter, or adjacent mat. When setting the screed-height control, keep in mind that the paving mix will shrink while being rolled and compacted; therefore, the depth for the screed setting must be greater than the actual depth that results after rolling. When paving a mat adjacent to one previously laid, a *joint matcher* (grade reference shoe) can be used to sense the elevation of the original mat (Photo 9-5).

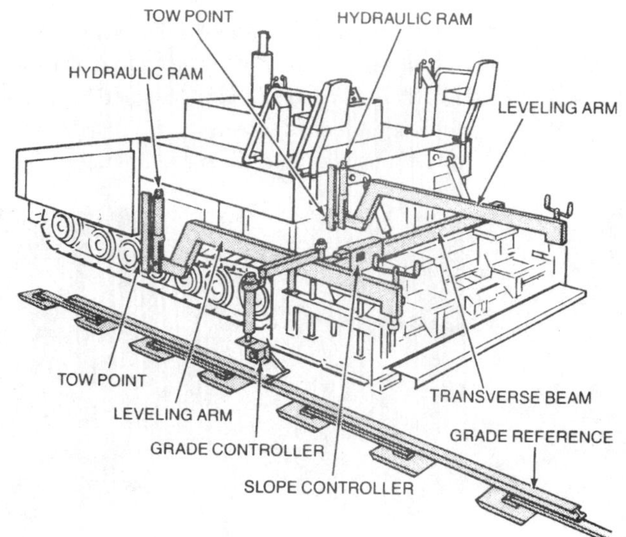


**Figure 9-8. Power leveling mechanisms.** (*Courtesy of Barber-Greene*)

**Photo 9-5. Power leveling mechanism used as a joint matcher.**
(*Photo by Dan Atcheson*)

Track-, or wheel-mounted pavers are available (Photos 9-6 and 9-7). The track-mounted paver is well suited for working over a loose aggregate base, while the wheel-mounted paver works well over firm aggregate, or any hard surface. The crawler paver has a low bearing pressure that minimizes damage to subbase material and asphalt coatings.

**Photo 9-6. Crawler paver.** (*Courtesy of Barber-Greene*)

**Photo 9-7. Wheel-mounted paver.** (*Courtesy of Barber-Greene*)

The operating speed of a paver is dependent primarily upon the width and thickness of the mat, and the workability of the paving mix. Crawler pavers can pave at a maximum speed of around 2 mph, and wheel pavers at about 3.5 mph.

Towed pavers (*spreader boxes*) are similar to self-propelled pavers; however, they do not have a vibrating screed (Figure 9-9). Because of this, the paving mix should be laid about 3/16 to 1/4 inch thicker than that laid by a paving machine. Make sure the dump truck pulling the spreader box does not travel too slow, or too fast. If the truck moves to slowly, the pavement will be too thick, and if it moves too fast, the pavement will be too thin.

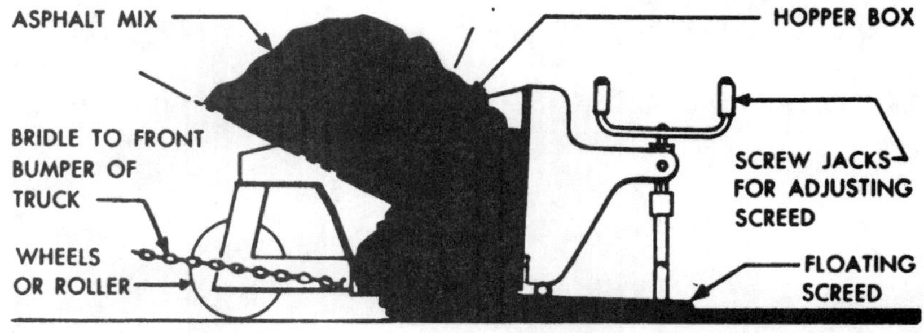


**Figure 9-9. Towed-type paver.** (*Courtesy of The Asphalt Institute*)

Another variety of paving machines is the *road widener*, which rides over an existing road and paves beyond the edge of the pavement (Photo 9-8). Asphalt mix, aggregate, stabilizing agents, or other material is fed into a receiving hopper and conveyed to the road through the side of the machine where it is leveled and sloped by a strike-off tool. Depending on the size of machine, the road widener can pave strips ranging from 1 to 14 feet wide. The road widener can also be equipped with an optional rear-mounted broom for cleanup work.

**Photo 9-8. Road widener.** (*Courtesy of Barber-Greene*)

It is desirable to maintain a constant rate of paving, which requires an even flow of materials to the paver without long delays while waiting for trucks, or hurrying to compensate for a temporary excess of materials cooling off in a waiting truck. This is costly to both the paving contractor and the trucking firm (assuming an independent contractor).

The use of trucks to deliver paving mix to the hopper often results in imperfections in the mat, due to the need to stop the paver while trucks are being changed. Changes in the forward speed of the paver and variations in the level of material in front of the screed are major contributors to changes in the screed force system, which results in variations in mat thickness. An uneven mat surface will cause vehicle axles to bounce, imposing more than twice the load of static axles. The relationship between axle loading and pavement damage is exponential, so a 10% load increase can result in 75 to 100% more damage. Therefore, some paving operations utilize a *windrow elevator* to pick up previously placed windrows (longitudinal piles) of paving material and continuously feed the mix into the paver's hopper (Photo 9-9). A windrow elevator works somewhat like a self-loading scraper. It picks up asphalt from a windrow dumped on the ground and paddles the asphalt into a hopper, using a slat conveyor which distributes the asphalt in front of the screed. The windrow elevator is also referred to as a *material transfer vehicle* (MTV).

**Photo 9-9. Windrow elevator.** (*Courtesy of Barber-Greene*)

One advantage of a windrow elevator is that it seldom needs to stop, as long as asphalt material is available in a windrow in front of the machine. On the other hand, a paving machine must stop each time a change in trucks is required. When a windrow elevator is used, trucks are unloaded quickly and can make more haul trips per day. Also, paving production is increased through greater job efficiency. An end-dump truck is required to supply asphalt to a paving machine, whereas, end-dump or bottom-dump trucks are used to dump asphalt windrows in front of a pickup machine (Photos 4-6 and 4-7). The windrow elevator shown in Photo 9-9 has a wide pick-up throat which allows it to ingest off-center and extra-wide windrows. It is also equipped with a surge-capacity mechanism which allows it to ingest, continuously feed and evenly distribute paving material into the hopper, regardless of the cross section of the windrow (Figure 9-10). Without surge capacity, windrows must be sized accurately, requiring the use of expensive bottom-dump wagons, and material must be added to, or taken away from the windrow in order to balance the intake of the windrow elevator with the output of the paver. Therefore, the volume of a windrow must remain fairly constant throughout its length. Referring to Figure 9-10, the volume of a windrow can be calculated as follows:

$$\text{Volume (Cu. Ft./L.F.)} = \frac{A + (A + C + C)}{2} \times B$$

$$= \frac{2A + 2C}{2} \times B$$

$$= (A + C) \times B \qquad \textbf{(Equation 9.4)}$$

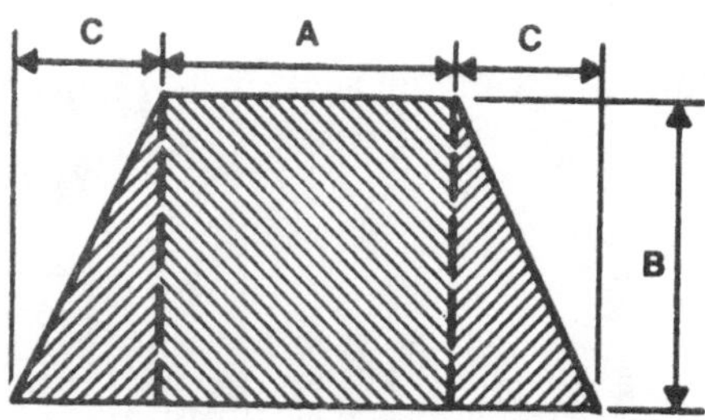


**Figure 9-10. Windrow cross section.**
(*Courtesy of Barber-Greene*)

**Example 9-5:** Assuming A = 2', B = 2' and C = 1', determine the volume of the windrow.
**Solution:** The volume per running foot of windrow is:

$$V = (2' + 1') \times 2' \qquad \text{(Eq. 9.4)}$$
$$= 6 \text{ cubic feet/lineal foot}$$

Another variety of the windrow elevator is the *material transfer paver* (MTP). With this machine, the windrow elevator is an integral part of the paving machine (Photo 9-10). The material transfer paver has a large on-board mix storage capacity to achieve a more continuous asphalt paving operation.

**Photo 9-10. Material transfer paver.** (*Courtesy of Barber-Greene*)

Regardless of the type of paving machine used, always spray an emulsified asphalt on every surface over which paving mix is to be installed, and do not install the mix until the color of the emulsion changes from brown to black.

An asphalt concrete pavement can be installed in one, two, three or more lifts, or courses. The first course can be subgrade material, or subbase material stabilized with asphalt or portland cement. The base course thickness will vary, depending on anticipated traffic loading.

On re-surfacing projects, a *leveling course* (scratch course) is often applied in order to smooth irregularities in the old pavement. This can be limited to the spot patching of depressions with "*leveling wedges*," or it can consist of a pavement layer installed throughout. A 2- to 3-inch binder course is often used instead of a leveling course. The binder-course mix is designed for stability, so the aggregate is normally larger than that used in the surface course which follows.

A 1- to 2-inch surface course is the last course installed in a multi-course pavement, or the only course installed in a single layer operation. The surface course should be smooth, yet skid-resistant, durable, water-resistant and weather-resistant.

When two or more lifts of asphalt paving are required, stagger the joints so that if the top joint cracks, the lower lift will remain intact (Figure 9-11). When a second strip of paving is placed adjacent to a paving strip previously installed, allow the screed to slightly overlap the first strip, since this will prevent a gap at the joint. A joint matching shoe is often installed at the side of a paver that rides over the first strip and senses grade information. Two pavers are often used, one ahead of another, to obtain a *hot joint* between the mats. This is referred to as *echelon paving*. A succession of paving units used to construct a pavement is referred to as a *paving train*.

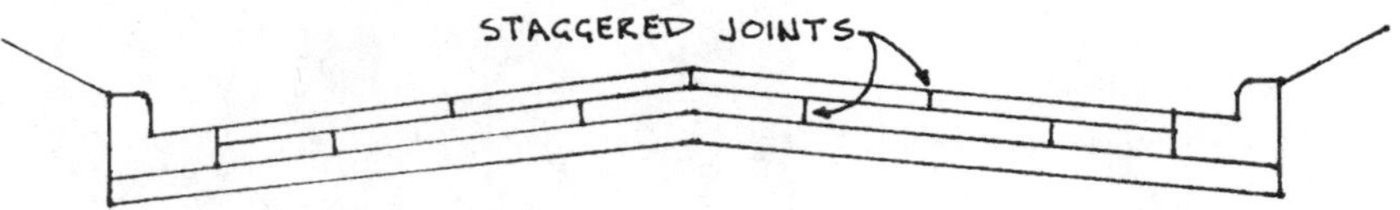


**Figure 9-11. Staggering joints.**

Constantly monitor the oil content, texture and temperature of the mix as it comes out from under the screed. A dull surface indicates too little oil in the mix, and a shiny surface indicates too much oil. If there is not enough fine material in the mix, the surface will have a coarse, rocky appearance. If the mix is too hot, it will often have a brownish color and will give off more smoke than usual. If the mix is too cool, it will become lumpy. Open-graded asphalt paving is usually spread at a temperature ranging from 200 to 250°F., and regular asphalt paving is spread at 250 to 325°F. Neither type of paving should be spread when the air temperature is less than 60°F.

## COMPACTING THE PAVEMENT

New paving must be allowed to cool to a certain extent, prior to rolling it. The cooling time depends on the type of mix, the weight of the roller, the temperature of the original mix, the temperature of the subgrade and air, and the amount of wind and sunshine. Some mixes can be rolled at about 250°F., while others must cool below 200°F. The ability to roll is often a matter of experimentation and judgement. If the mix tends to flow ahead of, or to the sides of the roller, the mix is too hot to roll. Other tell-tale signs of a mix too hot to roll include cracking at the roller edges, or blisters forming behind the roller. On the other hand, if the mix is too cool, it will become stiff and adequate compaction will be difficult, if not impossible. Never allow a roller remain stationary over a paving surface that is not thoroughly compacted and cooled, since this will cause the machine to sink into the paved surface.

There are three roller modes, including breakdown, intermediate and finish rolling. *Breakdown rolling* is used to achieve a passing density within the shortest amount of time. *Intermediate rolling* acts to supplement the breakdown rolling by improving smoothness and eliminating surface blemishes. *Finish rolling* is used to remove any surface marks left by previous rollers.

Transverse joints are rolled first, followed by longitudinal joints (Figure 9-12). Then the outside edges are rolled (a pneumatic-tired roller can be used to compact an unsupported edge), followed by initial (breakdown) rolling, intermediate rolling and finish rolling of the mat. When rolling the mat, make the first pass on the low side of the mat. If the mat is level, start rolling on the edge farthest from the next paver pass. If the next paver pass is to be made without delay, do not roll the last foot of the first pass adjacent to the next pass, because this strip can be used as a guide for the next paver pass.

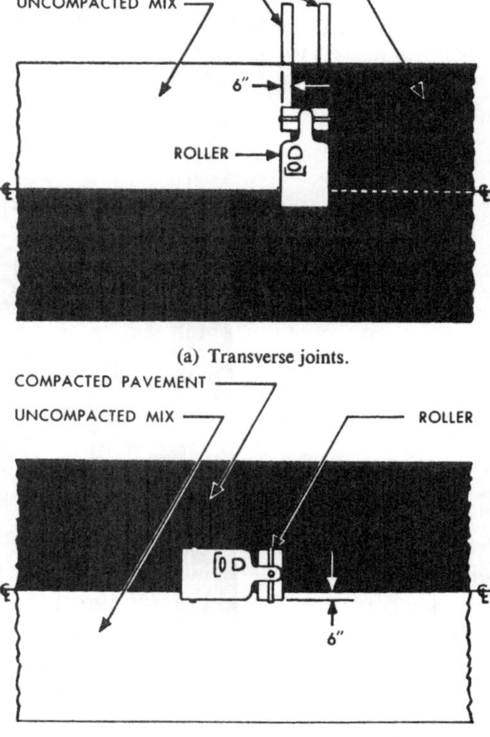


**Figure 9-12.**
**Proper method of rolling joints.**
(*Courtesy of The Asphalt Institute*)

Paving mix is rolled and compacted by smooth-wheel rollers, or pneumatic-tired rollers. Two-axle tandem smooth-wheel rollers are good for compacting joints (Photo 7-25), and three-wheel or tandem 12-ton smooth-wheel rollers are often used for initial (breakdown) rolling of the mat (Photo 7-26). Pneumatic-tired 8-ton rollers are often used for rolling outside edges, and for intermediate rolling because of their ability to seal the pavement surface (Photo 7-31). The initial tire inflation pressure should be set low to minimize tire tracks in the mix. For 15-inch-rim tire sizes, use a minimum 10-ply tire and a contact pressure ranging from 75 to 90 psi. Static tandem 8- to 12-ton smooth-wheel rollers, or 8- to 10-ton vibratory rollers are often used for finish rolling because of their leveling action.

Vibratory tandem steel rollers are often used for breakdown and intermediate rolling; however, some agencies prohibit their use over lifts less than 1-1/2 inches thick (Photo 7-27). The reason for this is due to the need for close impact spacing to avoid a washboard effect which becomes more pronounced as the lift thickness is decreased, or as the forward roller speed is increased (Figure 9-13).

Three-wheel smooth rollers are often preferred over tandem rollers because they are easier to steer (Photo 7-26). During initial rolling, be sure to move the roller with the drive wheels (bull wheels, or drive rolls) forward. The turning force of the drive-wheel rollers will prevent a buildup and displacement of the mix (*bow wave*) that would result from the tiller wheel (steering wheel, or guide roll) being pushed (Figure 9-14). Also, avoid sudden stops, since this can cause displacement of the paving mix. When rolling on an inclined surface, roll with the drive wheels on the down-slope end of the roller.

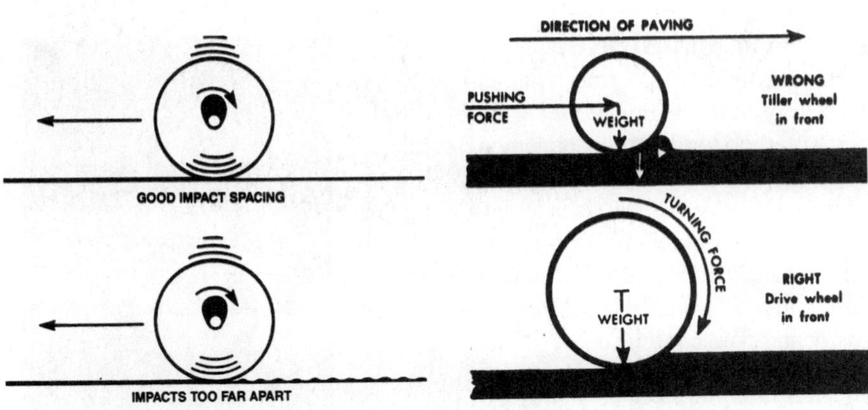


**Figure 9-13. Impact spacing.**
(*Courtesy of Barber-Greene*)

**Figure 9-14. Direction of rolling pavement.**
(*Courtesy of The Asphalt Institute*)

A smooth-wheel roller should be equipped with a roller wetting system to prevent the mix from sticking to the steel drums. Otherwise, the paving surface will be ruined and the drum will no longer be smooth. Roller marks are produced by the edges of a steel drum, and tension cracks are sometimes produced beneath the drum. These imperfections can be erased by a pneumatic-tired roller. Other mat imperfections can be caused by vibrating in place while the machine is stopped, or by a sudden stop while shifting into reverse gear. There are many variables that can cause problems and imperfections in the pavement finish, and Figure 9-15 can be used to help trouble-shoot these various problems.

| Insufficient or Non-Uniform Tack Coat | Improperly Cured Prime or Tack Coat | Mixture Too Coarse | Excess Fines in Mixture | Insufficient Asphalt | Excess Asphalt | Improperly Proportioned Mixture | Unsatisfactory Batches in Load | Excess Moisture in Mixture | Mixture Too Hot or Burned | Mixture Too Cold | Poor Spreader Operation | Spreader in Poor Condition | Inadequate Rolling | Rolling at Wrong Time | Over-Rolling | Rolling Mixture When Too Hot | Rolling Mixture When Too Cold | Roller Standing on Hot Pavement | Overweight Rollers | Roller Vibration | Unstable Base Course | Excessive Moisture in Subsoil | Excessive Prime Coat or Tack Coat | Poor Handwork Behind Spreader | Excessive Hand Raking | Labor Careless or Unskilled | Excessive Segregation in Laying | Faulty Allowance for Compaction | Operating Finishing Machine Too Fast | Mix Laid in Too Thick Course | Traffic Put On Mix While Too Hot | Types of Pavement Imperfections That May Be Encountered In Laying Plant Mix Paving Mixtures. |
|---|---|---|---|---|---|---|---|---|---|---|---|---|---|---|---|---|---|---|---|---|---|---|---|---|---|---|---|---|---|---|---|---|
| | | | | | X | X | X | | | | | | | | | | | | | | | | X | | | | | | | | | Bleeding |
| | | | | X | | | | X | X | | | | | | | | | | | | | | | | | | | | | | | Brown, Dead Appearance |
| | | | | | X | X | X | | | | | | | | | | | | | | | | X | | | | X | | | | | Rich or Fat Spots |
| | | X | X | | | X | X | | | X | X | X | X | X | X | X | X | | | | | | | X | X | X | X | | X | | | Poor Surface Texture |
| X | X | X | | | | X | X | | | X | X | X | X | X | | X | X | X | X | X | X | | | X | X | X | X | | X | | | Rough Uneven Surface |
| | | X | | X | | X | X | | | X | X | X | X | X | | | X | | | | | | | X | X | X | X | | | | | Honeycomb or Raveling |
| | | X | | | | | | | | X | X | X | X | X | | X | X | | | | X | | | X | X | X | X | X | | | | Uneven Joints |
| | | | X | | X | X | | | | X | | | X | X | | X | X | X | X | | | | | | | X | | | | | | Roller Marks |
| X | X | | X | | X | X | X | X | | | X | X | | X | | X | | | X | X | X | | | | X | | | | | X | X | Pushing or Waves |
| | | | X | X | | X | | | | | | | | X | X | X | | | X | | X | X | | | | | | | | | | Cracking (Many Fine Cracks) |
| | | | | | | | | | | | | | | | X | | | | X | | X | X | | | | | | | | | | Cracking (Large Long Cracks) |
| | | X | | | | X | | | | X | X | X | | X | X | X | | | X | | | | | | | | | | | | | Rocks Broken by Roller |
| | | X | | X | | X | | | X | X | X | X | | | | | | | | | X | | | | | | X | | X | | | Tearing of Surface During Laying |
| X | X | | X | | X | X | | X | | X | | | X | X | X | | X | | X | | X | X | X | | | | | | | | | Surface Slipping on Base |

**Figure 9-15. Possible Causes of Imperfections in Finished Pavement.**
(*Courtesy of The Asphalt Institute*)

## MARSHALL MIX DESIGN

Bruce Marshall, a former engineer with the Mississippi State Highway Department, developed a method for designing pavement mixtures, known as the *Marshall Test* (ASTM D-1559). The purpose of the Marshall Test is to determine the optimum asphalt content for a particular blend of aggregate. The test also establishes the maximum road resistance, and the *optimum pavement density* and void content.

There are a number of mix design and quality control tests available; however, in-place paving density as compared to optimum laboratory density is of primary concern to the paving contractor. To perform the density portion of Marshall Test, three paving mix specimens ("pills") approximately 2-1/2 inches high and 3-7/8 inches in diameter are compacted by a hammer weighing 10 pounds, dropped through a distance of 18 inches (Photo 9-11). The upper *and*

**Photo 9-11. Compacting a Marshall specimen.**
(*Photo by Dan Atcheson*)

lower surface of each specimen is struck as follows: 35 blows (light traffic), 50 blows (medium traffic), or 75 blows (heavy traffic). The number of blows required will be specified by the designer, according to the anticipated traffic load.

To determine the density of the compacted mix, the specimen is weighed dry (Dry Weight). The specimen is then submerged in a vessel (pycnometer) containing water, and the weight of water displaced by the sample is determined (Water Weight). The specimen is then taken from the water, the surfaces are allowed to dry, and the sample is weighed once more (Surface Dry Weight). The volume of the specimen can be determined by:

$$\text{Volume} = \text{Surface Dry Weight} - \text{Water Weight} \qquad \textbf{(Equation 9.5)}$$

The specific gravity of the specimen can be determined by:

$$\text{Specific Gravity} = \frac{\text{Dry Weight}}{\text{Volume}} \qquad \textbf{(Equation 9.6)}$$

The density of the specimen can then be determined by:

$$\text{Density (lb./Cu. Ft.)} = \text{Specific Gravity} \times 62.4 \text{ lb./Cu. Ft.}^{*} \qquad \textbf{(Equation 9.7)}$$

The in-place paving density is obtained by analyzing samples cored from the pavement, or by using a nuclear density gauge (Figure 7-18) (Photo 7-17). The density tests should be made at random locations, staying 1 to 2 feet from the edge of the pavement, and 20 feet from any a transverse joint. Many public agencies require one test per 1000 lane-feet of each completed pavement course. The density required usually ranges from 95 to 97% of optimum laboratory density.

## ESTIMATING ASPHALT COMPACTOR PRODUCTION

The production rate of an asphalt compactor is dependent upon the type of asphalt mix, the temperature of the mix, the ambient temperature, lift thickness, the density required and the skill of the operator. The production rates of asphalt compaction equipment can be determined by:

$$\text{Production (C.Y./Hr.)} = \frac{W \times S \times L \times 16.3 \times E}{P} \qquad \textbf{(Equation 9.8)}$$

where: W = Compacted width per pass (ft.)
S = Average speed (mph)
E = Job efficiency
P = Number of passes required
L = Lift thickness (in.)

**Density of fresh water (See Appendix B).*

and,

$$16.3 = \frac{5280 \text{ Ft.}}{12 \text{ In. x } 27 \text{ Cu. Ft.}}$$

**Example 9-6:** Determine the production rate of an asphalt compactor under the following conditions:

W = 5 feet
S = 4 mph
L = 6 inches
P = 5 each
E = 0.83

**Solution:** The probable production rate expressed in cubic yards per hour is:

$$\text{Production (CCY/Hr.)} = \frac{5 \text{ x } 4 \text{ x } 6 \text{ x } 16.3 \text{ x } 0.83}{5} \quad \text{(Eq. 9.8)}$$

$$= 325 \text{ C.Y./hour}$$

Production rates determined in cubic yards per hour can be converted to production rates expressed in terms of square yards per hour as follows:

$$\text{Production (S.Y./Hr.)} = \frac{3 \text{ x Production (C.Y./Hr.)}}{\text{Lift Thickness (Ft.)}} \quad \textbf{(Equation 9.9)}$$

**Example 9-7:** Determine the production rate of the compactor given in the previous example in terms of square yards per hour.

**Solution:** The production rate expressed in square yards of paving installed per hour is:

$$\text{Production (S.Y./Hr.)} = \frac{3 \text{ x } 325 \text{ C.Y./Hr.}}{0.5 \text{ Ft.}} \quad \text{(Eq. 9.9)}$$

$$= 1950 \text{ S.Y./hour}$$

The production rate of a pneumatic-tired roller can be estimated by referring to Figure 9-16. To use Figure 9-16, enter the chart from the left side and find the appropriate lift thickness. Project a horizontal line to the right until it intersects the diagonal line representing the roller speed. From this intersection, drop vertically to obtain the cubic yards of paving compacted per hour, *per pass*. The actual production rate can then be determined by:

$$\text{Production (C.Y./Hr.)} = \frac{\text{C.Y./Pass}}{\text{No. Passes Required}}$$ **(Equation 9.10)**

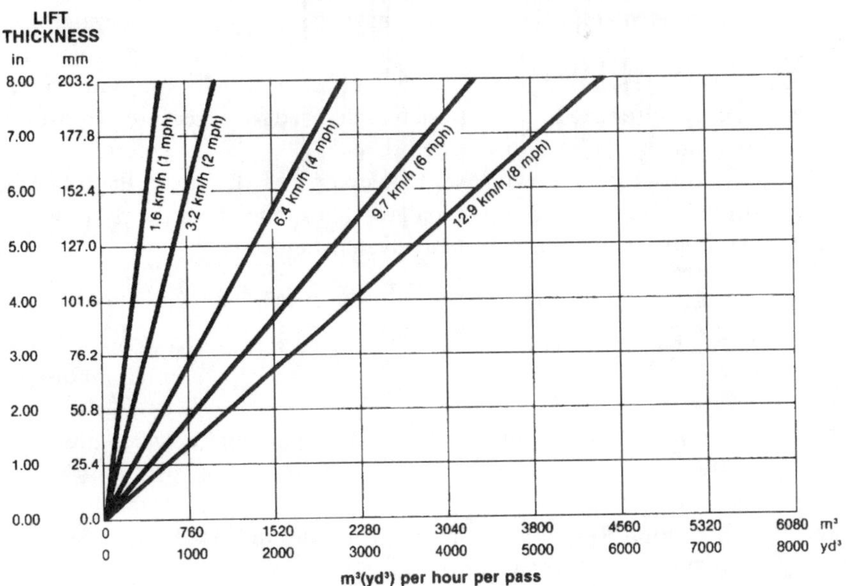


**Figure 9-16. PS-130 or PS-180 pneumatic roller production (C.Y./Hr./Pass)**
*(Courtesy of Caterpillar Inc.)*

**Example 9-8:** Assuming a 6-inch lift thickness, determine the production rate of a Caterpillar PS-130 pneumatic-tired roller traveling at 2 mph if four passes are required per lift.
**Solution:** Referring to Figure 9-16, the production rate per hour, per pass is:

Production (C.Y./Hr./Pass) = 1000 C.Y./hour/pass (Figure 9-16)

The actual production rate using four passes is:

$$\text{Production (C.Y./Hr.)} = \frac{\text{1000 C.Y./Hr./Pass}}{\text{4 Passes}} \quad \text{(Eq. 9.10)}$$

$$= \text{250 C.Y./hour}$$

## MATCHING PAVER AND ROLLER SPEEDS

The average forward travel speed of a paver is dependent upon the delivery rate of paving material to the paver, and the width and thickness of the mat. However, the net forward travel speed of the roller is dependent upon the rolling speed, the number of passes and laps required, plus extra passes for joints, etc. Since freshly laid material must be compacted before it is allowed to cool to a critical temperature, it is vital for the roller to keep up with the paver.

The number of passes required to achieve density is dependent primarily upon the target density, material behavior, lift thickness and the type and weight of the compactor used. Material behavior is dependent primarily upon the mix temperature, and aggregate size and gradation. Under average conditions, a static smooth-wheel roller requires 2 to 6 passes, a pneumatic-tired roller requires 4 passes, and a vibratory tandem roller, 2 to 4 passes.

In order for the breakdown roller to keep up with the paver, the roller must travel through the *rolling zone* a much greater distance than the paver within a given time frame. The required roller travel distance is:

Roller Travel Distance =
Paver Travel Distance x [(No. Laps x No. Passes/Lap) + 1] x 1.10
**(Equation 9.11)**

where: + 1 is placed in the equation due to the "make-up" pass required over a compacted section in order to reach fresh pavement (See Pass No. 19 in Figure 9-17)

and, 1.10 allows 10% non-productive travel distance while changing lanes (Some contractors suggest 15%, or a factor of 1.15).

**Example 9-9:** Assuming a paver travels 1000 feet per hour, determine the distance traveled by a breakdown roller required to make 6 passes over a 3-lap section of paving as shown in Figure 9-17.
**Solution:** The total roller travel distance required is:

Roller Travel Distance (Ft./Hr.) = 1000 x [(3 x 6) + 1] x 1.10 (Eq. 9.11)
= 1000 x 19 x 1.10
= 20,900 feet/hour

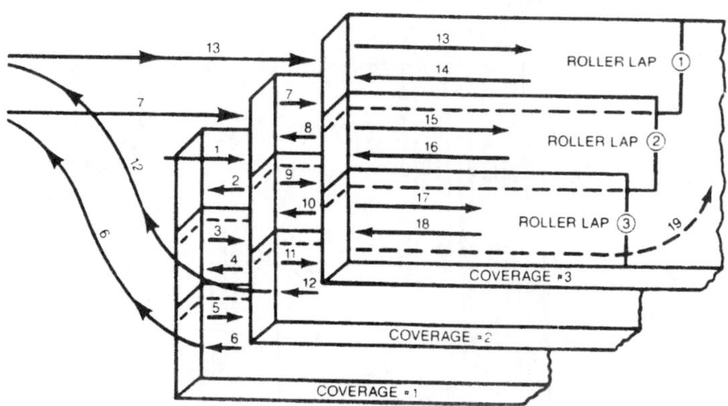


**Figure 9-17. Roller path for Example 9-9.** *(Courtesy of Barber-Greene)*

If the roller in the previous example works a 50-minute hour, it will be required to roll at an average travel speed of:

$$\text{Ave. Roller Speed (fpm)} = \frac{20{,}900 \text{ Ft./Hr.}}{50 \text{ Min./Hr.}}$$
$$= 418 \text{ feet/minute}$$

This is equivalent to:

$$\text{Ave. Roller Speed (mph)} = \frac{418 \text{ Ft./Min.}}{88 \text{ Ft./Min./mph}}$$
$$= 4.75 \text{ mph}$$

This is the average speed required to travel the 20,900 feet in 50 minutes; however, in reality, we would also be required to accelerate, decelerate and stop to reverse direction many times throughout the rolling pattern. Allowing 10% additional speed to make up for the starts and stops, we would actually be required to travel at the rate of:

$$\text{Required Roller Speed (mph)} = 4.75 \text{ mph} \times 1.10$$
$$= 5.23 \text{ mph}$$

This is too fast to achieve adequate compaction; therefore, to keep up with the paver, we must either add a second roller to reduce the number of passes required per roller, or use a wider roller that would require fewer laps, and perhaps, fewer passes per lap. Several rollers operating simultaneously is referred to as a *roller train*.

## PAVEMENT PATCHING

To avoid rough, uneven surfaces when patching a pavement, paving over a utility trench, or paving any small area, the edge of the new paving must be tapered (feathered) to the edge of the existing pavement. To feather properly, prime the existing pavement with an emulsion prior to installing the patchwork paving. A very cold pavement should be heated with a blow torch prior to priming. When the new paving is spread, use a *lute* to rake out all of the large aggregates at the feathered edge, then roll the area with a small roller before the mix cools. Since the edges along concrete curbs cannot be rolled, use a plate tamper or sledgehammer to compact the new pavement.

Install two lifts of material when paving over a trench, especially if more than 4 inches of paving is required. Install the first lift so that its surface is 1 inch below that of the existing asphalt after compaction, then spray the edges with an asphalt emulsion. Finally, dump the second lift, rake out all large aggregates at the edges and roll the lift level with the existing pavement. Apply sand over areas that cannot be properly compacted, to prevent traffic from loosening and picking up pieces of the new pavement, especially in cold weather. This will allow the material to stay in place and eventually harden.

## ASPHALT PLANTS

There are three types of asphalt hot-mix plants, including the batch plant, the continuous mix plant and the drum mix plant. In a *batch plant*, aggregate and asphalt is proportioned and mixed in batches, rather than continuously (Photo 9-12). Aggregate is drawn from a cold feed system, heated and dried in a rotary dryer, then discharged into a bucket elevator and lifted to the top of a batching tower. It is then poured over a vibrating screen where it is separated into a number of different sized materials that are stored separately in *hot bins*. Aggregate is drawn from the hot bins in predetermined amounts, weighed and dropped into a *pugmill* where it is mixed with asphalt. The resulting mix is dropped into a waiting truck, or conveyed to a holding bin (*surge bin*) until a truck is available.

**Photo 9-12. Batch-type asphalt mixing plant.** (*Courtesy of Cedarapids, Inc.*)

In a *continuous mix plant*, aggregate and asphalt are continuously proportioned and mixed. Aggregate is fed into a heated dryer and segregated according to size by screening into separate bins. The aggregate is heated again and fed into a pugmill where it is mixed with asphalt and discharged into trucks (Figure 9-18).

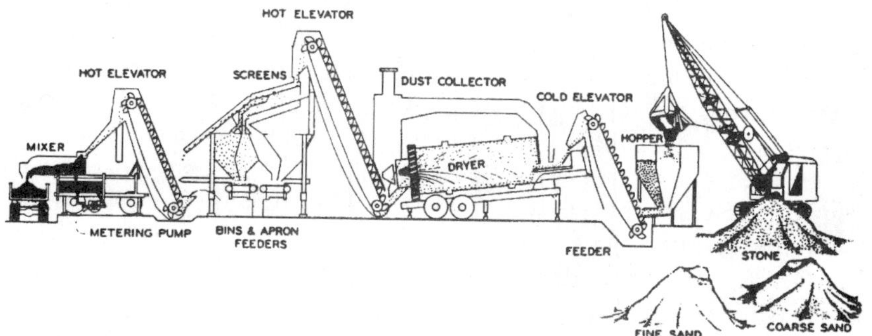


**Figure 9-18. Continuous mix plant.** (*Courtesy of Barber-Greene*)

A *drum mix plant* is similar to a continuous mix plant, except that the aggregate is dried and mixed simultaneously in a dryer drum (Figure 9-19) (Photo 9-13). The drum mixer removes moisture, heats the aggregate and simultaneously coats the aggregate with liquid asphalt (Figure 9-20). This eliminates the need for hot elevators, the gradation control unit and the pugmill. The recent trend has been toward the use of drum mix plants because they are less costly to build than batch and continuous mix plants, and they have fewer and less expensive wearing parts requiring maintenance and replacement. *Portable drum mixers* can be moved and placed into service within 2 or 3 days (Photo 9-14). However, drum mixers have no screening capability, and accurate aggregate sizing is dependent on the quarry crusher, proper stockpiling and cold feed compartments that blend the materials.

**Photo 9-13. Stationary drum mix asphalt plant.** (*Courtesy of Cedarapids, Inc.*)

**Photo 9-14. Portable drum mix asphalt plant.** (*Courtesy of Cedarapids, Inc.*)

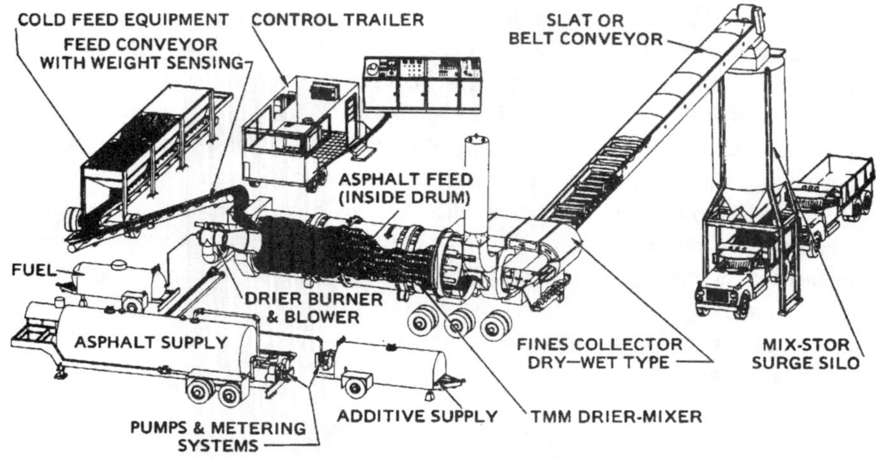


**Figure 9-19. Drum mix plant.** (*Courtesy of Cedarapids, Inc.*)

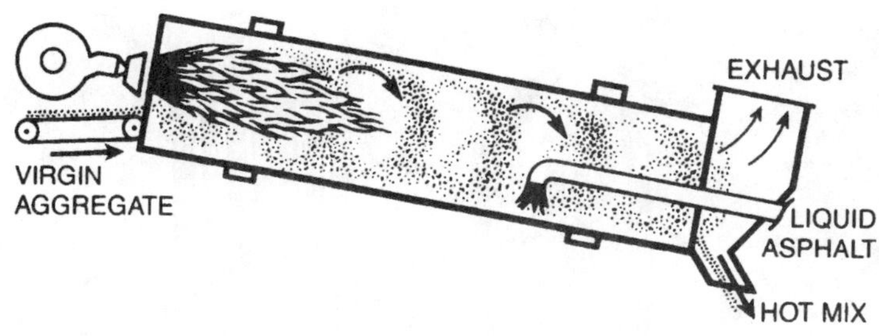


**Figure 9-20. Drum mixer.** (*Courtesy of Barber-Greene*)

## OPTIMUM SPACING OF SUPPLY CENTERS

Supply centers include borrow pits, quarries and batch plants supplying material that is hauled for uniform distribution along a roadway. Supply centers are usually set up in several locations, or they are moved as needed from one location to another, and they are usually installed as close to the road as possible.

Reducing the distance between supply centers increases the set-up costs, but reduces hauling costs, and vise-versa. We can use the following equation to determine the optimum spacing of supply centers:

$$\text{Opt. Distance} = 2 \times \sqrt{\frac{\text{Set-Up Cost}}{\text{Rate of Application} \times \text{Unit Hauling Cost}}}$$

**(Equation 9.12)**

The optimum number of supply centers can be determined by:

$$\text{Opt. Number} = \frac{\text{Length of Project}}{\text{Opt. Distance}} \qquad \textbf{(Equation 9.13)}$$

**Example 9-10:** Assuming that supply centers are set up, operated and moved as required, determine the optimum spacing and number of asphalt processing plants required to service a highway project under the following conditions:

| | |
|---|---|
| Highway Length | = 32 miles |
| Rate of Application | = 4000 tons per mile |
| Set-up Cost (excluding machinery cost) | = $20,000 |
| Hauling Cost | = $0.50/ton/mile |

**Solution:** The optimum spacing of supply centers is:

$$\text{Opt. Distance} = 2 \text{ x} \sqrt{\frac{20{,}000}{4000 \text{ x } 0.50}} \qquad \text{(Eq 9.12)}$$
$$= 6.32 \text{ miles}$$

The optimum number of supply centers is:

$$\text{Opt. Number} = \frac{32}{6.32} \qquad \text{(Eq. 9.13)}$$
$$= 5.06$$
$$= 5 \text{ each}$$

**PAVEMENT RECYCLING**

Deteriorated asphalt or concrete paving can be removed and reused as subbase material, or mixed with asphalt and recycled as paving material. Asphalt material removed from a roadway for recycling is referred to as *reclaimed asphalt pavement* (RAP). The original asphalt binder can be rejuvenated by heating, by adding softening agents, or by adding new asphalt. Concrete materials can be recycled to provide new aggregate, but the original cement binder becomes a part of the new aggregate and must be replaced with asphalt or new portland cement. Removed pavement material containing no reusable binding agent is referred to a *reclaimed aggregate material* (RAM).

There are two basic methods used for reclaiming pavements, including cold planing and hot-mix recycling. *Cold planing* (grinding, profiling, or milling) involves removing part, or all of an asphalt or concrete pavement at ambient temperature. Pavement is removed by a *cold planer* (profiler, milling machine, or grinder) that is equipped with a rotating drum outfitted with numerous replaceable tungsten-carbide cutting teeth (Photo 9-15). The bits are cooled with a water spray system that is also used for dust control. The cold planer is manufactured with a wheel-, or crawler-mounted chassis. The removed material can be left on the surface to be picked up later, or transferred to trucks by on-board conveyors. The material is often left in place, mixed with a binder (and additional aggregate, if necessary), spread, compacted, then sealed with either a surface treatment, or an overlay. This type of pavement is referred to as *in-place recycled pavement.*

Cold planers can also be used to alter the surface of a pavement by restoring specified grades or slopes, removing bumps, potholes, ruts and other surface imperfections, and providing a textured, skid-resistant surface with reduced hydroplaning characteristics.

**Photo 9-15. Cold planer.** (*Courtesy of Barber-Greene*)

The width of paving removed per pass depends on the size of the cold planer. Standard cutting mandrel widths vary from 4 feet to 12'-6", and large cold planers can be equipped with 1', 1'-6", or 2'-6" mandrel extensions.

The cold planer can be set to remove a specified pavement thickness. Production increases with the depth of cut, up to a certain point. As the depth of cut increases beyond the machine's peak production depth, the reduced forward speed begins to offset the production gains of the deeper cut. Asphalt concrete up to 3 to 4 inches thick, or portland cement concrete ranging from 1/2 to 1-1/2 inches thick can normally be removed during a single pass. Although thicker materials can be planed in a single pass (up to 10" on larger machines), it is usually more efficient to make two passes, removing less material per pass. Some contractors use two cold planers, one ahead of the other, each removing a portion of the total depth.

The forward speed of a cold planer (and thus, production) while removing any given type of paving is dependent primarily on the condition of the road, the hardness and extent of gradation of the aggregate in the mix, the strength of the asphalt bond, and the depth of material removed during a pass. Since the cutting teeth are essentially breaking the bond between the binder and the aggregate, and not actually fracturing the aggregate itself, a pavement with a high asphalt content and small aggregate is harder to mill than one with a high percentage of coarse aggregate. Forward travel speed can vary from 8 to 10 feet per minute when reclaiming several inches of high-quality asphalt concrete, and 100 to 150 feet per minute when removing less than one inch of deteriorated asphalt concrete. Caterpillar Inc. publishes cold planer production rates as shown in Table 9-2.

**Table 9-2: PR-750B Cold Planer Production Using a 120" Cutting Drum (Working 60 Min./Hr.)**

| Forward Speed (fpm) | S.Y./Min. | C.Y./Min.* | U.S. Tons/Min.* |
|---|---|---|---|
| 10 | 11.1 | 0.31 | 0.64 |
| 15 | 16.7 | 0.46 | 0.96 |
| 20 | 22.2 | 0.62 | 1.28 |
| 25 | 27.8 | 0.77 | 1.60 |
| 30 | 33.3 | 0.93 | 1.91 |
| 35 | 38.9 | 1.08 | 2.24 |
| 40 | 44.4 | 1.23 | 2.55 |
| 45 | 50.0 | 1.39 | 2.87 |
| 50 | 55.5 | 1.54 | 3.19 |
| 55 | 61.1 | 1.70 | 3.51 |
| 60 | 66.7 | 1.85 | 3.83 |

*Cubic yardage and tonnage figures are based on a one-inch depth of cut. For greater depths of cut, multiply the production rate by the cutting depth. Based on asphalt density of 115 pounds per square yard, one inch thick.
*(Courtesy of Caterpillar Inc.)*

The production rates in terms of square yards per minute given in Table 9-2 were obtained from the following equation:

$$\text{Production (S.Y./Min.)} = \text{Forward Speed (fpm)} \times \frac{\text{Cutting Edge Width (Ft.)}}{\text{9 Sq. Ft./S.Y.}} \times \text{Efficiency Factor}$$

**(Equation 9.14)**

where: Efficiency factor is taken from Table 1-4, or is estimated as productive time per clock hour.

**Example 9-11:** Determine the production rate of a Caterpillar PR-750B cold planer, cutting pavement with a 10-foot cutting drum while traveling at a forward speed of 20 feet per minute. Express the production rate in terms of square yards per minute while working a 50-minute hour.

**Solution:** The probable production rate is:

$$\text{Production (S.Y./Min.)} = 20\text{ fpm} \times \frac{10\text{ Ft.}}{9\text{ Sq. Ft./S.Y.}} \times 50/60 \quad \text{(Eq. 9.14)}$$

$$= 22.2\text{ S.Y./Min.} \times 50/60 \quad \text{(See Table 9-2)}$$

$$= 18.5\text{ S.Y./minute}$$

The production rates in terms of cubic yards per minute given in Table 9-2 were obtained from the following equation:

$$\text{Production (C.Y./Min.)} = \text{Production (S.Y./Min.)} \times \frac{\text{Cutting Depth (In.)}}{36\text{ In./Yd.}} \times \text{Efficiency Factor}$$

**(Equation 9.15)**

**Example 9-12:** Assuming a one-inch cutting depth, determine the production rate of the cold planer given in the previous example in terms of cubic yards per minute while working a 55-minute hour.
**Solution:** The probable production rate is:

$$\text{Production (C.Y./Hr.)} = 22.2\text{ S.Y./Min.} \times \frac{1''}{36''} \times 55/60$$

(Table 9-2; Eq. 9.15)

$$= 0.62\text{ C.Y./Min.} \times 55/60 \quad \text{(See Table 9-2)}$$

$$= 0.57\text{ C.Y./minute}$$

The production rates in terms of tons per minute given in Table 9-2 were obtained from the following equation:

$$\text{Production (Tons/Min.)} = \text{Production (C.Y./Min.)} \times \frac{\text{Pavement Density (lb./C.Y.)}}{2000\text{ lb./Ton}} \times \text{Efficiency}$$

**(Equation 9.16)**

**Example 9-13:** Assuming a pavement density of 115 pounds per square yard per inch of depth, determine the production rate of the cold planer given in the previous example in terms of U.S. tons per minute while working a 55-minute hour.

**Solution:** The density of the asphalt pavement per cubic yard is:

$$\text{Pavement Density (lb./C.Y.)} = 115 \text{ lb./S.Y./In.} \times 36 \text{ In./Yd.}$$
$$= 4140 \text{ lb./C.Y.}$$

The probable production rate is:

$$\text{Production (Tons/Min.)} = 0.62 \text{ C.Y./Min} \times \frac{4140 \text{ lb./C.Y.}}{2000 \text{ lb./Ton}} \times 55/60$$

(Table 9-2; Eq. 9.16)

$$= 1.28 \text{ Tons/Min.} \times 55/60$$ (See Table 9-2)

$$= 1.17 \text{ tons/minute}$$

Pavements can also be recycled using a machine called a *road reclaimer* (Photo 9-16). The reclaimer can also be used for mechanical stabilization of a road. The reclaimer can remove a cut 8 feet wide, and up to 18 inches deep. The machine cuts and pulverizes asphalt paving and base materials, and blends them with one, or a combination of the following materials: asphalt emulsions, binding agents, softening agents, stabilizers and/or water. After mixing, the material is re-deposited on the road to be re-laid, rolled and compacted.

**Photo 9-16. Road reclaimer.** (*Courtesy of Caterpillar Inc.*)

Caterpillar Inc. publishes reclaimer production rates as shown in Table 9-3. The production rates given in Table 9-3 were obtained from Equations 9.14 (S.Y./Min.) and 9.15 (C.Y./Min.) while using a cutting width of 8 feet, and cutting depths as indicated at the top of the table.

**Table 9-3: Stabilization/Reclamation Production**

| Travel Speed FPM | 6" | | 8" | | 10" | | 12" | | 13" | | 15" | | 16" | | 18" | |
|---|---|---|---|---|---|---|---|---|---|---|---|---|---|---|---|---|
| | yd²/min | yd³/min | yd²/min | yd³/min | yd²/min | yd³/min | yd²/min | yd³/min | yd²/min | yd³/min | yd²/min | yd³/min | yd²/min | yd³/min | yd²/min | yd³/min |
| 10 | 8.9 | 1.5 | 8.9 | 2.0 | 8.9 | 2.5 | 8.9 | 3.0 | 8.9 | 3.2 | 8.9 | 3.7 | 8.9 | 4.0 | 8.9 | 4.5 |
| 20 | 17.8 | 3.0 | 17.8 | 4.0 | 17.8 | 4.9 | 17.8 | 5.9 | 17.8 | 6.4 | 17.8 | 7.4 | 17.8 | 7.9 | 17.8 | 8.9 |
| 30 | 26.7 | 4.5 | 26.7 | 5.9 | 26.7 | 7.4 | 26.7 | 8.9 | 26.7 | 9.6 | 26.7 | 11.1 | 26.7 | 11.9 | 26.7 | 13.4 |
| 40 | 35.6 | 5.9 | 35.6 | 7.9 | 35.6 | 9.9 | 35.6 | 11.9 | 35.6 | 12.8 | 35.6 | 14.8 | 35.6 | 15.8 | 35.6 | 17.8 |
| 50 | 44.5 | 7.4 | 44.5 | 9.9 | 44.5 | 12.4 | 44.5 | 14.8 | 44.5 | 16.0 | 44.5 | 18.5 | 44.5 | 19.8 | 44.5 | 22.3 |
| 60 | 53.4 | 8.9 | 53.4 | 11.9 | 53.4 | 14.8 | 53.4 | 17.8 | 53.4 | 19.3 | 53.4 | 22.2 | 53.4 | 23.7 | 53.4 | 26.7 |
| 70 | 62.3 | 10.4 | 62.3 | 13.8 | 62.3 | 17.3 | 62.3 | 20.8 | 62.3 | 22.5 | 62.3 | 25.9 | 62.3 | 27.7 | 62.3 | 31.2 |
| 80 | 71.2 | 11.9 | 71.2 | 15.8 | 71.2 | 19.8 | 71.2 | 23.7 | 71.2 | 22.7 | 71.2 | 29.6 | 71.2 | 31.6 | 71.2 | 35.6 |
| 90 | 80.1 | 13.4 | 80.1 | 17.8 | 80.1 | 22.4 | 80.1 | 26.7 | 80.1 | 28.9 | 80.1 | 33.3 | 80.1 | 35.6 | 80.1 | 40.1 |

(*Courtesy of Caterpillar Inc.*)

*Hot-mix recycling* requires that the removed materials be transported to a central plant, where they are reheated and mixed with new asphalt and/or recycling agents, and/or new aggregate.

*Surface recycling* involves scarifying and removing only 3/4 to 1 inch of the asphalt road surface after it has been heated in place with an open flame, or infrared heater. An asphalt emulsion is mixed with the material, which is then relaid on the road to be rolled and compacted. A succession of mobile processing units used to recycle old pavement is referred to as a *recycling train* (Figure 9-21).

**Figure 9-21. Recycling train.** (*Courtesy of Barber-Greene*)

## CONCRETE PAVING

Most concrete paving is finished by large, specialized concrete spreaders that strike off, vibrate and finish the concrete as the machine moves forward over steel forms (Photo 9-17). Hand finishing is accomplished from a form-riding platform that follows the machine. The spreader is followed by a form-riding curing machine that spray-applies curing compound. Concrete spreaders can typically spread and finish slabs up to 10 inches thick and 24 feet wide, at the rate of 20 feet per minute.

**Photo 9-17. Concrete spreader.** (*Courtesy of CMI Corporation*)

# Chapter 10

# RIPPING

With the advent of heavy-duty tractors equipped with rippers, specialized scrapers and hauling units, rock that once required blasting can now be ripped, oftentimes at 50% of the cost of blasting.

Rippers are often installed on ripper racks mounted at the rear of many types of equipment, including motor graders (Photo 10-1), front-end loaders, compactors, tractors and bulldozers. The buckets of some backhoes or similar excavators can be removed and replaced with a ripper tooth. Ripper teeth are also installed on the back of an excavator bucket (Figure 1-31), or bulldozer blade. Teeth mounted on a bulldozer blade are designed to rip only when the bulldozer backs up. In forward motion, the teeth pivot to the rear and drag loosely along the ground.

**Photo 10-1. Rippers mounted on a motor grader.** (*Courtesy of Caterpillar Inc*)

Rippers are used primarily for ripping rock and asphalt; however, they can also be used to cut tree roots, remove tree stumps and pry up concrete slabs.

## TOWED RIPPERS

Although most rippers are mounted on self-propelled units, they are occasionally towed behind a tractor. *Towed rippers* are useful only in moderate materials such as hard ground, weak rock, or pavements. Since towed rippers are relatively lightweight, they tend to tip over when they hit large objects such as boulders.

## TRACTOR-MOUNTED RIPPERS

There are three basic types of tractor-mounted rippers, including the hinge-type or radial-lift ripper (Figure 10-1), the fixed parallelogram or parallel-lift ripper (Figure 10-2) and the adjustable parallelogram ripper (Figures 10-3 and 10-4). Any of these rippers can be attached to a tractor without interfering with the tractor's dozing ability. The type of ripper used will depend on the rock type and job conditions; however, the adjustable parallelogram ripper is the most versatile ripper and can be used for any application.

The *radial-lift ripper* (radial ripper) swings the shanks up and down through an arc, changing the tooth angle at various depths of ripper penetration (Figure 10-1). The radial-lift ripper is especially useful for ripping close to existing structures such as walls or footings, and for ripping soils containing boulders.

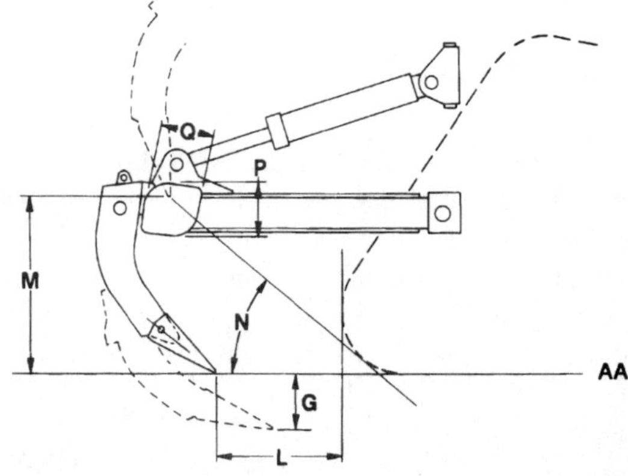


**Figure 10-1.**
**Radial-lift ripper.**
(*Courtesy of Caterpillar Inc.*)

The *fixed parallelogram ripper* (parallel ripper) raises and lowers the shanks through a vertical plane, maintaining a constant tooth angle at any depth of penetration, which keeps the tooth at its optimum cutting angle and reduces tip wear (Figure 10-2).

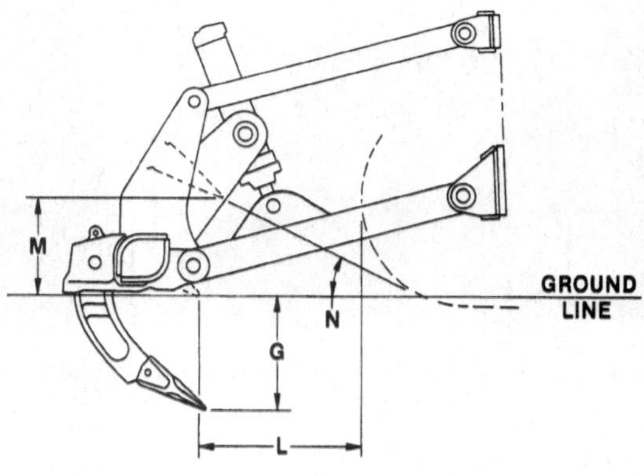


**Figure 10-2.**
**Fixed parallelogram ripper.**
(*Courtesy of Caterpillar Inc.*)

The *adjustable parallelogram ripper* (adjustable parallel ripper) is the most versatile ripper available, because the tooth angle can be varied according to prevailing conditions, even while the tractor is moving (Figure 10-3) (Photo 10-2). This type or ripper is also available with a hydraulically-controlled shank, and is referred to as an *impact ripper* (Figure 10-4). An impact ripper has a single shank and is used to rip extremely difficult rock.

All rippers (except the impact ripper) are available with single, or multiple shanks (Figure 10-5). The number of shanks used depends on the tractor power, depth of penetration and extent of breakage desired, and the resistance of the material being ripped. The extent of breakage required depends on the type of loading method used. Scrapers require smaller fragments than mass excavators,

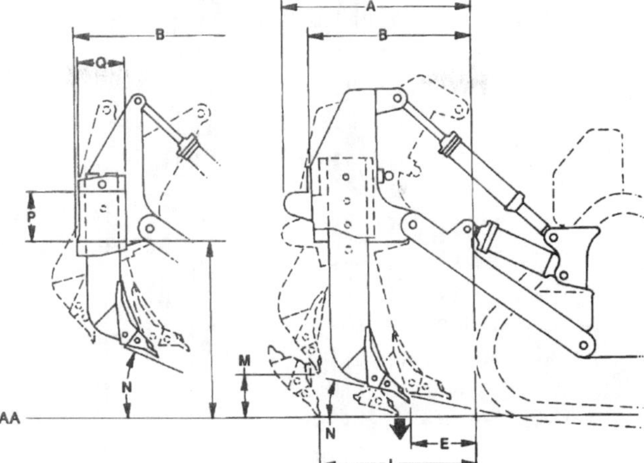


**Figure 10-3. Adjustable parallelogram ripper.**
(*Courtesy of Caterpillar Inc.*)

**Photo 10-2. Adjustable parallelogram ripper.**
(*Courtesy of Caterpillar Inc.*)

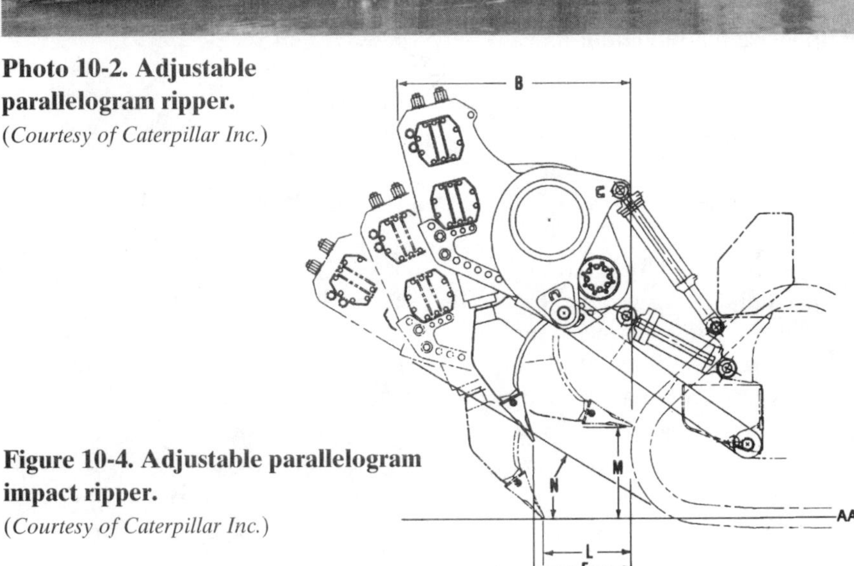


**Figure 10-4. Adjustable parallelogram impact ripper.**
(*Courtesy of Caterpillar Inc.*)

front-end loaders, or dozers. Ripping with only the outer teeth increases penetration and requires less power; however, this also results in larger rock sizes than that created when all three teeth are used. The size of fractured material depends primarily on the lateral spacing of the teeth. If the ripped pieces are too large, they can be reduced by rolling over them with the ripper tractor, by blasting, or by using a crane with a drop ball. When only two shanks are used, they should not be spaced too far apart, since this will cause the tractor to swerve to one side when one shank hits a hard spot. Only the center tooth is normally used for ripping extremely tough material.

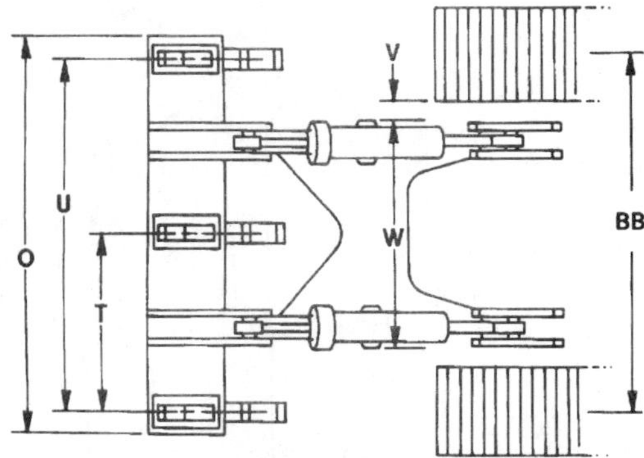


**Figure 10-5. Multiple ripper shanks.**
*(Courtesy of Caterpillar Inc.)*

There are three ripper tip (tooth) configurations available, including short, intermediate and long tips. The short tip is used in high-impact materials where tip breakage is a problem, and when ripping with tandem tractors, since the additional tractor pushing power exerts a tremendous amount of stress on the ripper teeth. The intermediate tip is used in moderate-impact materials where abrasion is not excessive, and the long tip is used in loose, abrasive materials where breakage is not a problem. Use the longest tip that will wear without excessive breakage. Some ripper teeth can be reversed, extending service life. Figure 10-6 shows ripper tooth details.

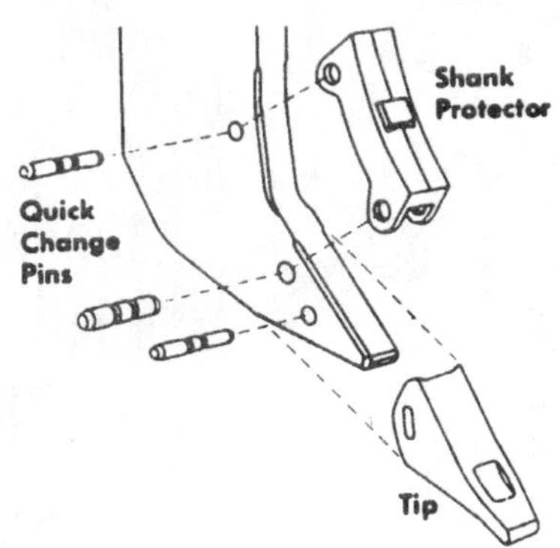


**Figure 10-6. Ripper tooth details.**
*(Courtesy of Caterpillar Inc.)*

Production is sometimes optimized by using a single shank with a tooth penetration that matches the thickness of the rock stratum (layer), but better production is often attained by using multiple shanks with shallow tooth penetration. However, rip as deeply as possible, because shallow penetration will cause tooth wear that is disproportionate to the production. Regardless of the tooth penetration used, avoid raising the rear end of the tractor over extended periods of time, since this reduces production and causes excessive undercarriage wear. Production is also increased by leaving a few inches of ripped material after each pass to increase traction. Teeth that slide into the same groove during subsequent passes without ripping usually indicate dull, worn teeth that should be replaced, or reversed. A ripper tractor is sometimes assisted by a pusher tractor equipped with a cushion blade (Figure 5-7) (Photo 5-15).

## DETERMINING RIPPABILITY

Prior to bidding a project, you must attempt to determine whether rock can be ripped, or whether it must be drilled and blasted. One of the best indicators of *rippability* is the velocity of sound waves traveling through the rock. Field studies using seismographic methods have shown that the velocity of sound waves transmitted through rock is inversely proportional to the rippability of the rock. That is, rock that transmits sound waves at slow velocities is more rippable than rock that transmits sound waves at high velocities. Sound velocity can range from 1000 fps (feet per second) in loose soil to 20,000 fps in hard, solid rock.

*Seismic velocity charts* such as Figure 10-7 can be compared with field seismic information and used as a guide for determining the rippability of various types of rock. The rippability of various rock types given in a seismic velocity chart is unique to the type of tractor and ripper used, but the actual rippability is dependent upon many variables, including the extent of stratification (layering), fractures, planes of weakness, the extent of softening due to weathering, rock brittleness and grain size. Many rocks become more resistant to ripping as depth increases, due to less exposure to weathering. Some rocks cannot be ripped until they have been weakened by light blasting.

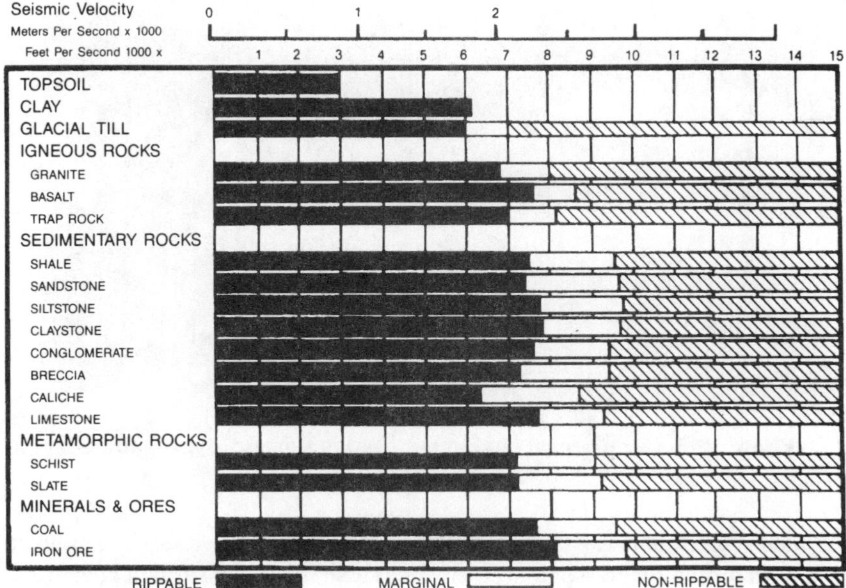


**Figure 10-7. D9N ripper performance, estimated by seismic wave velocities.**
(*Courtesy of Caterpillar Inc.*)

Tooth penetration is usually the key to ripping success, regardless of seismic velocity. Thus, in materials with low seismic velocities, rippability will be seriously impaired if there are no bedding joints, or fractures. This is especially true in homogeneous materials such as mudstone, claystone and fine-grained caliche, or tightly cemented conglomerates and glacial till.

## SEISMIC SUBSURFACE INVESTIGATIONS

To perform a seismic investigation, a sound sensor (*geophone*) and *timer* is set up at station 0 (Figure 10-8). The geophone is wired to the timer, which is wired to a *hammer* (Photo 10-3). The hammer is used to strike steel plates placed at equally spaced *impact stations* (Stations 1,2,3, etc.). The hammer striking a plate starts the timer and sends sound waves, and the geophone stops the timer when the first sound wave is received. As the distance between the geophone and impact stations is increased, the deeper the sound waves will penetrate the earth. Knowing the distance and sound-wave travel time, the velocity of the sound wave can be determined. Sound velocity is normally measured in terms of feet per second (fps) as follows:

$$V\text{ (fps)} = \frac{\text{Distance From Phone to Plate (Ft.)}}{\text{Sound Wave Travel Time (Sec.)}} \qquad \textbf{(Equation 10.1)}$$

**Photo 10-3. Geophysical equipment used in seismic investigations.**
(*Soiltest, Inc.*)

Sound velocity data can then be plotted as shown in Figure 10-9. For any given rock or soil formation, the sound velocity plots essentially as a straight line; therefore, a slope change in sound velocity lines indicates a different type of rock or soil formation. The point at which velocity lines of different slopes intersect is defined as the *critical distance* ($C_1$ and $C_2$ of Figure 10-9).

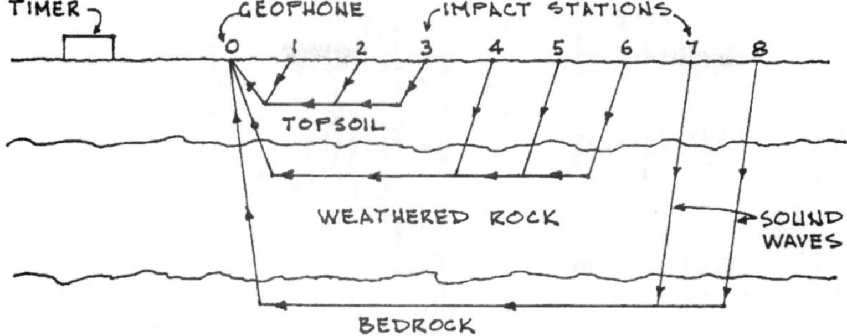


**Figure 10-8. Paths of sound waves through formations.**

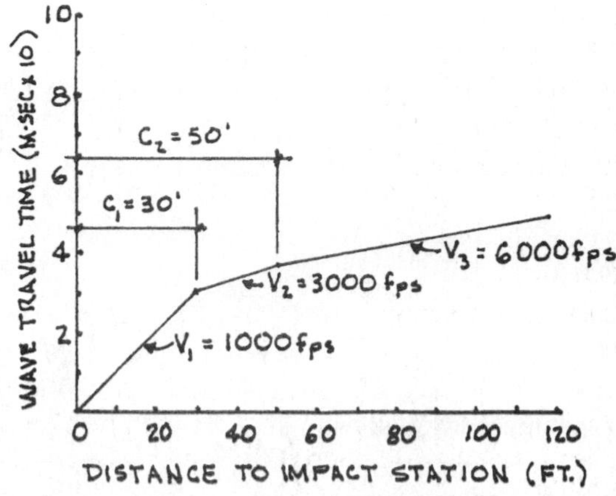


**Figure 10-9. Relationship between distance, travel time and velocity of sound waves in soil.**

Seismic data can be used to determine the hardness and thickness of a rock, or soil formation. The following equations can be used to determine the thickness of the top two formations, *provided* the sound velocity increases with the depth. The thickness of the upper layer can be determined by:

$$T_1 = \frac{C_1}{2}\sqrt{\frac{V_2 - V_1}{V_2 + V_1}} \qquad \textbf{(Equation 10.2)}$$

where: $T_1$ = Thickness of upper layer (ft.)
$C_1$ = Critical distance for upper layer (ft.)
$V_1$ = Sound velocity in upper layer (fps)
and, $V_2$ = Sound velocity in second layer (fps).

The thickness of the second layer can be determined by:

$$T_2 = \frac{C_2}{2}\sqrt{\frac{V_3 - V_2}{V_3 + V_2}} + T_1 \frac{V_3\sqrt{V_2^2 - V_1^2} - V_2\sqrt{V_3^2 - V_1^2}}{V_1\sqrt{V_3^2 - V_2^2}}$$

**(Equation 10.3)**

where: $T_1$ = Thickness of upper layer (ft.)
$T_2$ = Thickness of lower layer (ft.)
$C_2$ = Critical distance for second layer (ft.)
$V_1$ = Sound velocity in upper layer (fps)
$V_2$ = Sound velocity in second layer (fps)
and, $V_3$ = Sound velocity in third layer (fps).

**Example 10-1:** Use data given in Figure 10-9 to determine the thickness of the first and second layers of rock.
**Solution:** Information from Figure 10-9 is:
$C_1$ = 30 feet
$C_2$ = 50 feet
$V_1$ = 1000 fps
$V_2$ = 3000 fps
$V_3$ = 6000 fps

The thickness of the upper layer is:

$$T_1 = \frac{30'}{2} \times \sqrt{\frac{3000 - 1000}{3000 + 1000}} \qquad \text{(Eq. 10.2)}$$
$$= 10.61 \text{ feet}$$

The thickness of the second layer is:

$$T_2 = \frac{50'}{2} \times \sqrt{\frac{6000 - 3000}{6000 + 3000}}$$
$$+ 10.61' \times \frac{6000 \times \sqrt{(3000)^2 - (1000)^2} - 3000 \times \sqrt{(6000)^2 - (1000)^2}}{1000 \times \sqrt{(6000)^2 - (3000)^2}} \qquad \text{(Eq. 10.3)}$$
$$= 12.85 \text{ feet}$$

**Example 10-2:** Assuming the layers of materials given in the previous example are topsoil, limestone and caliche, respectively, determine their rippability while using a Caterpillar D9N tractor.

**Solution:** The sound-wave velocities are:

Topsoil = 1000 fps (Example 10-1)
Limestone = 3000 fps
Caliche = 6000 fps

Referring to Figure 10-7, we can see that the topsoil and limestone are easily rippable. The caliche is also rippable; however, it is almost marginal.

## CORE-DRILLED SUBSURFACE INVESTIGATIONS

Subsurface investigations are sometimes made by core-drilling test holes at random locations throughout the site, and removing and analyzing the soil samples.

**Example 10-3:** Given the following test hole data, determine the volumes of each type of material to be excavated in the basement diagrammed in Figure 10-10. Assume the soils above and below the layer of rock are common earth and clay, respectively.

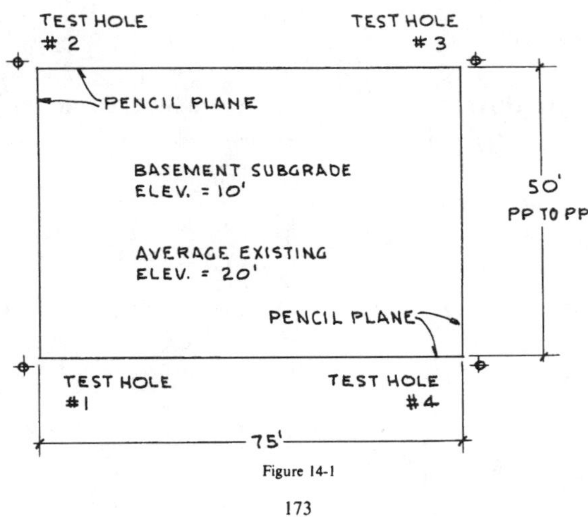


**Figure 10-10. Plan view of basement excavation.**
*(Illustrated by James Edward Atcheson)*

**Test Hole Data:**

| Test Hole | Depth to Top of Rock | Thickness of Rock |
|---|---|---|
| 1 | 4 feet | 3 feet |
| 2 | 1 foot | 4 feet |
| 3 | 5 feet | 3 feet |
| 4 | 8 feet | 2 feet |

**Solution:** Test hole data indicate an excavation with a three-dimensional view as diagrammed in Figure 10-11.

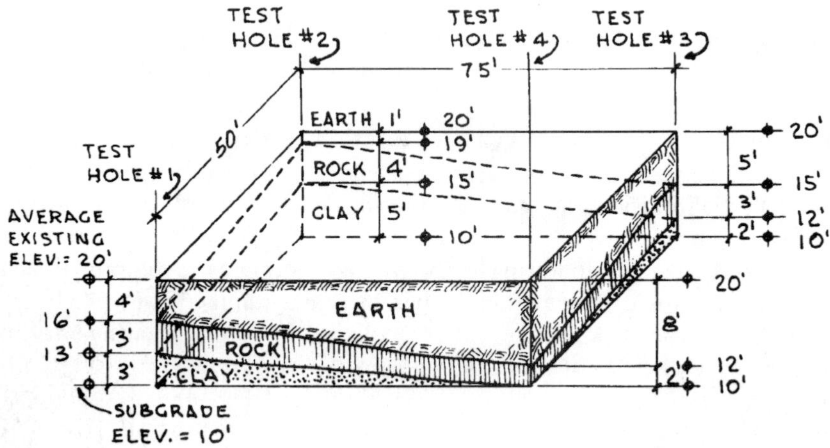


**Figure 10-11. Oblique view of basement excavation.**
*(Illustrated by James Edward Atcheson)*

The volume of any given bed of material is:

$$\text{Volume} = \text{Length x Width x Ave. Bed Thickness} \quad \textbf{(Equation 10.4)}$$

The volume of clay is:

$$\text{Volume (Clay)} = 75' \text{ x } 50' \text{ x } \left( \frac{3' + 5' + 2' + 0'}{4} \right) \text{ (Eq. 10.4; Figure 10-11)}$$

$= 3750$ Sq. Ft. x 2.5'
$= 9375$ Cu. Ft. ÷ 27 Cu. Ft./C.Y.
$= 347$ cubic yards

The volume of rock is:

$$\text{Volume (Rock)} = 3750 \text{ Sq. Ft. x } \left( \frac{3' + 4' + 3' + 2'}{4} \right)$$

$= 3750$ Sq. Ft. x 3'
$= 11{,}250$ Cu. Ft. ÷ 27
$= 417$ cubic yards

The volume of earth is:

$$\text{Volume (Earth)} = 3750 \text{ Sq. Ft. x} \left( \frac{4' + 1' + 5' + 8'}{4} \right)$$
$$= 3750 \text{ Sq. Ft. x } 4.5'$$
$$= 16{,}875 \text{ Cu. Ft.} \div 27$$
$$= 625 \text{ cubic yards}$$

## ESTIMATING RIPPER PRODUCTION

Most ripping is done in first gear at approximately 1-1/2 mph; however, the speed is so variable that it is usually not a reliable factor when estimating ripper production.

There are three basic methods that can be used to determine ripper production, and all three methods require that measurements be made at the job site. The most accurate production estimating method is to record the time spent ripping (including turn-around, or back-up time), remove and weigh the ripped material, taking care to remove ripped material, only. The ripper production rate can be estimated by:

$$\text{Production (Tons/Hr.)} = \frac{\text{Weight of Material (Tons)}}{\text{Ripping Time (Hr.)}} \qquad \textbf{(Equation 10.5)}$$

If the density of the ripped material is known, the following equation can be used to express ripper production rates in terms of bank cubic yards per hour:

$$\text{Production (BCY/Hr.)} = \frac{\text{Production (Tons/Hr.)}}{\text{Density (Tons/BCY)}} \qquad \textbf{(Equation 10.6)}$$

**Example 10-4:** After 20 minutes of ripping, caliche is removed and weighed at 312 tons. Use data from Table 1-1 to determine the hourly production rate of the ripper.

**Solution:** Based on tonnage, the production rate is:

$$\text{Production (Tons/Hr.)} = \frac{312 \text{ Tons}}{0.33 \text{ Hr.}} \qquad \text{(Eq. 10.5)}$$
$$= 945 \text{ tons/hour}$$

This production rate is based on the weight of material loaded into trucks or scrapers in a loose condition. Referring to Table 1-1, the density of a bank cubic yard of caliche is:

$$\text{Density (Tons/BCY)} = \frac{3800 \text{ lb./BCY}}{2000 \text{ lb./Ton}} \qquad \text{(Table 1-1)}$$

$$= 1.9 \text{ tons/BCY}$$

Expressed in terms of bank cubic yards per hour, the production rate is:

$$\text{Production (BCY/Hr.)} = \frac{945 \text{ Tons/Hr.}}{1.9 \text{ Tons/BCY}} \qquad \text{(Eq. 10.6)}$$

$$= 497 \text{ BCY/hour}$$

Another less accurate production estimating method is to record the time spent ripping (including turn-around, or back-up time), remove the material and use the *average end area method* to determine the volume of ripped material. Using the average end area method, areas of adjacent cross-sections are averaged and multiplied by the perpendicular distance between the cross-sections (Figure 10-12); i.e.,

$$V = \frac{A_1 + A_2}{2} \times L \qquad \textbf{(Equation 10.7)}$$

where: $A_1$ = End area of one end of the solid
$A_2$ = End are at the opposite end of the solid
and, L = Distance between End Areas $A_1$ and $A_2$.

Note: $A_1$ and $A_2$ must be parallel and not equal to zero*

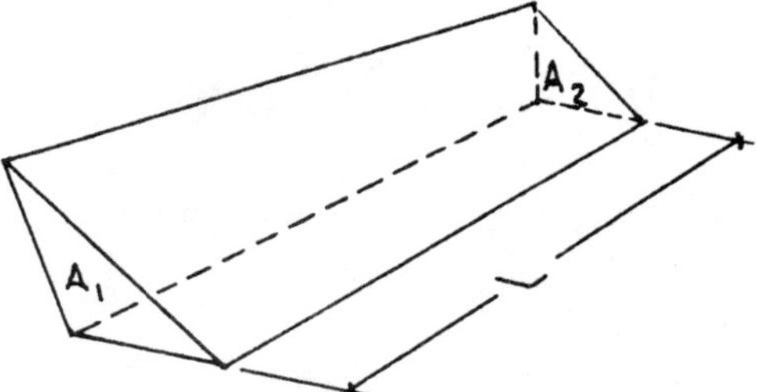


**Figure 10-12.**
**Average end area method.**
(*Illustrated by James Edward Atcheson*)

**Example 10-5:** After 2 minutes of ripping, a tractor cuts a trench 200 feet long. After the ripped material has been removed, it is determined that the end areas of the trench are 6 square feet and 8 square feet. Determine the production rate of the ripper.

**The equations for determining the volume of a pyramid and wedge are given in Appendix D.*

**Solution:** Expressed in hours, the cycle time is:

$$\text{Cycle Time (Hr.)} = \frac{\text{2 Min.}}{\text{60 Min./Hr.}}$$
$$= \text{0.03 hours}$$

The volume of ripped material is:

$$\text{Volume} = \frac{\text{6 Sq. Ft.} + \text{8 Sq. Ft.}}{2} \text{ x } 200' \qquad \text{(Eq. 10.7)}$$
$$= \text{1400 Cu. Ft.} \div \text{27 Cu. Ft./C.Y.}$$
$$= \text{52 BCY}$$

The ripper production rate is:

$$\text{Production (BCY/Hr.)} = \frac{\text{52 BCY}}{\text{0.03 Hr.}}$$
$$= \text{1733 BCY/hour}$$

The quickest, but least accurate ripper production estimating method is to record the ripping time, distance and width, and the depth of tooth penetration. The following equation can then be used to estimate the ripper production rate:

$$\text{Production (BCY/Hr.)} = \frac{1.89 \text{ x D x W x L x E}}{\text{T}} \qquad \textbf{(Equation 10.8)}$$

where: D = Average depth of tooth penetration (ft.)
W = Average width of loosened material (ft.)
L = Length of ripping distance (ft.)
E = Job efficiency
and, T = Ripper cycle time, including backing, turns, etc. (min.).

Note: For those who thoroughly study equations and endeavor to understand their development, we will examine where the number 1.89 in Equation 10.8 came from. The original equation would actually be:

$$\text{Production (BCY/Hr.)} = \frac{\text{(D x W x L)} \div \text{27 Cu. Ft./C.Y.}}{\text{T (Min.)} \div \text{60 Min./Hr.}} \text{ x E}$$

$$= \frac{2.22 \times D \times W \times L \times E}{T}$$

However, experience has shown that results from this estimating method are about 10 to 20% greater than those obtained from the more accurate method of cross-sectioning. By reducing the results by an average of 15%, we obtain fairly accurate production rates as follows:

$$\text{Production (BCY/Hr.)} = 0.85 \times \frac{2.22 \times D \times W \times L \times E}{T}$$

$$= \frac{1.89 \times D \times W \times L \times E}{T} \qquad \text{(Eq. 10.8)}$$

**Example 10-6:** Determine a ripper's production rate under the following conditions:

D = 2 feet
W = 3 feet
L = 300 feet
T = 3.4 minutes
E = 0.83

**Solution:** The probable production rate is:

$$\text{Production (BCY/Hr.)} = \frac{1.89 \times 2 \times 3 \times 300 \times 0.83}{3.4} \qquad \text{(Eq. 10.8)}$$

$$= 830 \text{ BCY/hour}$$

**Example 10-7:** Assuming the ripper tractor given in the previous example costs $180.00/hour, including operator's wages, determine the unit production cost.

**Solution:** The unit production cost is:

$$\text{Unit Production Cost} = \frac{\$180.00/\text{Hr.}}{830 \text{ BCY/Hr.}} \qquad \text{(Eq. 1.18)}$$

$$= \$0.22/\text{BCY}$$

## RIPPER PRODUCTION GRAPHS

The production estimating methods mentioned previously must be conducted at the job site. Prior to bidding a job, you must rely on previous experience, or you can use a production graph as that shown in Figure 10-13. The data given in production graphs are based on the following assumptions:

1) The tractor rips full-time, without dozing.
2) The machine is a power-shift tractor with a single-shank ripper.
3) Efficiency is 100% (60-minute hour).
4) In igneous rock with a seismic velocity of 8000 fps and higher for the D11N, or 6000 fps and higher for the D10N or D9N, reduce the production rate by 25%.
5) The upper production limits of the chart reflect ripping under ideal conditions, only. If conditions such as thick lamination, vertical lamination, or any factor which would adversely affect production are present, the lower production limit should be used.

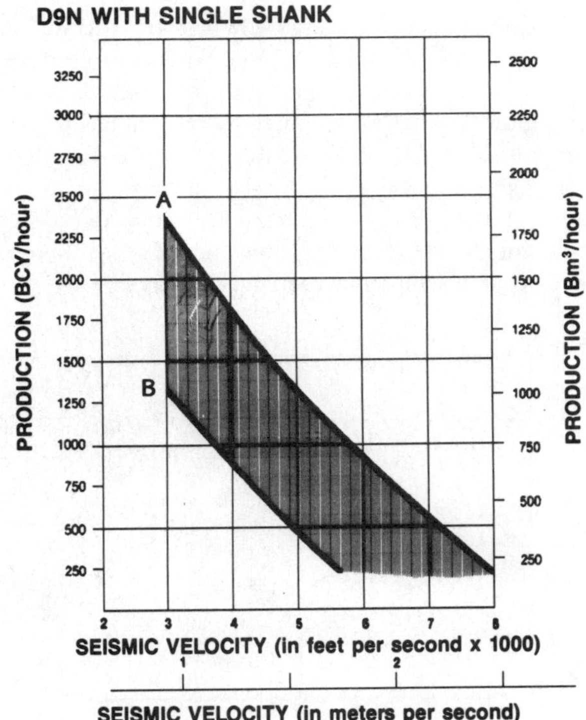


**Figure 10-13. D9N ripper production rates.**
(*Courtesy of Caterpillar Inc.*)

**Example 10-8:** Assuming a seismic velocity of 5000 feet per second and extreme ripping conditions, use Figure 10-13 to determine the ripping production rate of a Caterpillar D9N tractor equipped with a single-shank ripper while working a 50-minute hour.

**Solution:** The probable ripper production rate is:

Production (BCY/Hr.) = 500 BCY/Hr. x 0.83 (Figure 10-13)
= 415 BCY/hour

Production rates can often be increased by ripping down slope. However, in stratified material, the orientation of bedding planes often dictate the direction of ripping, since the shank tip must be placed beneath the laminations for the most efficient production. On level ground where bedding planes are perpendicular to the surface, rip at right angles to the direction of the beds, and if material cannot be effectively removed by ripping in one direction, *cross-ripping* at right angles to the original ripping direction will often improve production. Also, setting off small blast charges to loosen the rock prior to ripping (*pre-blasting*) will increase production rates.

## TYPES OF ROCK

The rippability of rock is dependent upon many factors; however, the primary factor is the type of rock. Therefore, we need to have a basic understanding of the types of rock listed in Figure 10-7. There are three fundamental types of rock, including igneous, sedimentary and metamorphic rock (Figure 10-14). *Igneous rock* is formed by the cooling of molten material, and because of its method of formation, it is basically homogeneous and has few planes of weakness; therefore, it is usually the most difficult type of rock to rip. *Sedimentary rock* is formed from deposition of material from water, ice or air. Due to stratification, sedimentary rock is relatively easy to rip. *Metamorphic rock* is formed from igneous or sedimentary rock altered by heat and pressure; therefore, its rippability lies between that of sedimentary and igneous rock.

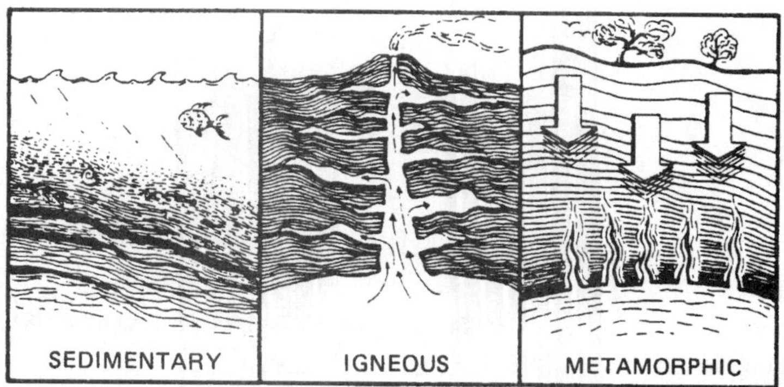


**Figure 10-14. Classification of rocks.** (*Courtesy of Barber-Greene*)

## IGNEOUS ROCKS

Igneous rock is formed from the cooling of molten material, and there are two types of igneous rock, including extrusive and intrusive rock (Figure 10-15). *Extrusive* igneous rock is molten rock that cools at the earth's surface (lava) (Photo 10-4). *Intrusive* igneous rock is molten rock that cools deep within the earth (*plutonic rock*), and is eventually exposed by erosion and/or movement of the earth's surface (Photo 10-5).

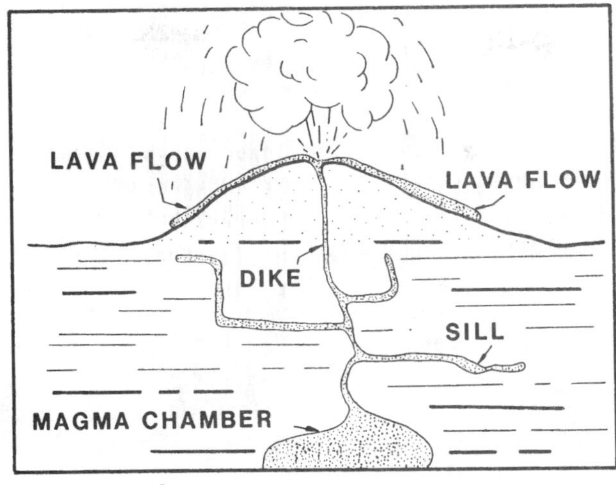


**Figure 10-15. Formation of volcanic and plutonic rocks.**
(*Courtesy of John A. Ciciarelli*)

**Photo 10-4.**
**Volcano near Roswell, New Mexico.**
(*Photo by Dan Atcheson*)

**Photo 10-5.**
**Plutonic rock now at the earth's surface.**
(*Photo by Dan Atcheson*)

Igneous rocks are classified according to composition and texture. *Composition* is defined by the mineral constituents in the rock, and *texture* is defined by the size of the mineral grains. As a general rule, the faster an igneous rock cools, the finer will be the grain. Thus, extrusive igneous rocks are *fine-grained*, and intrusive igneous rocks are *coarse-grained*. Coarse-grained igneous rock such as granite is easier to rip than fine-grained igneous rock such as basalt (Figure 10-7). Some igneous rock cannot be ripped unless it is weakened by weathering, and some igneous rock does not even respond to light blasting unless it has a closely-jointed structure.

*Granite* is a coarse-grained plutonic rock composed primarily of orthoclase feldspar, quartz, and small amounts of dark minerals such as hornblende, magnetite, biotite and muscovite. The primary color of granite may be gray, white, pink or yellow-brown. Granite is the most common coarse-grained igneous rock in the world.

*Rhyolite* is the fine-grained equivalent of granite. Rhyolite is found abundantly in Yellowstone National Park, and in the San Francisco district of Arizona.

*Gabbro* is a coarse-grained igneous rock composed primarily of feldspar, hornblende and other dark minerals. Except for its darker color, gabbro has properties similar to granite. It is found predominantly in the Adirondack Mountains of New York.

*Basalt* is the volcanic equivalent of gabbro. It is dark and contains gas bubbles. Basalt is the most abundant volcanic rock, and it is found throughout the Northwestern United States. Slightly coarser sheets of partially weathered and altered basalt in known as "*trap rock*."

*Diorite* is a coarse-grained igneous rock that is intermediate in composition between granite and basalt. It has a "salt-and-pepper" appearance and is found predominantly in Vermont and New York.

*Pumice* is a light-colored volcanic equivalent of granite. It contains gas bubbles and is very lightweight. *Scoria* is similar to pumice in texture, but basaltic in composition. Like pumice, scoria resembles a sponge.

*Obsidian* is a glassy volcanic rock with no crystalline texture. It is normally black, or dark brown and is similar to granite or rhyolite in chemical composition. Dull, weathered and partially altered obsidian is known as "*pitchstone*."

## SEDIMENTARY ROCKS

Sedimentary rocks are the most abundantly exposed rocks on the earth's surface. Most sedimentary rocks are the products of the disintegration and decomposition of other rocks (igneous, sedimentary, or metamorphic), and are deposited in layers by the action of water, ice and wind.

*Textures* of sedimentary rocks can be classified into two groups, including clastic and non-clastic. *Clastic* (fragmental) sedimentary rocks are made up of fragments of other rocks. Examples of clastic sediments are gravel, sand and silt.

*Non-clastic* (chemical) sedimentary rocks contain calcium and magnesium, and are formed by material precipitated from solution in water, and by evaporation. Chemical sedimentary rocks often contain shell fragments of marine animals. Examples of chemical sediments are limestone, dolomite, gypsum and salt (halite).

Most sediments are moved from a source area and deposited in a sedimentary, or *depositional basin* (Figure 10-16). As the sediments enter the basin, the largest particles are the first to precipitate out of the water, followed by smaller particles farther out in the basin. The non-clastic particles are the last to be precipitated and deposited farthest from the source area. Given time, loose sediments are consolidated into solid rock, primarily through compaction and cementation. Due to the method of deposition, sedimentary rocks are formed in basically horizontal layers of various thicknesses. These layers are referred to as beds, or *strata*.

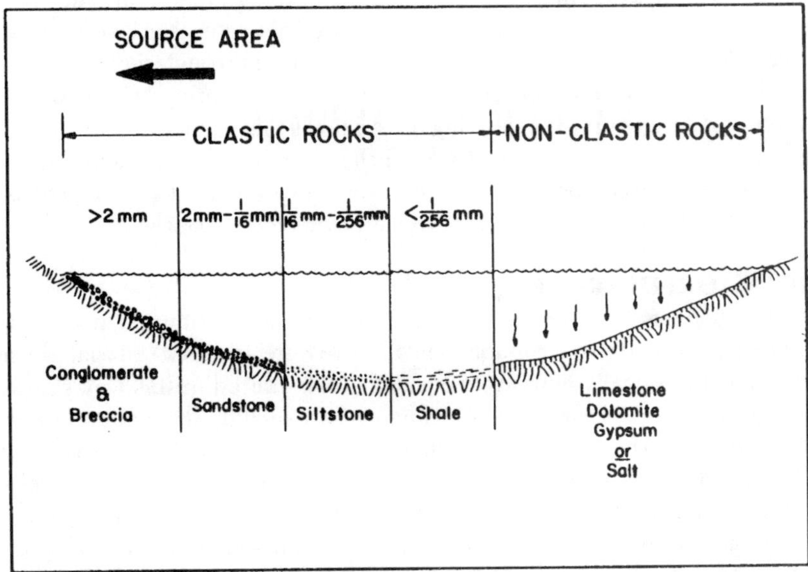


**Figure 10-16. Details of a depositional basin.** (*Courtesy of John A. Ciciarelli*)

Conglomerates and breccias have grain sizes greater than 2 mm and are deposited closest to the source area. A *conglomerate* is composed of rounded water-worn fragments, and a *breccia* is composed of broken, angular fragments. Both types of deposits are made up primarily of quartz cemented by a mass of finer material filling spaces between the larger fragments. *Glacial till* is a cemented mixture of clay and boulders deposited by the movement and melting of glaciers.

*Sandstone* is composed primarily of cemented quartz sand. Sandstone has grain sizes ranging from 1/16 to 1 mm and is deposited farther from the source area. Depending on the grain size, sandstone is classified as, fine-, medium-, or coarse-grained, and the grains can be seen with the unaided eye. Depending primarily on the color of the cement, sandstone may be white, gray, yellowish, or dark red.

The next deposit to be precipitated is *siltstone* which has a grain size ranging from 1/256 mm to 1/16 mm. Like sandstone, siltstone feels gritty; however, the individual grains are too small to be seen with the unaided eye.

*Shale* is composed principally of clay particles, often intermixed with a small amount of sand. Shale has a grain size smaller than 1/256 mm and represents the finest fraction of the clastic sediments. Shale is thinly bedded and fractures into thin chips. Shale feels smooth and the grains are too small to be seen, even with a microscope. Shale and slate are often indistinguishable. To determine which is which, breathe on the rock and smell of it. If it smells like mud, it is shale; if not, it is slate.

*Limestone*, dolomite, gypsum, salt and chert are non-clastic sedimentary rocks and are chemically precipitated out of water and deposited farthest from the source area. Limestone often contains fossilized shell animals.

*Caliche* is a sedimentary rock with a unique method of deposition. During rainy seasons in an arid or semiarid climate, rain water sinks into the earth due to gravity and raises the level of the water table. During dry seasons, the level of the water table is lowered as ground water rises to the surface through capillary action to replace water that has evaporated. As the water rises, it brings with it substances in solution derived from underlying material. As this rising water evaporates at the surface, a material referred to as caliche is deposited. Although caliche can be made of many compounds, calcium carbonate is the most abundant constituent. Most caliches are excellent for use as paving base material.

## METAMORPHIC ROCKS

If an igneous or sedimentary rock is formed, and later buried at great depth where it is exposed to tremendous heat and pressure, the rock will change (metamorphose) physically and/or chemically. Rock altered in this way is called a *metamorphic rock.* Examples of *metamorphism* include shale altering to *slate*, limestone to *marble*, sandstone to *quartzite*, granite to *gneiss*, slate to *phyllite* or *schist*, and bituminous coal to *anthracite coal.* Massive metamorphic rocks such as gneiss, marble and quartzite are not rippable unless they are fractured.

Metamorphism can also be produced by an intrusion of hot magma, accompanied by mineralized gases that soak through the enclosing rock. This type of metamorphism is referred to as *contact metamorphism.* In addition to the metamorphism of the parent rock near the magma intrusion, the cooling gases deposit minerals (sometimes precious) within voids and fissures in the surrounding rock.

Metamorphic rocks are classified according to texture, grain alignment and mineral composition. During metamorphism, flat minerals are often reoriented to a parallel alignment, referred to as a *foliated* texture. This parallel structure allows foliated metamorphic rock to split easily in the direction of the foliation. In other cases, pore spaces are diminished, and the texture is referred to as *non-foliated.* Non-foliated metamorphic rock does not split easily. Table 10-1 can be used to categorize various types of metamorphic rocks.

## ROCK STRUCTURES

*Structural geology* deals with the orientation and configuration of rocks in the earth's crust. Your knowledge of rock structures will help you to interpret geologic maps and predict subsurface features. Standard symbols used on geologic maps are shown in Figure 10-17.

## FOLDING

Under compression forces, rocks frequently buckle and fold (Photo 10-6). In a series of folds, the portion that is arched upward is called an *anticline*, and the portion arched downward is called a *syncline* (Figure 10-18).

**Table 10-1: Metamorphic Rock Identification Chart**

| Texture | Composition | Rock Name |
|---|---|---|
| **Foliated** | | |
| Fine-grained | Clay, hornblende, serpentine, quartz, talc | Slate |
| Medium-grained | Clay, hornblende, serpentine, quartz, mica, garnet | Schist |
| Coarse-grained | Feldspar, quartz, hornblende, mica | Gneiss |
| **Non-Foliated** | | |
| Fine-grained | 90-98% carbon | Anthracite coal |
| Medium and coarse-grained | Calcite with some quartz | Marble |
| | Almost all quartz | Quartzite |

(*Courtesy of John A. Ciciarelli*)

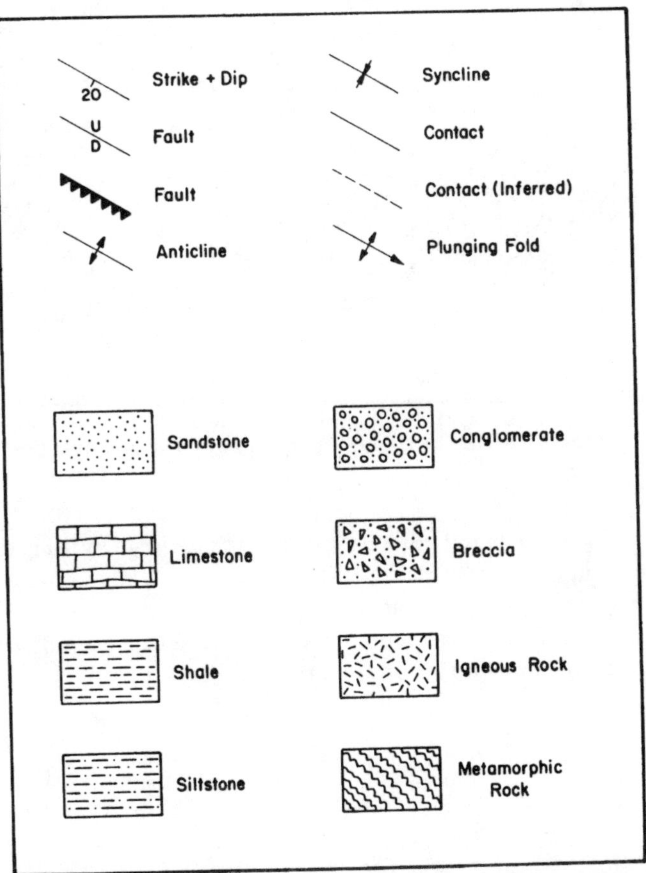


**Figure 10-17. Geologic map symbols.** (*Courtesy of John A. Ciciarelli*)

**Photo 10-6. Folded rock strata.** (*Photo by Dan Atcheson*)

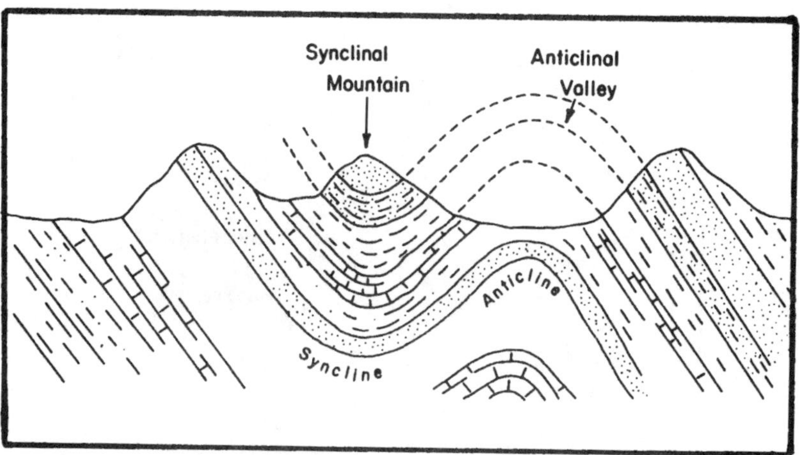


**Figure 10-18. Cross-section of an anticline and syncline.**
(*Courtesy of John A. Ciciarelli*)

The *attitude* of folds and other rock strata are described by using terms known as strike and dip. The *strike* of a bed is the direction of a line formed by the intersection of the bed surface and an imaginary horizontal reference plane (Figure 10-19). The *dip* is the amount of tilt of the inclined bed, measured in degrees perpendicular to the strike direction (Figures 10-19 and 10-20).

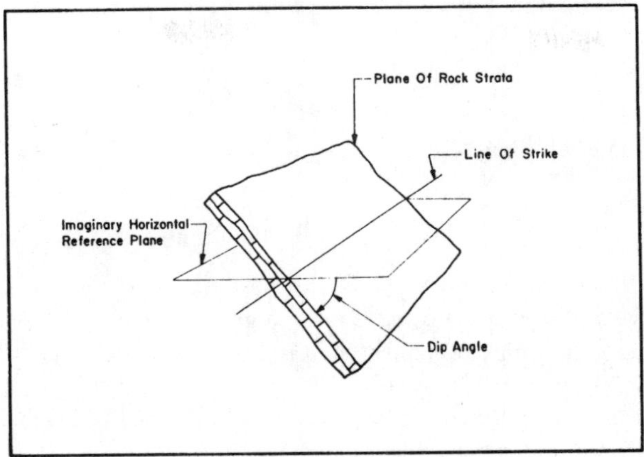


**Figure 10-19. Line of strike.** (*Courtesy of John A. Ciciarelli*)

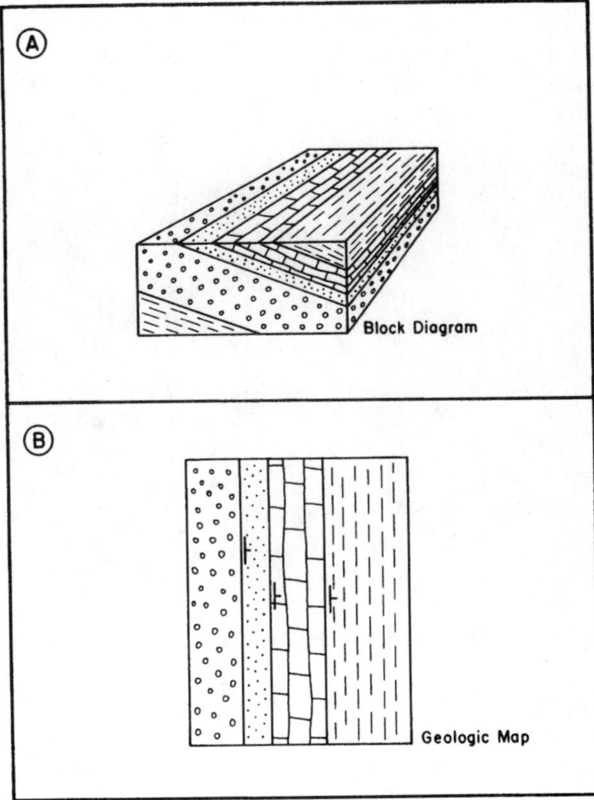


**Figure 10-20. Tilted sedimentary strata.** (*Courtesy of John A. Ciciarelli*)

The actual thickness of a tilted bed can be determined by:

$$T = E \times \sin \emptyset$$ **(Equation 10.9)**

where: T = Actual thickness of bed
E = Thickness of exposed bed
and, Ø = Dip angle.

**Example 10-9:** Determine the actual thickness of a bed of limestone if its exposed thickness is 30 feet and the dip angle is 45 degrees.
**Solution:** The actual thickness of the bed is:

$$\begin{aligned} T &= 30' \times \sin 45° \\ &= 30' \times 0.707 \\ &= 21.2 \text{ feet} \end{aligned}$$ (Eq. 10.9)

## FAULTING

A *fault* is a break in the earth's crust where there has been lateral displacement, and the plane that separates the moving rock is called a *fault plane* (Photo 10-7). The block of rock that the fault plane dips toward is called the *hanging wall*, while the other block is called the *foot wall* (Figure 10-21).

**Photo 10-7. Series of faults.** (*Photo by Dan Atcheson*)

Faults are classified on the basis of the type of rock movement on either side of the fault plane. Tension in the earth's crust can cause the hanging wall to move down relative to the foot wall, creating a *normal fault* (Figure 10-22). Compression can cause the hanging wall to move up relative to the foot wall, creating a *reverse fault*, or *thrust fault* (Figure 10-23).

It is possible for the rock on both sides of a fault line to be eroded to a common level, and your knowledge regarding the location of a fault can be of vital importance in bidding a project, since rock being ripped on one side of the fault line can be different from the rock on the opposite side. However, rock adjacent to a fault line is often in a weakened condition, since the forces involved in faulting tend to fracture the surrounding rock.

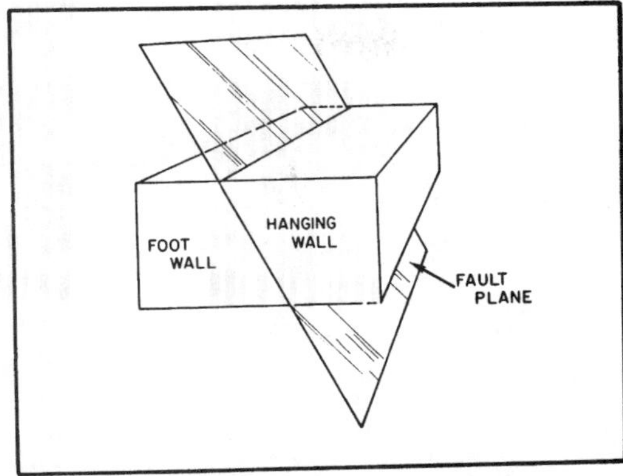


**Figure 10-21.**
**Hanging wall, foot wall and fault plane.**
(*Courtesy of John A. Ciciarelli*)

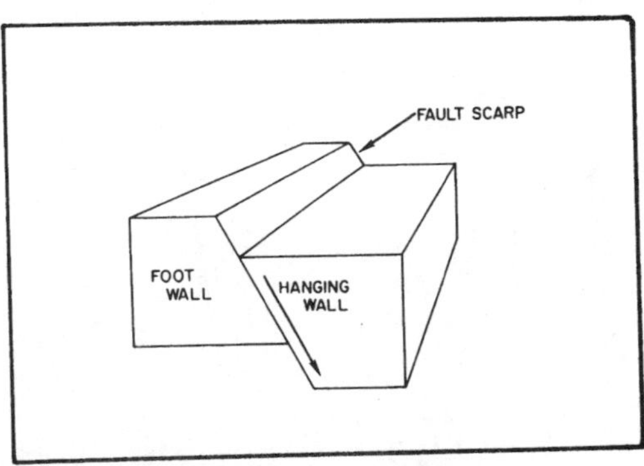


**Figure 10-22. Normal fault.**
(*Courtesy of John A. Ciciarelli*)

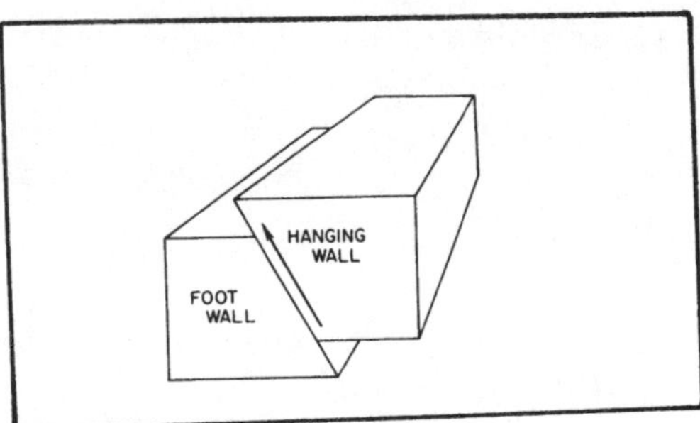


**Figure 10-23. Reverse fault.**
(*Courtesy of John A. Ciciarelli*)

*Chapter 11*

# EQUIPMENT COSTS

We have learned in previous chapters to determine hourly production rates of various types of equipment under a variety of job conditions. In this chapter, we will learn how to determine the hourly equipment costs. By knowing the hourly production and cost rates, we can determine the unit production cost for handling earth by using the following equation:

$$\text{Unit Production Cost} = \frac{\text{Hourly Production Cost}}{\text{Hourly Production Rate}} \qquad \textbf{(Equation 11.1)}$$

Equipment costs are separated into two major categories, including owning costs and operating costs. *Owning costs* are the costs incurred whether or not the machine is in operation, and *operating costs* are costs incurred only when the machine is actually in operation.

## OWNING COSTS

Owning costs are fixed and are directly related to the length of time the equipment is owned; therefore, owning costs are often referred to a *fixed costs*. Owning costs include equipment depreciation, interest, insurance, taxes, storage and miscellaneous costs. Owning costs are usually determined on an hourly basis as follows:

$$\text{Hourly Owning Costs} = \frac{\text{Total Owning Costs}}{\text{Service Life (Hr.)}} \qquad \textbf{(Equation 11.2)}$$

The service life is normally assumed to be 2000 hours per year of ownership. This is based on working 40 hours per week at 50 weeks per year. Thus, a 5-year service life is equivalent to: 5 Yr. x 2000 Hr./Yr. = 10,000 hours. If the equipment is idle for some of these hours, the owning costs can be accounted for in the general operating overhead, and when the equipment is in use, the hourly owning costs can be charged to the current project. In either case, the owning costs must be recaptured if the construction company is to remain solvent.

The actual machine operating hours is dependent upon the weather, the availability of work, the type of equipment required and the down time. Time lost due to bad weather is unique to the local climate; however, the Bureau of Public Roads claims a nation-wide average loss of 20% of the total available work days. Some equipment such as draglines and track-mounted machines are less affected by weather, than are wheel-mounted equipment.

No contractor is busy 100% of the time, and even when he has work, he might have to rent specialized equipment to work the project efficiently, leaving some of his personal equipment lying idle in the yard. As an average, a successful and capable contractor bidding during good economic conditions will work all of his equipment only 75 to 80% of the available good-weather time.

*Down time* is time lost during good weather, due to equipment failure, strikes, or shortages of men and materials. Down time will vary with equipment age and upkeep, and quality of management. The Bureau of Public Roads claims down-time losses ranging from 20 to 65% of the available good-weather working time.

If we apply what is known as the *Rule of the Three Twenties* to the 2000 hours of work time ideally available each year, we will have an average annual productive work time of 1024 hours based on the following tabulation:

| Cause Of Loss | Time Lost (Hr.) | Time Available (Hr.) |
|---|---|---|
| Bad Weather | 0.20 x 2000 = 400 | 2000 – 400 = 1600 |
| No Work | 0.20 x 1600 = 320 | 1600 – 320 = 1280 |
| Down Time | 0.20 x 1280 = 256 | 1280 – 256 = 1024 |

## DEPRECIATION

*Depreciation* is the decrease in equipment value due to wear and obsolescence through use. Depreciation is not just some figure you place on an income tax return to reduce your tax liability; it is a physical reality, and it must be included in the hourly machine cost charged to a client. The funds received should then be placed into an account entitled "*depreciation reserve*." Whether this account is maintained in a separate bank account, or on a ledger sheet, the funds must be available for the purchase of new equipment when the old one wears out. Otherwise, there might be many a sleepless night, due to worry about where the funds for even a down payment will come from, which is normally around 25% of the total purchase price.

We will discuss five basic depreciation methods, but first, we should understand the definitions of delivered cost, initial cost, salvage value and equipment service life. The *delivered cost* is the total cost of the equipment, including options, tires, sales tax, transportation, unloading, assembly, initial servicing, etc. (Figure 11-1).

The *initial cost* (also referred to as initial net cost, net first cost, base value or basis) is the delivered cost less the cost of tires (Figure 11-1). Tires are related to the equipment, as a ribbon is related to the typewriter. Since tires are normally replaced several times over the lifetime of the equipment, they are considered to be a *current period expense*, and therefore, are not depreciable unless it can be proved that they are replaced at least once every year. Tire costs will be accounted for in the operating costs of the equipment.

The equipment *salvage value* (residual value) is a component in two of the five depreciation methods used. The salvage value is the value of the equipment at the end of the depreciation period (Figure 11-1). It can be determined by using historical data, by referring to equipment auction prices, or by consulting local equipment dealerships. The actual salvage value of a machine is dependent upon many variables, including the age, physical condition and type of equipment, the demand for the equipment, its productivity and obsolescence relative to newer models, and the current heavy construction economic climate. Salvage value can vary from scrap value to 60% of the original cost; however, the average salvage value ranges from 5 to 20% of the original cost. If it can be shown that the salvage value is 10% or less of the initial cost, it can be ignored when calculating depreciation. If it exceeds 10%, the salvage value must be deducted from the initial cost when determining depreciation. The amount that can be depreciated is the difference between the initial cost and the salvage value, and is referred to as the *depreciable amount* (Figure 11-1).

The *book value* (B.V.) of a machine at any given time is the difference between the initial cost and the amount of depreciation taken up to that point in time (Figure 11-1); i.e.,

$$\text{Book Value} = \text{Initial Cost} - \text{Depreciation} \qquad \textbf{(Equation 11.3)}$$

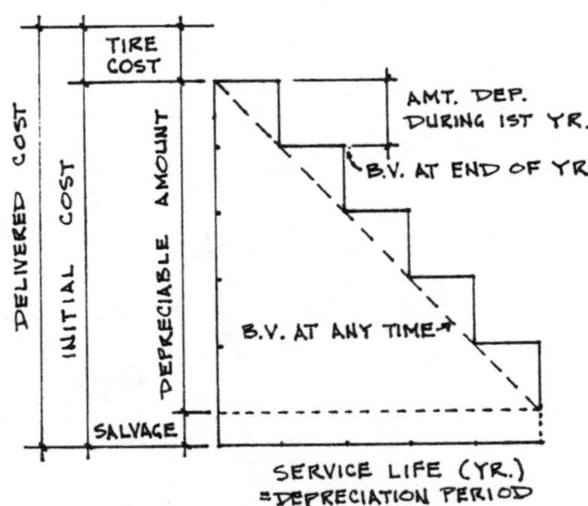


**Figure 11-1. Factors used in depreciation.**

Equipment *service life* (depreciation period) is another component in four of the five depreciation methods used (Figure 11-1). The IRS has established guidelines ("Bulletin F") for determining realistic equipment life. In general, it is 5 years (or 10,000 hours) for general construction equipment; however, the IRS will allow shorter periods if it can be shown that unique circumstances, such as operating in continuous sub-zero temperature zones, dictate a shorter equipment life. Some equipment manufacturers publish guides for determining useful equipment life based on the work application and operating conditions (Table 11-1).

## Table 11-1: Guide for Selecting Ownership Period Based on Application and Operating Conditions

| | ZONE A Moderate | ZONE B Average | ZONE C Severe |
|---|---|---|---|
| **TRACK-TYPE TRACTORS** | Pulling scrapers, most agricultural drawbar, stockpile, coalpile. No impact. Intermittent full throttle operation. | Production dozing in clays, sands, gravels. Pushloading scrapers, borrow pit ripping, most landclearing and skidding applications. Medium impact conditions. Production landfill work. | Heavy rock ripping. Tandem ripping. Pushloading and dozing in hard rock. Work on rock surfaces. Continuous high impact conditions. |
| D3-D7 | 12,000 Hr | 10,000 Hr | 8,000 Hr |
| D8-D11 | 22,000 Hr | 18,000 Hr | 15,000 Hr |
| **MOTOR GRADERS** | Light road maintenance. Finishing. Plant and road mix work. Light snowplowing. Large amounts of traveling. | Haul road maintenance. Road construction, ditching. Loose fill spreading. Landforming, landleveling. Summer road maintenance with medium to heavy winter snow removal. Elevating grader use. | Maintenance of hard packed roads with embedded rock. Heavy fill spreading. Ripping-scarifying of asphalt or concrete. Continuous high load factor. High impact. |
| | 20,000 Hr | 15,000 Hr | 12,000 Hr |
| **EXCAVATORS** | Utility construction, low density material, rehandling and scrap handling applications. | Continuous digging in sandy clay / sandy gravel, site development and lumber yard applications. | Continuous digging in rock / natural bed clay, high impact, using hammer, working in forests or quarries. |
| 206B FT, 211B LC<br>212B, 213B LC,<br>214B / 214B FT, 224B<br>E70B, E110B, E120B, E140 | 8,000 Hr | 6,000 Hr | 4,000 Hr |
| **EXCAVATORS** | Shallow depth utility construction where excavator sets pipe and digs only 3 or 4 hours / shift. Free flowing, low density material and little or no impact. Most scrap handling arrangements. | Mass excavation or trenching where machine digs all the time in natural bed clay soils. Some traveling and steady, full throttle operation. Most log loading applications. | Continuous trenching or truck loading in rock or shot rock soils. Large amount of travel over rough ground. Machine continuously working on rock floor with constant high load factor and high impact. |
| 231D-245D<br>320, E240C, 325<br>330, E450, E650 | 12,000 Hr | 10,000 Hr | 8,000 Hr |
| **FRONT SHOVELS** | Continuous loading in loose banks or stockpile. Good underfoot conditions. (Might be considered similar to "normal" wheel loader conditions.) | Continuous loading in well-shot rock or fairly tight bank. Good underfoot conditions; dry floor, little impact or sliding on undercarriage. | Continuous loading in poorly-shot rock, virgin or lightly-blasted tight banks, e.g., shales, cemented gravels, caliches, etc. Adverse underfoot conditions; rough floors; high impact sliding on undercarriage. |
| 235D, 245D<br>E450, E650 | 18,000 Hr | 15,000 Hr | 10,000 Hr |
| **WHEEL TRACTOR-SCRAPERS** | Level or favorable hauls on good haul roads. No impact. Easy-loading materials. | Varying loading and haul road conditions. Long and short hauls. Adverse and favorable grades. Some impact. Typical road-building use on a variety of jobs. | High impact condition, such as loading ripped rock. Overloading. Continuous high total resistance conditions. Rough haul roads. |
| 613C, 615C | 12,000 Hr | 10,000 Hr | 8,000 Hr |
| E and E Series II | 22,000 Hr | 17,000 Hr | 12,000 Hr |
| **ARTICULATED TRUCKS** | Earthmoving and stockpile use with well matched loading equipment. Short to medium hauls on well-maintained level haul roads. Free flowing material. Few impact loads. | Varying load and haul road conditions. High rolling resistance and poor traction during part of the job. Some adverse grades. Some impact loads. Typical use in road-building, dam construction, open-pit mining, etc. | Continuous use on very poorly maintained haul roads, high rolling resistance and poor traction. Frequent adverse grades and high impact loads. Poorly-matched loading equipment with continuous over-loading. |
| | 15,000 Hr | 10,000 Hr | 8,000 Hr |
| **WHEEL TRACTORS & COMPACTORS** | Light utility work. Stockpile work. Pulling compactors. Dozing loose fill. No impact. | Production dozing, pushloading in clays, sands, silts, loose gravels. Shovel clean-up. | Production dozing in rock. Pushloading in rocky, bouldery borrow pits. High impact conditions. Landfill compactor work. |
| | 15,000 Hr | 12,000 Hr | 8,000 Hr |
| **WHEEL LOADERS** | Intermittent truck loading from stockpile, hopper charging on firm, smooth surfaces. Free flowing, low density materials. Utility work in governmental and industrial applications. Light snowplowing. Load and carry on good surface for short distances with no grades. | Continuous truck loading from stockpile. Low to medium density materials in properly sized bucket. Hopper charging in low to medium rolling resistance. Loading from bank in good digging. Load and carry on poor surfaces and slight adverse grades. | Loading shot rock (large loaders). Handling high density materials with counterweighted machine. Steady loading from very tight banks. Continuous work on rough or very soft surfaces. Load and carry in hard digging; travel longer distances on poor surfaces with adverse grades. |
| 910E-966F | 12,000 Hr | 10,000 Hr | 8,000 Hr |
| 960F-992C | 15,000 Hr | 12,000 Hr | 10,000 Hr |
| 994 | 60,000 Hr | 50,000 Hr | 40,000 Hr |

*(Courtesy of Caterpillar Inc.)*

Because of its sturdy construction, a diesel engine will have a longer service life than a gasoline engine, all other conditions being equal. Although the initial cost of a diesel engine is more than that of a gasoline engine, the diesel engine is more economical in the long run, due to reduced fuel consumption and lower fuel prices. Also, routine maintenance costs are lower, since diesel engines have no spark plugs, carburetor, or electrical ignition components to service.

**Straight-Line (SL) Depreciation Method**

The *straight-line method* spreads the depreciation uniformly over the depreciation period (Figure 11-2), and depreciation for any given year can be calculated by:

$$D_{Annual} = \frac{\text{Delivered Cost} - \text{Salvage Value} - \text{Tire Cost}}{\text{Service Life (Yrs.)}}$$

**(Equation 11.4)**

**Example 11-1:** Use the straight-line depreciation method to determine the annual depreciation, the book value at the end of each year, and the hourly depreciation cost for a wheel tractor under the following conditions:

Delivered Cost = \$65,000.00
Salvage Value = \$5000.00
Tire Cost = \$15,000.00
Service Life = 5 years (2000 Hr./Yr.)

**Solution:** The depreciation during any given year is:

$$D_{Annual} = \frac{\$65{,}000 - \$5000 - \$15{,}000}{5\text{ Yr.}} \quad \text{(Eq. 11.4)}$$
$$= \$9000/\text{year}$$

The hourly depreciation cost is:

$$\text{Hourly Depreciation Cost} = \frac{\$9000/\text{Yr.}}{2000\text{ Hr./Yr.}} \quad \text{(Eq. 11.2)}$$
$$= \$4.50/\text{hour}$$

The book value of the tractor at the end of the first year is:

$$\text{Book Value (End of Year 1)} = \$50{,}000 - \$9000 \quad \text{(Eq. 11.3)}$$
$$= \$41{,}000$$

The depreciation schedule is tabulated as follows:

| Year | Depreciation | Book Value at End of Year (Excl. Tires) |
|---|---|---|
| 0 | 0 | 50,000 |
| 1 | 9000 | 41,000 |
| 2 | 9000 | 32,000 |
| 3 | 9000 | 23,000 |
| 4 | 9000 | 14,000 |
| 5 | 9000 | 5,000 |

The straight line method automatically ensures that you do not break the IRS cardinal rule of depreciating beyond the salvage value (Figure 11-2).

All of the graphic illustrations of depreciation in this book are shown in a stepwise fashion, where the "riser" indicates the depreciation for an entire year and the "tread" indicates the book value at the end of the year. In reality, the depreciation and book value at any given point in time is prorated throughout the service life, as indicated by a dashed diagonal line in all of the illustrations (Figure 11-1), and this proration must be taken into consideration if the machine is sold or traded at mid-year.

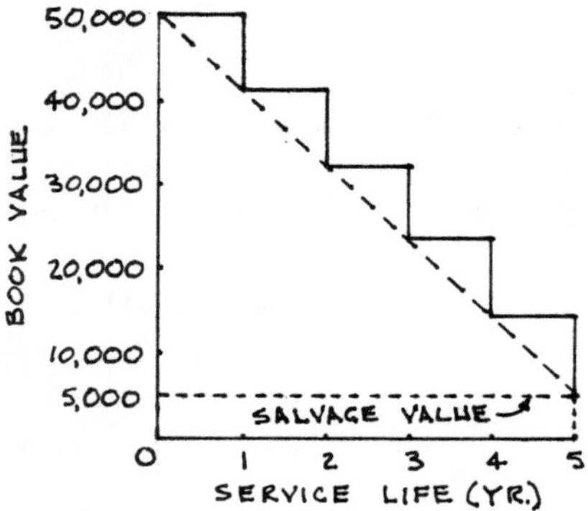


**Figure 11-2.**
**Straight-line depreciation.**

**Example 11-2:** Assuming the tractor given in the previous example is sold at the end of the third month in the second year of the service life, determine the depreciation taken and the book value of the tractor at the time of sale.
**Solution:** The original service life is:

$$\text{Original Service Life} = 5 \text{ Yr.} \times 12 \text{ Mo./Yr.}$$
$$= 60 \text{ months}$$

However, since the machine is sold after 2 years, 3 months of the service life, the ownership period is:

$$\text{Ownership Period} = 2.25 \text{ Yr.} \times 12 \text{ Mo./Yr.}$$
$$= 27 \text{ months}$$

The total depreciation taken up to the time of sale is:

$$\text{Total Depreciation} = \frac{27 \text{ Mo.}}{60 \text{ Mo.}} \times \$50{,}000$$
$$= \$22{,}500$$

Therefore, the book value at the time of sale is:

$$\text{B.V.} = \$50{,}000 - \$22{,}500 \qquad \text{(Eq. 11.3)}$$
$$= \$27{,}500$$

A piece of equipment is often overhauled or modified during its lifetime. A modification is considered a *capital improvement* that increases the base value, and often extends to service life of the equipment. The following examples illustrate how to depreciate equipment that is modified during its service life (Also refer to Figure 11-3).

**Example 11-3:** Assuming a $13,000 improvement is made during the second year, that extends service life to 6 years, use the straight-line depreciation method to determine the annual depreciation and book value at the end of each year for the tractor described in Example 11-1.
**Solution:** The depreciation during each of the first two years is:

$$D_1 = D_2 = \$9000 \qquad \text{(Example 11-1)}$$

Since the $13,000 improvement is make during the second year, the book value at the end of the second year is:

$$\text{B.V. (End of Year 2)} = \$50{,}000 - \$9000 - \$9000 + \$13{,}000$$
$$= \$45{,}000$$

This value is the new depreciable amount that must be depreciated through the end of the sixth year, with an annual depreciation of:

$$D_{Annual} = \frac{\$45{,}000 - \$5000}{4} \qquad \text{(Eq. 11.4)}$$
$$= \$10{,}000/\text{year}$$

The depreciation schedule is tabulated as follows:

| Year | Depreciation | Improvement | Book Value at End of Year (Excl. Tires) |
|---|---|---|---|
| 0 | 0 | 0 | 50,000 |
| 1 | 9,000 | 0 | 41,000 |
| 2 | 9,000 | 13,000 | 45,000 |
| 3 | 10,000 | 0 | 35,000 |
| 4 | 10,000 | 0 | 25,000 |
| 5 | 10,000 | 0 | 15,000 |
| 6 | 10,000 | 0 | 5,000 |

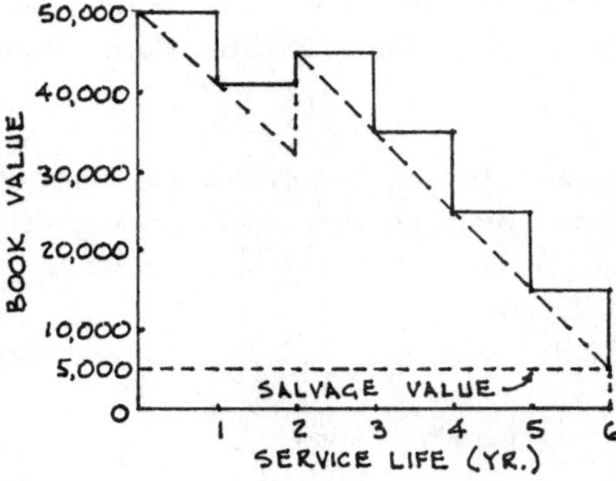


**Figure 11-3. Depreciating a modification.**

**Example 11-4:** Work the previous example, assuming the improvement increases the salvage value to $7000.

**Solution:** The depreciation during each of the first two years is:

$$D_1 = D_2 = \$9000 \qquad \text{(Example 11-3)}$$

The book value at the end of the second year is:

$$\text{B.V. (End of Year 2)} = \$45,000 \qquad \text{(Example 11-3)}$$

The annual depreciation throughout the remainder of the service life is:

$$D_{Annual} = \frac{\$45,000 - \$7000}{4} \qquad \text{(Eq. 11.4)}$$

$$= \$9500/\text{year}$$

The depreciation schedule is tabulated as follows:

| Year | Depreciation | Improvement | Book Value at End of Year (Excl. Tires) |
|---|---|---|---|
| 0 | 0 | 0 | 50,000 |
| 1 | 9000 | 0 | 41,000 |
| 2 | 9000 | 13,000 | 45,000 |
| 3 | 9500 | 0 | 35,500 |
| 4 | 9500 | 0 | 26,000 |
| 5 | 9500 | 0 | 16,500 |
| 6 | 9500 | 0 | 7,000 |

**Sum-Of-Year-Digits (SOYD) Depreciation Method**

The *sum-of-the-year-digits* (SOYD) *method* is referred to as an *accelerated* form of depreciation, since it allows the greatest depreciation to occur during the first year, and reduces the annual depreciation each year until the book value equals the salvage value (Figure 11-4). The depreciation during any given year can be determined by:

$$D_n = \frac{\text{Year Digit}}{\text{Sum Of Years Digit}} \times (\text{Delivered Cost} - \text{Salvage Value} - \text{Tire Costs})$$

**(Equation 11.5)**

where: $D_n$ = Depreciation for the current (n'th) year
Year Digit = Year taken in inverse order (For the first year, use 5; for the second year, use 4, etc.)

and, Sum of Years Digit = Sum of years' digits for the depreciation period (For a 5-year service life, use $1 + 2 + 3 + 4 + 5 = 15$).

**Example 11-5:** Use the sum-of-the-year-digits depreciation method to determine the annual depreciation and book value at the end of each year for the tractor described in Example 11-1.

**Solution:** The depreciation during the first year is:

$$D_1 = \frac{5}{15} \text{ x } (\$65{,}000 - \$5{,}000 - \$15{,}000) \qquad \text{(Eq. 11.5)}$$

$$= \frac{5}{15} \text{ x } \$45{,}000$$

$$= \$15{,}000$$

The annual depreciation during the remaining years is:

$$D_2 = \frac{4}{15} \text{ x } \$45{,}000 = \$12{,}000$$

$$D_3 = \frac{3}{15} \text{ x } \$45{,}000 = \$9{,}000$$

$$D_4 = \frac{2}{15} \text{ x } \$45{,}000 = \$6{,}000$$

$$D_5 = \frac{1}{15} \text{ x } \$45{,}000 = \$3{,}000$$

The depreciation schedule is tabulated as follows:

| Year | Depreciation | Book Value at End of Year (Excl. Tires) |
|---|---|---|
| 0 | 0 | 50,000 |
| 1 | 15,000 | 35,000 |
| 2 | 12,000 | 23,000 |
| 3 | 9,000 | 14,000 |
| 4 | 6,000 | 8,000 |
| 5 | 3,000 | 5,000 |

The SOYD depreciation method also ensures that the book value is equal to the salvage value at the end of the depreciation period (Figure 11-4). Figure 11-5 shows a comparison of the SOYD and straight-line depreciation methods.

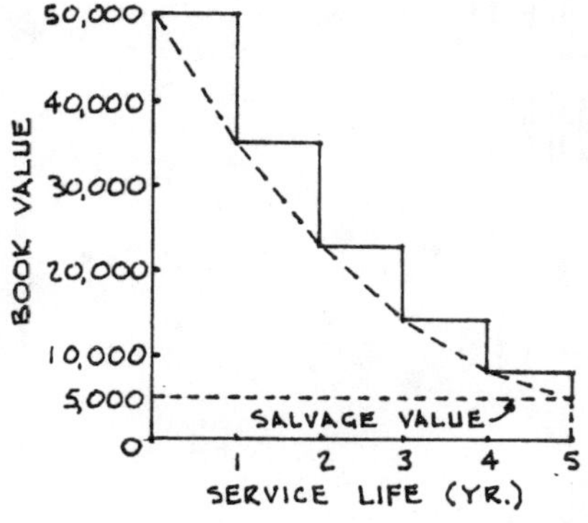


**Figure 11-4.**
**SOYD depreciation method.**

**Figure 11-5. Comparison of the SOYD and straight-line depreciation methods.**

**Declining Balance Method**

The *declining balance* (DB) *method* is also an accelerated depreciation method, since it allows greater depreciation in the early portion of equipment life. When this method is applied to new equipment with a useful life of at least 3 years, the rate of depreciation is about double that of the straight-line rate; therefore, this method is often referred to as the *double-declining balance* (DDB) *method* (Figure 11-6). The depreciation of *new* equipment during any given year can be determined by:

$$D_n = \frac{2}{\text{Life (Yrs.)}} \text{ x B.V. at Start of Year (Excl. Tire Cost)}$$

**(Equation 11.6)**

Notice that salvage value is not a component of Equation 11.6, since this depreciation method automatically leaves a small salvage value; however, take care not to depreciate the equipment beyond the allowable salvage value.

**Example 11-6:** Use the declining balance method to determine the annual depreciation and book value at the end of each year for the tractor described in Example 11-1.

**Solution:** The depreciation during the first year is:

$$D_1 = \frac{2}{5} \text{ x } \$50{,}000 \qquad \text{(Eq. 11.6)}$$

$$= 0.4 \text{ x } \$50{,}000$$
$$= \$20{,}000$$

The annual depreciation during the remaining years is:

$$D_2 = 0.4 \text{ x } (\$50{,}000 - \$20{,}000)$$
$$= 0.4 \text{ x } \$30{,}000$$
$$= \$12{,}000$$

$$D_3 = 0.4 \text{ x } (\$30{,}000 - \$12{,}000)$$
$$= 0.4 \text{ x } \$18{,}000$$
$$= \$7200$$

$$D_4 = 0.4 \text{ x } (\$18{,}000 - \$7200)$$
$$= 0.4 \text{ x } \$10{,}800$$
$$= \$4320$$

$$D_5 = 0.4 \text{ x } (\$10{,}800 - \$4320)$$
$$= 0.4 \text{ x } \$6480$$
$$= \$2592 \text{ (Use 1480)*}$$

The depreciation schedule is tabulated is follows:

| Year | Depreciation | Book Value At End Of Year (Excl. Tires) |
|---|---|---|
| 0 | 0 | 50,000 |
| 1 | 20,000 | 30,000 |
| 2 | 12,000 | 18,000 |
| 3 | 7,200 | 10,800 |
| 4 | 4,320 | 6,480 |
| 5 | 2,592 (Use 1,480)* | 3,888 (Use 5,000)* |

* *Keep in mind that the IRS will not allow you to depreciate below the salvage value.*

If the salvage value is set too low while using the declining balance method, the amount of depreciation taken during the service life will be less than the depreciable amount. To compensate for this problem, the depreciation method is usually changed to the straight-line method during the fourth or fifth year, to ensure closure on the salvage value. Figure 11-7 shows a comparison of the declining-balance and straight-line depreciation methods.

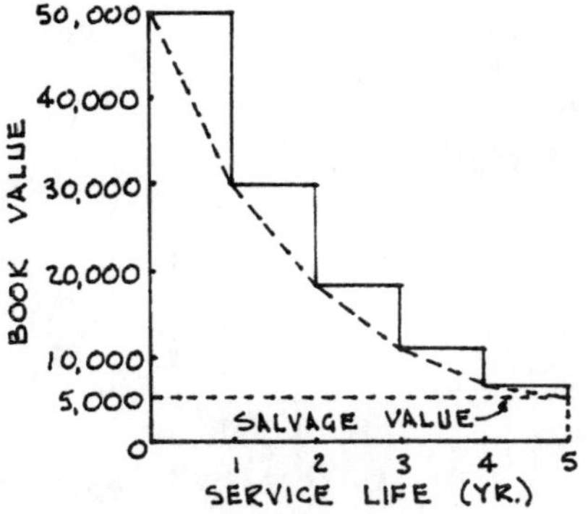


**Figure 11-6. Declining-balance depreciation method.**

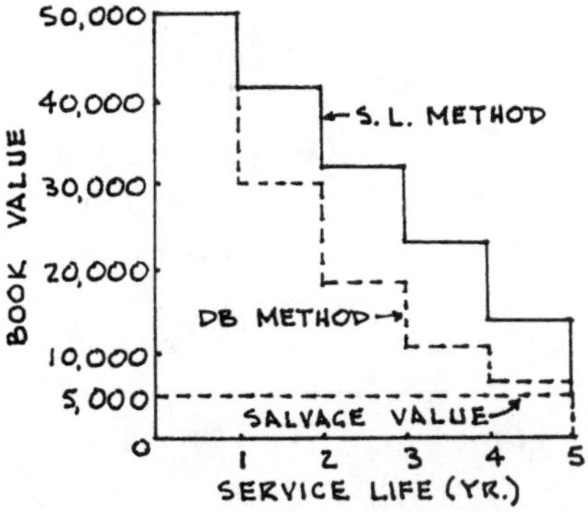

**Figure 11-7. Comparison of the declining-balance and straight-line depreciation methods.**

The depreciation rate for *second-hand* equipment for any given year can be determined by:

$$D_n = \frac{1.5}{\text{Life (Years)}} \text{ x B.V. at Start of Year (Excl. Tire Cost)}$$

**(Equation 11.7)**

**Example 11-7:** Use Equation 11.7 to determine the annual depreciation and book value at the end of each year for a second-hand tractor under the following conditions:

Delivered Cost = \$20,000
Salvage Value = \$2000
Tire Value = \$5000
Service Life = 3 years

**Solution:** The depreciation during the first year is:

$$D_1 = \frac{1.5}{3} \text{ x (\$20,000 – \$5000)} \qquad \text{(Eq. 11.7)}$$

$$= 0.5 \text{ x \$15,000}$$

$$= \$7500$$

The annual depreciation during the remaining years is:

$$D_2 = 0.5 \times (\$15{,}000 - \$7500)$$
$$= 0.5 \times \$7500$$
$$= \$3750$$

$$D_3 = 0.5 \times (\$7500 - \$3750)$$
$$= 0.5 \times \$3750$$
$$= \$1875 \text{ (Use \$1750)*}$$

The depreciation schedule is tabulated as follows:

| Year | Depreciation | Book Value at End of Year (Excl. Tires) |
|---|---|---|
| 0 | 0 | 15,000 |
| 1 | 7500 | 7,500 |
| 2 | 3750 | 3,750 |
| 3 | 1875 (Use 1750)* | 1,875 (Use 2000)* |

**You cannot depreciate beyond the salvage value.*

**Hours-of-Use Depreciation Method**

Many earthmoving contractors use *the hours-of-use depreciation method.* The amount of depreciation is based on the number of hours the unit was in production for a given year. Thus, the depreciable amount is prorated over the hours of actual use, rather than over the number of years of life. As an example, if a machine costs $30,000 and its service life is 10,000 hours, the depreciation would be charged at: $30,000 ÷ 10,000 Hr. = $3.00 per hour of use. This method ensures that depreciation will be recovered at the same time the machine is generating profit; however, there are instances when the hours-of-use depreciation method is not advantageous. For instance, if a machine is used only 500 hours per year, the production method would extend the period of recovery out over 20 years, and if the machine were sold after 5 years, you would recover only 25% of the available depreciation.

**Production Method**

The *production method* of depreciation is similar to the hours-of-use method, except that production units (cubic yards, tons, etc.) are used. This method is often used in mining operations. For example, if a machine costs $100,000 and is expected to produce 2,000,000 tons of coal during its lifetime, the depreciation would be charged at: $100,000 ÷ 2,000,000 Tons = $0.05 per ton of coal that the machine produces.

**Comparing Depreciation Methods**

Any given piece of equipment is defined as an *asset*, which can be viewed as a "profit center." That is, any legal depreciation strategy can be applied to the machine to reduce or alter the taxable income that it will generate. Each individual asset can be depreciated separately, and the contractor may establish as many depreciation accounts as he desires. He can use a separate account for each asset, or two or more assets can be combined into one account. Assets can also be grouped by common useful lives, or by common uses. During an asset's lifetime, you can make one change from an accelerated depreciation method to a straight-line method; however, you are not allowed to then change from a straight-line to an accelerated depreciation method.

Regardless of the depreciation method used, the depreciable amount always remains constant; therefore, larger deductions taken in the early years using an accelerated method are offset by smaller deductions in the later years.

When using any depreciation method, you are not allowed to depreciate beyond the salvage value. Also, any cash or trade-in allowance above the equipment's current book value when the machine is sold or traded must be treated as a *capital gain* and is subject to taxation. However, capital gain is taxed at a lesser rate than that of ordinary income.

The straight-line depreciation method is not the most realistic depreciation method, since the actual resale value of a given piece of equipment decreases at a greater rate during the first year than during subsequent years. However, when determining hourly ownership costs to be charged to a project, this method is preferred over the accelerated depreciation methods, because it is unrealistic and noncompetitive to bid a project using high ownership costs, just because you are currently using new equipment whose depreciation has been accelerated.

There is financial advantage to using an accelerated depreciation for tax purposes. Increasing the depreciation early on also reduces your tax liability early on. The savings is equivalent to a short-term interest-free loan. However, to avoid future financial problems, you must be aware that your subsequent depreciation deductions will be reduced, and your tax liability will increase. The problem will be compounded if accelerated depreciation is taken during unprofitable years, followed by profitable years. Equipment rental firms that intend to sell the equipment after the first two years of ownership often capitalize on the features of the accelerated depreciation methods.

If a piece of equipment is kept beyond its depreciation period, continue to charge depreciation when billing a client on an hourly basis. This is a legitimate, but "hidden" profit. However, don't keep a machine to the point where major repairs are required, or the profit may turn into a loss.

Some types of equipment are purchased for a specific project, to be sold at the end of the project. Under these circumstances, the depreciation should be based on the difference between the original cost and the estimated resale price. For instance, if a machine that costs $50,000 is expected to work 2000 hours and be resold for $35,000, the depreciation would be charged at:

$$\text{Depreciation} = \frac{(\$50{,}000 - \$35{,}000)}{2000 \text{ Hr.}}$$
$$= \$7.50/\text{hour}$$

**INTEREST, INSURANCE AND TAXES (IIT COSTS)**

The contractor must carry liability and loss *insurance* on his equipment, and he must pay local personal property *taxes* on the assessed value of the equipment. He must also pay *interest* on the note used to purchase the equipment, or lose interest on the money invested in the equipment, if the unit was paid for in cash. These expenses are accounted for under the heading of IIT costs. Some contractors also include the cost for equipment storage with IIT costs, but others include this cost in the general operating overhead. These IIT costs are also ownership costs, and are usually charged off as a percentage of a constant value referred to as the *average annual value* (AAV); i.e.,

Yearly IIT Cost = IIT % x AAV **(Equation 11.8)**

The IIT percentages used are developed from accounting records, and they are unique to the contractor and locality, especially with regard to insurance costs, which depend to a great extent on the contractor's safety record. However, average IIT percentages are:

| | |
|---|---|
| Taxes | 1.5% - 2.0% |
| Insurance | 1.0% - 3.0% |
| Interest | Too variable; check with lender |
| Storage | 0.5% - 5.0% |

IIT costs can be separated into individual values as follows:

Yearly Interest Cost = Interest % x AAV **(Equation 11.9)**
Yearly Insurance Cost = Insurance % x AAV **(Equation 11.10)**
Yearly Tax Cost = Tax % x AAV **(Equation 11.11)**

There are two formulas that can be used to determine the average annual value. *If salvage value is not considered* when calculating depreciation, the average annual value can be determined by:

$$\text{AAV} = \frac{\text{Delivered Cost} \times (n + 1)}{2n}$$ **(Equation 11.12)**

where: n = Service life (years).

**Example 11-8:** Assuming no salvage value, determine the average annual value and IIT costs for the tractor given in Example 11-1 under the following conditions:

$$\text{Interest} = 8\% \text{ of AAV}$$
$$\text{Insurance} = 3\% \text{ of AAV}$$
$$\text{Taxes} = 2\% \text{ of AAV}$$

**Solution:** The total IIT percentage to be applied to the average annual value is:

$$\text{Total IIT \%} = 8\% + 3\% + 2\%$$
$$= 13\%$$

The average annual value is:

$$\text{AAV} = \frac{\$65{,}000 \text{ x } (5 + 1)}{2 \text{ x } 5} \qquad \text{(Eq. 11.12; Example 11-1)}$$
$$= \$39{,}000/\text{year}$$

The average yearly IIT cost is:

$$\text{Average Yearly IIT Cost} = 0.13 \text{ x } \$39{,}000/\text{Yr.} \qquad \text{(Eq. 11.8)}$$
$$= \$5070$$

Assuming 2000 hours of operation per year, the average hourly ITT cost is:

$$\text{Average Hourly IIT Costs} = \frac{\$5070/\text{Yr.}}{2000 \text{ Hr./Yr.}} \qquad \text{(Eq. 11.2)}$$
$$= \$2.54/\text{hour}$$

**Example 11-9:** Determine the average individual hourly cost for interest, insurance and taxes in the previous example.
**Solution:** The average annual value is:

$$\text{AAV} = \$39{,}000/\text{year} \qquad \text{(Example 11-8)}$$

The average yearly interest cost is:

$$\text{Average Yearly Interest Cost} = 0.08 \text{ x } \$39{,}000/\text{Yr.}$$
$$= \$3120/\text{year}$$

(Eq. 11.9; Example 11-8)

The average hourly interest cost is:

$$\text{Average Hourly Interest Cost} = \frac{\$3120/\text{Yr.}}{2000\ \text{Hr./Yr.}} \qquad \text{(Eq. 11.2)}$$

$$= \$1.56/\text{hour}$$

The average hourly insurance cost is:

$$\text{Average Hourly Insurance Cost} = \frac{0.03 \times \$39{,}000/\text{Yr.}}{2000\ \text{Hr./Yr.}}$$

$$= \$0.59/\text{hour}$$

The average hourly tax cost is:

$$\text{Average Hourly Tax Cost} = \frac{0.02 \times \$39{,}000/\text{Yr.}}{2000\ \text{Hr./Yr.}}$$

$$= \$0.39/\text{hour}$$

Figure 11-8 illustrates a graphic representation of the average annual value when using the straight-line depreciation method, a delivered cost of $65,000, and no salvage value. The area within the "dashed" rectangle representing the average annual value equals the area under the stepwise figure representing the straight-line depreciation method, and yields a constant value to be used for estimating IIT costs throughout the service life of the equipment.

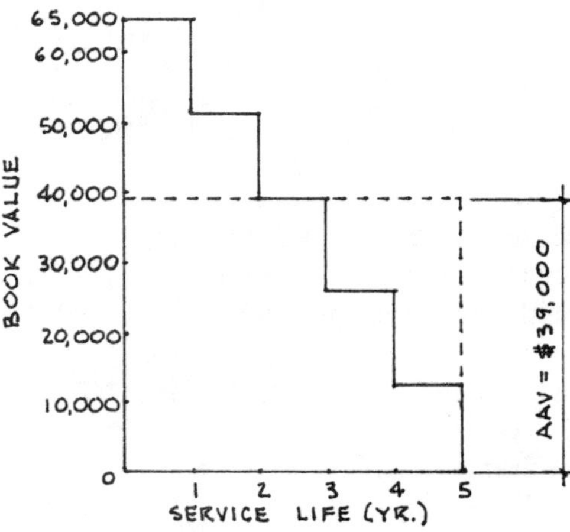


**Figure 11-8.**
**Average annual value without salvage value included.**

*If salvage value is considered* in depreciation, the average annual value can be determined by:

$$AAV = \frac{[\text{Delivered Cost x (n + 1)}] + [\text{Salvage Value x (n - 1)}]}{2n}$$

**(Equation 11.13)**

where: n = Service life (years).

**Example 11-10:** Work Example 11-8 assuming a salvage value of $5000.
**Solution:** The average annual value is:

$$AAV = \frac{[\$65{,}000 \text{ x } (5 + 1)] + [\$5000 \text{ x } (5 - 1)]}{2 \text{ x } 5} \quad \text{(Eq. 11.13; Example 11-1)}$$

$$= \$41{,}000/\text{year}$$

The average yearly IIT cost is:

$$\text{Average Yearly IIT Costs} = 0.13 \text{ x } \$41{,}000/\text{Yr.} \quad \text{(Eq. 11.8; Example 11-8)}$$
$$= \$5330/\text{year}$$

Assuming 2000 hours of operation per year, the average hourly IIT cost is:

$$\text{Average Hourly IIT Costs} = \frac{\$5330/\text{Yr.}}{2000 \text{ Hr./Yr.}} \quad \text{(Eq. 11.2)}$$
$$= \$2.67/\text{hour}$$

Figure 11-9 illustrates a graphic representation of the average annual value when using the straight-line depreciation method, a delivered cost of $65,000, and a salvage value of $5000. The area within the "dashed" rectangle representing the average annual value equals the area under the stepwise figure representing the straight-line depreciation method. The AAV method works equally well for any depreciation method, and yields a constant value to be used for estimating IIT costs throughout the service life of the equipment.

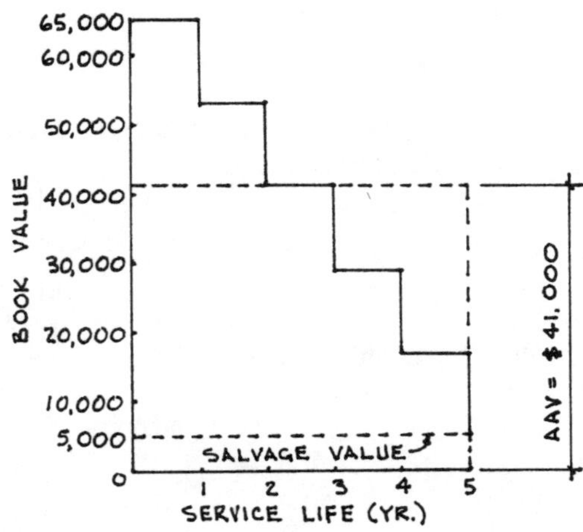


**Figure 11-9. Average annual value with salvage value included.**

Some contractors determine IIT costs by applying a percentage to an *average investment*, calculated on an individual yearly basis, where the average investment for a particular year is determined by:

$$\text{Ave. Yearly Investment (Year n)} = \frac{\text{B.V. (Start of Yr.)} + \text{B.V. (End of Yr.)}}{2}$$

**(Equation 11.14)**

When determining the average yearly investment, depreciation is based on the delivered cost, which *includes the cost for tires*. Using the straight-line method, the yearly depreciation for the tractor given in Example 11-1 would be:

$$D_{Annual} = \frac{\$65{,}000 - \$5000}{5}$$
$$= \$12{,}000$$

The depreciation schedule is tabulated as follows:

| Year | Depreciation | Book Value at End of Year (Incl. Tires) |
|---|---|---|
| 0 | 0 | 65,000 |
| 1 | 12,000 | 53,000 |
| 2 | 12,000 | 41,000 |
| 3 | 12,000 | 29,000 |
| 4 | 12,000 | 17,000 |
| 5 | 12,000 | 5,000 |

Assuming an IIT percentage of 13% of the average yearly investment, the yearly IIT costs are determined as follows:

The average investment during the first year is:

$$\text{Ave. Investment (Year 1)} = \frac{\$65{,}000 + \$53{,}000}{2} \qquad \text{(Eq. 11.14)}$$
$$= \$59{,}000$$

The hourly IIT cost for the first year is:

$$\text{Hourly IIT Cost (Year 1)} = \frac{0.13 \times \$59{,}000}{2000 \text{ Hr.}} \qquad \text{(Eq. 11.2)}$$

$$= \$3.84/\text{hour}$$

Continuing in this fashion, the hourly IIT costs for the subsequent years are:

Hourly IIT Costs (Year 2) = $3.06/hour
Hourly IIT Costs (Year 3) = $2.28/hour
Hourly IIT Costs (Year 4) = $1.50/hour
Hourly IIT Costs (Year 5) = $0.72/hour

**Comparing Investment Methods**

There is no tax liability advantage to using the average-annual-value method, or the average-yearly-investment method when determining the interest, insurance and local taxes paid, because the IRS will only allow deductions for the expenses actually paid during any given year. These methods are used only as a guide for you to pre-determine this portion of the owning costs of the equipment. Since you want to establish realistic costs that allow you to be equally competitive each year throughout the service life of the equipment, the average-annual-value method is preferred when determining hourly ownership costs to be charged to a project.

## OPERATING COSTS

*Operating costs* are costs incurred while the equipment is actually in operation. Operating costs include fuel, lubricants, hydraulic fluid, filters, periodic service, major repairs and tires. Some contractors also include the operator's wages in the operating costs.

## FUEL COSTS

The hourly *fuel cost* depends on the price of fuel and the rate of fuel consumption. The rate of consumption depends on the engine horsepower, engine type, fuel type and the operating conditions. Operating conditions (also referred to as *load factors*) are separated into three categories, including low (favorable), medium (average) and high (unfavorable). Some equipment manufacturers publish hourly fuel consumption rates as shown in Table 11-2.

**Table 11-2: 816B Wheel Tractor Fuel Consumption Guide (Gal./Hr.)**

| | Load Factor* | |
|---|---|---|
| Low | Medium | High |
| 7.0 - 8.0 | 9.5 - 11.0 | 11.5 - 12.5 |

*Load Factor Guide

High: Heavy dozing, compacting heavy material, heavy landfill work.
Medium: Production dozing, most push loading, shovel cleanup, normal compaction.
Low: Considerable idling, or travel with no load.

(*Courtesy of Caterpillar Inc.*)

If you have no published fuel consumption tables, measurement can be made in the field, or you can estimate the fuel consumption rate by using the following equation:

Fuel Consumption (Gal./Hr.) = 0.057 x hp x Fuel Consumption Factor
**(Equation 11.15)**

where: hp = Engine horsepower (Equation 11.15 is valid for engines up to 700 hp)

Fuel Consumption Factor = Percentage of time that the engine is delivering maximum output (Use Table 11-3 as a guide for determining the fuel consumption factor for various types of equipment)

and, 0.057 = Gallons of fuel consumed per horsepower, per hour. (This is based on consuming 0.4 pounds of fuel per horsepower per hour at maximum engine output. Since fuel weighs 7.01 pounds per gallon, this is equivalent to: 0.4 lb./hp/Hr.÷ 7.01 lb./Gal. = 0.057 gallons/hp/hour).

Once the fuel consumption rate has been calculated, the hourly fuel cost can be determined by:

Hourly Fuel Cost = Fuel Consumption (Gal./Hr.) x Fuel Cost/Gal.
**(Equation 11.16)**

**Table 11-3: Fuel Consumption Factors**

| Vehicle | Working Conditions Favorable | Average | Unfavorable |
|---|---|---|---|
| Crawler | 0.60 | 0.70 | 0.80 |
| Scraper, Self-Loading | 0.48 | 0.55 | 0.65 |
| Scraper, Push-Loaded | 0.40 | 0.50 | 0.60 |
| Front-End Loaders | 0.35 | 0.45 | 0.55 |
| Bottom Dumps | 0.30 | 0.40 | 0.50 |
| Haulers | 0.25 | 0.35 | 0.45 |

(*Courtesy of Terex Corporation*)

**Example 11-11:** Assuming low load conditions, determine the hourly fuel costs for a Caterpillar 816B wheel tractor if fuel costs $1.00 per gallon.
**Solution:** The hourly fuel consumption rate is:

Fuel Consumption (Gal./Hr.) = 7.5 gallons/hour (Table 11-2)

The hourly fuel cost is:

Hourly Fuel Cost = 7.5 Gal./Hr. x $1.00/Gal. (Eq. 11.16)
= $7.50/hour

**Example 11-12:** Assuming unfavorable load conditions, use Equation 11.15 and data from Table 11-3 to determine the hourly fuel costs for a 375-hp front-end loader if fuel costs $1.00 per gallon.
**Solution:** The hourly fuel consumption rate is:

Fuel Consumption (Gal./Hr.) = 0.057 x 375 x 0.55 (Eq. 11.15; Table 11-3)
= 11.8 gallons/hour

The hourly fuel cost is:

Hourly Fuel Cost = 11.8 Gal./Hr. x $1.00/Gal. (Eq. 11.16)
= $11.80/hour

## ROUTINE MAINTENANCE COSTS

*Routine maintenance* includes the material and labor costs for lubricating oil, grease, hydraulic fluid, filters, air cleaners, minor adjustments and other routine service. Historical cost records can be of great assistance when determining routine maintenance costs, but when data is not available, you can estimate the routine service costs as follows:

Routine Service Costs (Favorable Conditions) = 1/5 x Hourly Fuel Costs
**(Equation 11.17)**

Routine Service Costs (Average Conditions) = 1/3 x Hourly Fuel Costs
**(Equation 11.18)**

Routine Service Costs (Unfavorable Conditions) = 1/2 x Hourly Fuel Costs
**(Equation 11.19)**

Some routine maintenance, such as checking pressures, fluid levels, hoses and connections, and greasing fittings can be performed while the engine is warming up before use, or during the cooling-off period before the engine is shut down. Warming and cooling the engine will greatly extend the service life of the machine.

**Example 11-13:** Determine the hourly routine service costs for the front-end loader working under the conditions given in the previous example.
**Solution:** Since we are working under unfavorable load conditions, the routine service cost is:

Routine Service Cost = 1/2 x $11.80/Hr. (Eq. 11.19)
= $5.90/hour

Some equipment manufacturers publish hourly lubricant consumption and grease-fitting data (Table 11-4), and filter data (Table 11-5). The filter-change data given in Table 11-5 comes from the filter change-interval recommendations given in Table 11-6. Labor for lubrication and filter changes is recommended as follows:

Oil Change = 25 minutes = 0.42 hours
Filter Change = 5 minutes = 0.08 hours
Grease Fitting = 1 minute = 0.02 hours

**Table 11-4:**
**Approximate Hourly Consumption of Lubricants for E240 Excavator** *

| Crankcase | Final Drives | Hydraulic Control | Lubricant Changes | Grease Fittings |
|---|---|---|---|---|
| 0.021 Gal. | 0.004 Gal. | 0.04 Gal. | 10 Ea. | 600 Ea. |

**When operating in heavy dust, deep mud or water, increase the quantities by 25%.* *(Courtesy of Caterpillar Inc.)*

**Table 11-5: Total Number of Filters Changed Over 2000 Hr. for E240 Excavator**

| Filter | No. Changes |
|---|---|
| Engine | |
| Fuel | 4 |
| Oil | 4 |
| Air Cleaner | 2 |
| Hydraulic | |
| Return | 4 |
| Drain | 4 |
| Pilot | 4 |
| Total | 22 |

(*Courtesy of Caterpillar Inc.*)

**Table 11-6: Recommended Filter Change Intervals**

| Filters | Change Interval (Hr.)* |
|---|---|
| Engine | 250 |
| Transmission | 500 |
| Hydraulic | 500 |
| Fuel | |
| Primary | 2000 |
| Final | 500 |
| Air | |
| Primary | 2000 |
| Secondary | 1000 |

**Recommended change interval may vary with machine and sulfur content of diesel fuel. Always consult Lube & Maintenance Guide.*
(*Courtesy of Caterpillar Inc.*)

**Example 11-14:** Assuming the following material and labor costs, determine the hourly costs for lubrication and filter changes for a Caterpillar E240 hydraulic excavator operated 2000 hours per year.

Filters and Air Cleaners = \$175.00 per complete set
Lube Oil = \$4.00/gallon
Grease = \$0.03/fitting
Labor = \$15.00/hour (incl. labor burden)

**Solution:** The total hourly lubricant consumption rate for the crankcase, transmission and final drives is:

Hourly Lubricant Consumption (Gal./Hr.) = 0.021 + 0.004 + 0.04 (Table 11-4)
= 0.065 gallons/hour

The hourly material cost for lubricants is:

Lubricant Cost (Material) = 0.065 Gal./Hr. x \$4.00/Gal (Given)
= \$0.26/hour

The hourly labor cost for lubricant changes is:

Lubricant Cost (Labor) = 10 Changes/Yr. x 0.42 Hr./Change x \$15.00/Hr. (Table 11-4; Given)
= \$63.00/Yr. ÷ 2000 Hr./Yr. (Eq. 11.2)
= \$0.03/hour

The hourly material cost for grease fittings is:

Grease Cost (Material) = 600 Ea./Yr. x \$0.03/Ea. (Table 11-4; Given)
= \$18.00/Yr. ÷ 2000 Hr./Yr.
= \$0.01/hour

The hourly labor cost for packing the grease fittings is:

Grease Cost (Labor) = 600 Ea./Yr. x 0.02 Hr./Ea. x \$15.00/Hr. (Given)
= \$180.00/Yr. ÷ 2000 Hr./Yr.
= \$0.09/hour

The hourly material cost for filters is:

Filter Cost (Material) = 4 Sets/Yr. x \$175.00/Set (Table 11-5; Given)
= \$700.00/Yr. ÷ 2000 Hr./Yr.
= \$0.35/hour

The hourly labor cost for changing filters is:

Filter Cost (Labor) = 22 Ea./Yr. x 0.08 Hr./Ea. x \$15.00/Hr. (Table 11-5; Given)
= \$26.40/Yr. ÷ 2000 Hr./Yr.
= \$0.01/hour

The total hourly lubrication and filter costs are tabulated as follows:

| Item | Material Cost/Hour | Labor Cost/Hour |
|---|---|---|
| Lubricating Oil | \$0.26 | \$0.03 |
| Grease Fittings | \$0.01 | \$0.09 |
| Filters | \$0.35 | \$0.01 |
| Totals | \$0.62 | \$0.13 |

The total cost for lubricants, grease and filters is:

$$\text{Total Cost} = \$0.62/\text{Hr.} + \$0.13/\text{Hr.}$$
$$= \$0.75/\text{hour}$$

**MAJOR REPAIR COSTS**

With the exception of the operator's wages (and on occasion, tires) major repair cost (referred to as *repair cost*, hereafter) normally constitutes the single highest operating cost of heavy construction equipment. Repair cost is dependent upon equipment type (equipment application), operating conditions and maintenance standards. Repair cost does not include routine service, or the replacement cost for high-wear items such as tires, cutting edges, ripper teeth and shanks. Some manufacturers determine the repair cost on an hourly basis as follows:

$$\text{Average Hourly Repair Cost} = \frac{\text{Repair Factor x (Delivered Price} - \text{Tires)}}{1000}$$

**(Equation 11.20)**

where: Repair factors are taken from Table 11-7.

**Table 11-7: Repair Factors for Various Types of Equipment**

| Equipment Type | Operating Conditions | | |
|---|---|---|---|
| | Excellent | Average | Severe |
| Track-Mounted Tractors | 0.08 | 0.09 | 0.13 |
| Wheel Tractors | 0.05 | 0.06 | 0.09 |
| Wheel Tractor-Scrapers | 0.07 | 0.10 | 0.13 |
| Off-Highway Trucks | 0.06 | 0.09 | 0.12 |
| Track Loaders | 0.07 | 0.10 | 0.13 |
| Wheel Loaders | 0.05 | 0.06 | 0.09 |
| Motor Graders | 0.03 | 0.06 | 0.08 |

**Example 11-15:** Determine the average hourly repair cost for the tractor given in Example 11-1 if the tractor is working under average job conditions.
**Solution:** The average hourly repair cost is:

$$\text{Average Hourly Repair Cost} = \frac{0.06 \text{ x } (\$65{,}000 - \$15{,}000)}{1000}$$

$$= \$3.00/\text{hour}$$

(Eq. 11.20; Table 11-7)

Other manufacturers determine repair cost on an hourly basis as follows:

$$\text{Average Hourly Repair Cost} = \text{Repair Factor x Hourly Depreciation Cost} \text{ x } \frac{\text{Depreciation Period (Hrs.)}}{10{,}000 \text{ Hr.}}$$

**(Equation 11.21)**

where: Repair factors are taken from Table 11-8
and, Hourly depreciation cost *excludes* salvage value.

**Table 11-8: Repair Factors (Based on a Depreciation Period of 10,000 Hours)**

| Equipment Type | Operating Conditions Favorable | Average | Unfavorable |
|---|---|---|---|
| Scrapers-All Types | 42% | 50% | 62% |
| Front-End Loaders—Rubber-Tired | 45% | 55% | 70% |
| Haulers | 37% | 45% | 60% |
| Bottom Dumps | 30% | 35% | 45% |
| Crawler Tractors (by Application) | | | |
| Industrial | 10% | 25% | 75% |
| General Contracting | 40% | 60% | 80% |
| Quarrying | 50% | 85% | 115% |
| Mining | 70% | 110% | 150% |
| Logging | 55% | 135% | 215% |

(*Courtesy of Terex Corporation*)

**Example 11-16:** Determine the average hourly repair cost for a crawler tractor used in general construction under the following conditions.

Delivered Cost = $50,000
Depreciation Method = Straight-line
Service Life = 5 years (2000 Hr./Yr.)
Job Conditions = Average

**Solution:** Excluding salvage value, the annual depreciation is:

$$D_{Annual} = \frac{\$50,000}{5\ Yr.} = \$10,000/year \qquad \text{(Eq. 11.4)}$$

The hourly depreciation cost is:

$$\text{Hourly Depreciation Cost} = \frac{\$10,000/Yr.}{2000\ Hr./Yr.} = \$5.00/hour \qquad \text{(Eq. 11.2)}$$

The average hourly repair cost is:

$$\text{Average Hourly Repair Cost} = 0.60 \times \$5.00 \times \frac{10,000}{10,000} \qquad \text{(Eq. 11.21; Table 11-8)}$$

$$= \$3.00/hour$$

Repair costs are relatively low early on, and escalate as the equipment ages (Figure 11-10); therefore, when an average hourly repair cost is determined, it is more than that which is required early on. However, using an average hourly cost produces extra funds early on, which are reserved to cover higher repair costs that will be experienced later on. The extra funds are often referred to as *repair reserve*. The actual repair costs for any given year can be estimated by:

$$\text{Year's Repair Cost} = \frac{\text{Year Digit}}{\text{Sum of Years Digit}} \times \text{Total Repair Cost}$$

**(Equation 11.22)**

where: Year Digit = Year taken in ascending order (For the first year, use 1; for the second year, use 2, etc.)

Sum of Years Digit = Sum of the years' digits for the depreciation period (For a 5-year service life, use $1 + 2 + 3 + 4 + 5 = 15$)

and, Total Repair Cost = Repair Factor x (Delivered Cost – Tire Cost)

**(Equation 11.23)**

where: Repair factor is taken from Table 11-9.

**Table 11-9: Typical Lifetime Repair Costs (% of Delivered Cost Less Tires)**

| Equipment Type | Operating Conditions | | |
|---|---|---|---|
| | Favorable | Average | Unfavorable |
| Crawler Tractors | 85 | 90 | 95 |
| Wheel Tractors | 55 | 65 | 75 |
| Motor Graders | 50 | 55 | 60 |
| Off-Highway Trucks | 75 | 85 | 95 |
| Wagons | 45 | 55 | 65 |
| Scrapers | 85 | 95 | 105 |
| Track Loaders | 80 | 90 | 100 |
| Wheel Loaders | 50 | 60 | 70 |

**Example 11-17:** Determine the repair cost for each year of operation for the tractor given in Example 11-1 while working under unfavorable conditions.
**Solution:** The total (lifetime) repair cost is:

$$\text{Total Repair Cost} = 0.75 \times (\$65{,}000 - \$15{,}000) = \$37{,}500 \quad \text{(Eq. 11.23; Table 11-9)}$$

The repair cost for the first year is:

$$\text{Repair Cost (Year 1)} = \frac{1}{15} \times \$37{,}500 = \$2500 \quad \text{(Eq. 11.22)}$$

The hourly repair cost for the first year is:

$$\text{Hourly Repair Cost (Year 1)} = \frac{\$2500/\text{Yr.}}{2000\ \text{Hr./Yr.}} = \$1.25/\text{hour} \quad \text{(Eq. 11.2)}$$

The repair cost for the second year is:

$$\text{Repair Cost (Year 2)} = \frac{2}{15} \times \$37{,}500 = \$5000$$

The hourly repair cost for the second year is:

$$\text{Hourly Repair Cost (Year 2)} = \frac{\$5000/\text{Yr.}}{2000\ \text{Hr./Yr.}}$$
$$= \$2.50/\text{hour}$$

The hourly repair costs for subsequent years are tabulated as follows:

| Year | Hourly Repair Cost |
|---|---|
| 3 | 3.75 |
| 4 | 5.00 |
| 5 | 6.25 |

The cumulative repair cost during the lifetime of the tractor is tabulated as follows:

| Year | Year's Repair Cost | Cumulative Repair Cost |
|---|---|---|
| 0 | 0 | 0 |
| 1 | 2,500 | 2,500 |
| 2 | 5,000 | 7,500 |
| 3 | 7,500 | 15,000 |
| 4 | 10,000 | 25,000 |
| 5 | 12,500 | 37,500 |

This data is plotted on a *repair cost profile* as diagrammed in Figure 11-10.

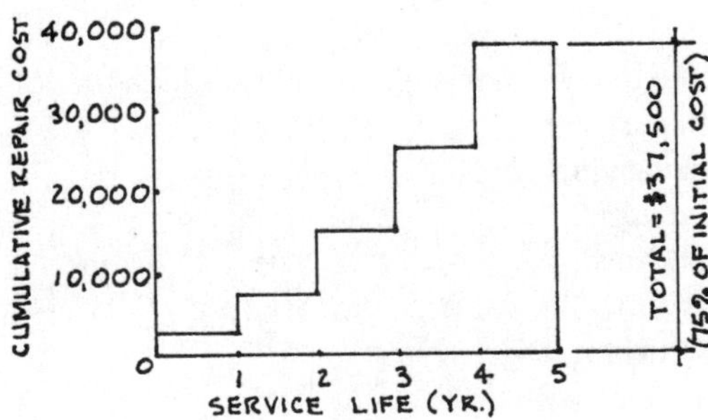


**Figure 11-10. Repair cost profile.**

Caterpillar Inc., publishes *repair reserve charts* as shown in Figure 11-11. The *basic hourly repair cost* (basic repair factor) is based on the first 10,000 hours of service (parts at current published U.S. Consumers' List Prices, and labor at a total cost of $40.00 per hour). *Extended-life multipliers* are given for those cases where the machine is used beyond 10,000 hours. When an extended-life multiplier is used, the adjusted cost per hour applies to the entire service life, and not just to the additional hours beyond 10,000.

**TRACK-TYPE TRACTORS**

| **Cost distribution** | | **Extended-life Multipliers** | |
|---|---|---|---|
| D3 to D7 — | 60% Parts | 0-10,000 hours | 1.0 |
| | 40% Labor | 0-15,000 | 1.1 |
| D8 to D11 — | 70% Parts | 0-20,000 | 1.3 |
| | 30% Labor | | |

Includes basic tractor equipped with ROPS canopy, straight bulldozer and hydraulic control.

**Note:** Repair time may be less on Elevated Sprocket Tractors due to modular design of power train components.

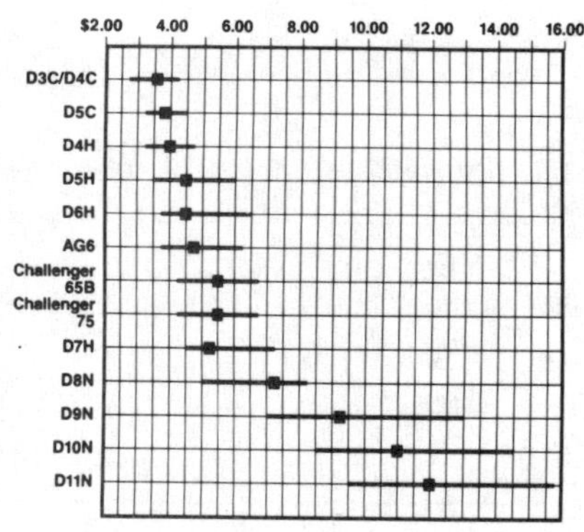


**Figure 11-11. Repair reserve.** (*Courtesy of Caterpillar Inc.*)

To estimate hourly repair costs, enter the chart for the given machine and determine the basic cost (repair factor) for the applicable job conditions. Operating conditions are placed in zones along the bar as shown in Figure 11-12. Zone A applies to favorable conditions, Zone B to average conditions, and Zone C to unfavorable conditions.

**Figure 11-12. Operating condition zones.**
(*Courtesy of Caterpillar Inc.*)

The basic repair factors given in Figure 11-11 exclude the following:

- Dozer blade
- Ground engaging tools
- Undercarriage
- Tires and rims
- Material and labor for periodic service
- Service truck mileage costs
- Serviceman's travel costs
- Machine transportation to and from the shop
- Shop and tool overhead

**Example 11-18:** Determine the average hourly repair cost for a Caterpillar D5H dozer working under average conditions, if the service life is anticipated to be 7 years (14,000 hours).
**Solution:** The basic hourly repair cost is:

Basic Hourly Repair Cost = \$4.50/hour (Figure 11-11)

However, this is based on a service life through 10,000 hours. Since we are anticipating a 14,000-hour service life, the estimated basic hourly repair cost is:

Estimated Hourly Repair Cost = \$4.50/Hr. x 1.1 (Figure 11-11)
= \$4.95/hour

## OPTIMUM REPAIR TIME

The repair cost determined by using various methods previously discussed will be widely varied for any specific machine operating under a given set of working conditions. This is due to a variety of reasons, including whether a repair was performed before, or after a catastrophic failure. A before-failure repair of a major component can cost one-third of that required for an after-failure repair (Figure 11-13). However, to achieve maximum economy, a repair should not be made prematurely, since this will result in the sacrifice of part service life, and an increase in hourly repair cost (Figure 11-14). Component life can be anticipated by oil analysis, maintenance inspections, observing service meter units (SMU), and by studying operator notes.

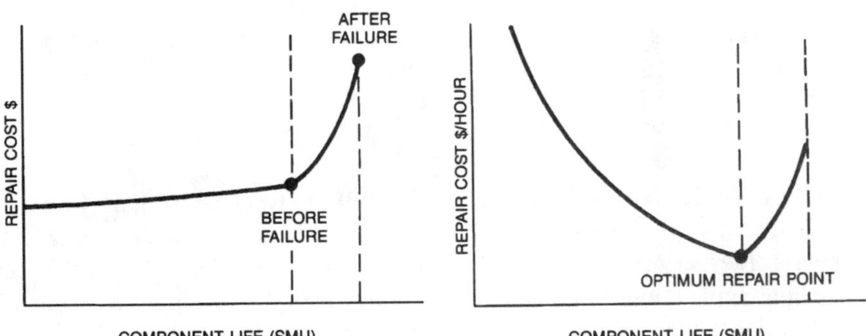


**Figure 11-13. Repair cost before, versus after failure.**
*(Courtesy of Caterpillar Inc.)*

**Figure 11-14. Optimum repair point.**
*(Courtesy of Caterpillar Inc.)*

## CUTTING EDGE REPLACEMENT COSTS

The hourly cost for replacing cutting edges on scraper bowls and loader buckets can be determined by:

$$\text{Hourly Cutting Edge Cost} = \frac{\text{Cutting Edge Cost}}{\text{Edge Service Life}} \qquad \textbf{(Equation 11.24)}$$

where: Edge service life is taken from Table 11-10.

**Table 11-10: Cutting Edge Service Life for Scrapers and Loaders (Hours)**

| Operating Conditions | Scrapers | Loaders |
|---|---|---|
| Favorable | 750 | 3500 |
| Average | 500 | 2000 |
| Unfavorable | 250 | 500 |
| Coal Stock and Reclaim | 1000 | — |

(*Courtesy of Terex Corporation*)

**Example 11-19:** Assuming a scraper requires four cutting edge sections costing \$175.00 per section, determine the hourly cutting edge replacement cost if the scraper is operated in unfavorable conditions.
**Solution:** The hourly cutting edge replacement cost is:

$$\text{Hourly Cutting Edge Cost} = \frac{4 \text{ Ea. x } \$175.00/\text{Ea.}}{250 \text{ Hr.}} \qquad \text{(Eq. 11.24; Table 11-10)}$$

$$= \$2.80/\text{hour}$$

## UNDERCARRIAGE COSTS (TRACK-MOUNTED EQUIPMENT)

Undercarriage replacement and maintenance costs for a track-mounted machine is often very expensive. Undercarriage wear can be independent of the basic machine application. For example, if a machine is worked in an easy digging application while traveling over a high-impact and abrasive working surface, the basic equipment components will experience relatively low wear, while the undercarriage encounters high wear. For this reason, the hourly undercarriage cost for track-mounted equipment should be calculated independently of other major repair costs.

Caterpillar Inc. has established three primary job conditions that affect track-mounted undercarriage wear, including impact, abrasiveness and the "Z" factor.

**Impact**

*Impact* is induced primarily by a hard, uneven working surface. The result of impact is structural damage to the undercarriage, including bending, chipping, cracking, spalling and hardware problems, such as a lack of pin and bushing retention. Impact is rated as follows:

- High: Non-penetrable hard surfaces with 6-inch (or higher) bumps.
- Moderate: Partially penetrable surface with bumps ranging from 3 to 6 inches in height.
- Low: Completely penetrable surface providing full shoe plate support, with few bumps.

**Abrasiveness**

*Abrasion* is the tendency of the underfoot materials to grind away the wear surfaces of the track components. Abrasiveness is rated as follows:

- High: Saturated wet soil containing a high proportion of hard, angular, or sharp sand, or rock particles.
- Moderate: Slightly damp soils containing a low proportion of hard, angular, or sharp particles.
- Low: Dry soils, or rock containing a low proportion of hard, angular, or sharp sand, or rock chip particles.

**"Z" Factor**

The *"Z" factor* represents the combined effect on component life of the many intangible environmental, operation and maintenance considerations on a given job. These considerations are rated as follows:

- Environmental and Terrain: Non-abrasive soil can become packed into sprocket teeth. Also, corrosive chemicals in the materials being moved can increase wear rates, while moisture and temperature can exaggerate the effect. Side-hill work can increase wear on the downhill sides of components.
- Operation: Some operator practices increase track wear. Examples include high-speed operation, especially in reverse. Tight turns and constant corrections in direction, and/or stalling the tractor under load can cause the tracks to slip, resulting in increased wear.
- Maintenance: Good maintenance extends undercarriage life and reduces costs. Good maintenance includes proper track tension, daily cleaning, periodic wear measurement and timely recommended services.

**Estimating Undercarriage Cost**

Impact and abrasion is not difficult to determine; however, the selection of an appropriate "Z" factor requires good judgement, common sense and careful analysis of job conditions.

To determine undercarriage costs, select the appropriate machine and its corresponding *basic factor* from Table 11-11. Next, determine and sum the appropriate *condition multipliers* taken from Table 11-12. The hourly undercarriage cost can then be determined by:

Hourly Undercarriage Costs = Basic Factor x Sum of Condition Multipliers
**(Equation 11.25)**

**Table 11-11: Undercarriage Basic Factors**

| Model | Basic Factor |
|---|---|
| D11N | 17.5 |
| D10N | 12.0 |
| D9N | 9.5 |
| D8N | 8.5 |

(*Courtesy of Caterpillar Inc.*)

**Table 11-12: Conditions Multipliers**

| Job Conditions | Impact | Abrasiveness | "Z" |
|---|---|---|---|
| High | 0.3 | 0.4 | 1.0 |
| Moderate | 0.2 | 0.2 | 0.5 |
| Low | 0.1 | 0.1 | 0.2 |

(*Courtesy of Caterpillar Inc.*)

**Example 11-20:** Determine the hourly undercarriage cost for a Caterpillar D9N tractor working over a high impact, low-abrasive surface with a moderate "Z" factor.
**Solution:** The hourly undercarriage cost is:

Hourly Undercarriage Cost = 9.5 x (0.3 + 0.1 + 0.5) (Eq. 11.25;
= $8.55/hour Tables 11-11 and 11-12)

**TIRE REPLACEMENT COST**

Tire *replacement* cost is a major operating expense, often exceeded only by operator's wages and the major repair cost. The hourly tire replacement cost can be determined as follows:

$$\text{Hourly Tire Replacement Cost} = \frac{\text{Tire Cost}}{\text{Tire Service Life (Hr.)}}$$

**(Equation 11.26)**

Tire service life is unique to each vehicle, and it can be influenced by approximately 50 variables. The Goodyear Tire and Rubber Co. has developed some guidelines which are presented in Table 11-13, with their permission. The guidelines are prefaced with the following statements (*READ THE PREAMBLE CAREFULLY*):

"... there is no accurate, fool-proof method of forecasting tire life. Tire engineers have many theoretical methods for calculating tire life, but these are so involved and time-consuming that they are impractical for field use."

"However, the tire industry has made many surveys of tire performance, coming up with a "ball park" estimate of tire life. Studies done by at least two major equipment manufacturers are in close agreement."

With regard to using the system, some fundamental cautions must be observed:

1. The system applies basically to off-the-road dump trucks and scrapers.
2. The system is only as good as the user's accuracy in determining appropriate factors.
3. The system does not include the important effect of "operator attitude."
4. The system has no legal or warranty status.

**Table 11-13: Estimated Tire Service Life Of Hauling Units (Trucks & Scrapers)**

| No. | Condition | Factor |
|---|---|---|
| I | **Maintenance** | |
| | Excellent | 1.090 |
| | Average | .981 |
| | Poor | .763 |
| II | **Speeds (Maximum)** | |
| | 10 mph ~ 16 km/h | 1.090 |
| | 20 mph ~ 32 km/h | .872 |
| | 30 mph ~ 48 km/h | .763 |
| III | **Surface Conditions** | |
| | Soft Earth — No Rock | 1.090 |
| | Soft Earth — Some Rock | .981 |
| | Well Maintained — Gravel Road | .981 |
| | Poorly Maintained — Gravel Road | .763 |
| | Blasted — Sharp Rock | .654 |
| IV | **Wheel Positions** | |
| | Trailing | 1.090 |
| | Front | .981 |
| | Driver (Rear Dump) | .872 |
| | (Bottom Dump) | .763 |
| | (Self Propelled Scraper) | .654 |

| No. | Condition | Factor |
|---|---|---|
| V | **Loads** (See No. VIII note) | |
| | T&RA/ETRTO* Recommended Loading | 1.090 |
| | 20% Overload | .872 |
| | 40% Overload | .545 |
| VI | **Curves** | |
| | None | 1.090 |
| | Medium | .981 |
| | Severe | .872 |
| VII | **Grades** (Drive Tires Only) | |
| | Level | 1.090 |
| | 5% Max. | .981 |
| | 15% Max. | .763 |
| VIII | **Other Miscellaneous Combinations** (See note below) | |
| | None | 1.090 |
| | Medium | .981 |
| | Severe | .872 |

**Condition VIII** is to be used when overloading is present in combination with one or more of the primary conditions of maintenance, speeds, surface conditions and curves. The combination of severe levels in these conditions, together with an overload, will create a new and more serious condition which will contribute to early tire failure to a larger extent than will the individual factors of each condition.

(*Courtesy of Goodyear Tire and Rubber Company*)

The estimated service life of a given type of tire can be determined by multiplying the *base average tire life* (Table 11-14) by the appropriate tire life factors taken from Table 11-13.

**Table 11-14: Base Average Tire Life**

| Type Of Tire | Base Average Life Hours | Base Average Life Miles |
|---|---|---|
| E-3 Std. Bias Belted | 2510 | 25,100 |
| E-4 Xtra Tread | 3510 | 35,100 |
| Radial RL4 Xtra Tread | 4200 | 42,000 |

(*Courtesy of Caterpillar Inc.*)

**Example 11-21:** Determine the probable tire service life for a single-engine overhung scraper with E-3 tires working under the following conditions:

Maintenance = Average
Maximum Speed = 20 mph
Surface Conditions = Soft earth (no rock)
Load = Recommended
Curves = Medium
Grades = Level
Miscellaneous Conditions = None

**Solution:** Referring to Table 11-13, we obtain the following tire life factors:

| Condition | Factor |
|---|---|
| Maintenance | 0.981 |
| Speed | 0.872 |
| Surface | 1.090 |
| Wheel Positions | 0.872 * |
| Load | 1.090 |
| Curves | 0.981 |
| Grades | 1.090 |
| Misc. Conditions | 1.090 |

**Average of 1.090 for trailing and 0.654 for scraper driving.*

The estimated tire life is:

Tire Life = (0.981 x 0.872 x 1.090 x 0.872 x 1.090
x 0.981 x 1.090 x 1.090) x 2510 Hrs.
= 1.03 x 2510 Hrs. (Table 11-14)
= 2585 hours

Data given in Tables 11-13 and 11-14 was restricted to off-highway trucks and scrapers, and only three types of tires were discussed. You can use Table 11-15 as a tire service-life guideline for other types of equipment.

**Table 11-15: Tire Life for Various Equipment Types (Hours)**

| Equipment Type | Job Conditions | | |
|---|---|---|---|
| | Favorable | Average | Unfavorable* |
| Scrapers, Twin | 4000 | 3000 | 2500 |
| Scrapers With Twin Hitch | 3600 | 2700 | 2250 |
| Scrapers, Single | | | |
| Tractor | 4000 | 3000 | 2500 |
| Scraper | 5500 | 3500 | 2500 |
| Loaders | 4000 | 3000-3500 | 1000-2500 |
| Haulers | 4000 | 3000-3500 | 2000-2500 |
| Bottom Dumps | 8000 | 5000 | 3500 |

**Poor haul-road maintenance can result in even lower tire life figures.*
(*Courtesy of Terex Corporation*)

Caterpillar Inc., has also developed *tire life curves* such as that shown in Figure 11-15. The data given in the curve is based on the following conditions:

- Curves do not allow for additional life from recapping. They assume new tires run to destruction, but this is not necessarily recommended.
- Curves are based on standard machine tires. Optional tires will shift these curves either up, or down.
- Sudden failure (blow out) due to exceeding the ton-mph limitations is not considered, nor are premature failures due to puncture by stumps, sharp tree limbs, or rocks.

Application Zones:

- Zone A: Almost all tires actually wear through the tread from abrasion.
- Zone B: Tires wear out, but others fail prematurely due to rock cuts, rips and non-repairable punctures.
- Zone C: Few, if any tires ever wear through the tread before having to be discarded, usually from rock cuts.

Note: Tire life can often be increased by using extra tread and extra deep tread tires.

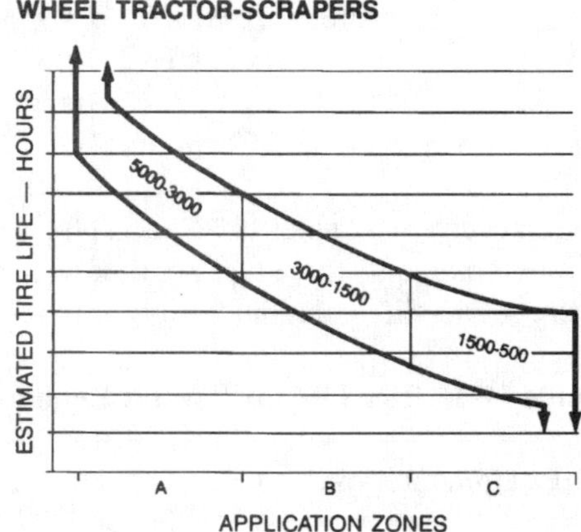


**Figure 11-15. Tire life curve (wheel tractor-scrapers).**
(*Courtesy of Caterpillar Inc.*)

Approximately 75% of all tires can be recapped, provided they are recapped before the tread is worn down to the casing. Recapping costs average about 50% of the original cost of the tire, but tire life can be extended by 75% to 85% of the original tire life. When tires are recapped, the hourly tire replacement costs can be determined by:

$$\text{Hourly Tire Replacement Cost} = \frac{\text{Replacement Cost} + \text{Recap Cost}}{\text{Original Life} + \text{Recap Life}}$$

**(Equation 11.27)**

By substituting the average cost and service life values given above into Equation 11.27, we have:

$$\text{Hourly Tire Replacement Cost} = \frac{1.50 \text{ x Replacement Cost}}{1.80 \text{ x Original Life}}$$

$$= \frac{0.83 \text{ x Replacement Cost}}{\text{Original Life}}$$

**(Equation 11.28)**

**Example 11-22:** Assuming recapping tires that originally cost $15,000, determine the hourly tire replacement cost for the tires given in the previous example.
**Solution:** The hourly tire replacement cost is:

$$\text{Hourly Tire Replacement Cost} = \frac{0.83 \times \$15{,}000}{2585 \text{ Hr.}} \qquad \text{(Eq. 11.28)}$$

$$= \$4.82/\text{hour}$$

## TIRE REPAIR COST

Tires often require repairs during their service life, and the repair costs can be estimated by:

Hourly Tire Repair Cost =Tire Repair Factor x Hourly Tire Replacement Cost **(Equation 11.29)**

where: Tire repair factor is taken from Table 11-16.

**Table 11-16: Tire Repair Factors**

| General Working Conditions | Tire Repair Factor With or Without Recapping |
|---|---|
| Favorable | 12% |
| Average | 15% |
| Unfavorable | 17% |

(*Courtesy of Terex Corporation*)

**Example 11-23:** Assuming favorable job conditions, determine the hourly tire repair costs for the tires given in the previous example.
**Solution:** The hourly tire repair cost is:

Hourly Tire Repair Cost = 0.12 x \$4.82/Hr. (Eq. 11.29; Table 11-16)
= \$0.58/hour

## OPERATOR'S WAGES

Some contractors include the operator's wages in the operating cost of the equipment. Be sure to include the cost for workman's compensation, social security, and fringe benefits when calculating the operator cost.

*Taxable fringe benefits* include paid time allowed for vacation and sick leave. To determine the total taxable labor hours, convert taxable fringe benefits into *taxable hours* and add them to the employee's regular hours. For example, if you offer 40 hours per year for vacation and 40 hours for sick leave, your total taxable hours would be:

$$\text{Total Taxable Hours} = 2000 \text{ Hr.} + 40 \text{ Hr.} + 40 \text{ Hr.}$$
$$= 2080 \text{ hours/year}$$

Next, multiply the total taxable hours by the employee's hourly wages (base wages) to determine the *total taxable wages*. For example, if an employee is paid \$10.00 per hour, the total taxable wages would be:

$$\text{Total Taxable Wages} = 2080 \text{ Hr./Yr.} \times \$10.00\text{/Hr.}$$
$$= \$20{,}800\text{/year}$$

The hourly taxable wages would be:

$$\text{Hourly Taxable Wages} = \frac{\$20{,}800\text{/Yr.}}{2000 \text{ Hr./Yr.}}$$
$$= \$10.40\text{/hour}$$

You must now multiply the total taxable wages by a factor referred to as *labor burden*. Labor burden (*indirect labor costs*) normally consists of FICA, state employment tax and federal unemployment tax. Most contractors also include the cost of general liability insurance and public liability insurance with labor burden. Liability insurance costs vary widely from state to state, and they depend on the contractor's loss experience. Taxes and liability insurance are presently about 25 to 30% of the total taxable wages for most trades. Assuming 30% in our present example, the yearly labor burden would be determined as follows:

$$\text{Annual Labor Burden (Taxes \& Ins.)} = 0.30 \times \$20{,}800\text{/Yr.}$$
$$= \$6240\text{/year}$$

The hourly cost for labor burden would be:

$$\text{Hourly Labor Burden (Taxes \& Insurance)} = \frac{\$6240\text{/Yr.}}{2000 \text{ Hr./Yr.}}$$
$$= \$3.12\text{/hour}$$

Some companies offer *non-taxable fringe benefits* such as medical coverage, tax-deferred pension plans and profit sharing plans. If all three benefits are in place, the cost is presently about 4 to 5% of the base wages. However, the cost for an individual non-taxable fringe benefit can be determined as follows. Assume that it costs \$126.00 per month to provide medical coverage for an employee. Then, the yearly cost for the plan would be:

$$\text{Annual Medical Coverage} = 12 \text{ Mo./Yr.} \times \$126\text{/Mo.}$$
$$= \$1512\text{/year}$$

The hourly cost for medical coverage is:

$$\text{Hourly Medical Coverage} = \frac{\$1512/\text{Yr.}}{2000\ \text{Hr./Yr.}}$$
$$= \$0.76/\text{hour}$$

The total hourly labor cost would be:

$$\text{Total Hourly Labor Cost} = \$10.40/\text{Hr.} + \$3.12/\text{Hr.} + \$0.76/\text{Hr.}$$
$$= \$14.28/\text{hour}$$

This is 42.8% greater than the base hourly wage!

Keep in mind that overtime will often be required when a project deadline dictates, and overtime wages are at least 1-1/2 times that of regular wages. Another option is to work more than one shift per day. In fact, many contractors choose to work overtime, or add extra shifts when renting high-cost equipment, if it will increase profits.

**Example 11-24:** Use the following information to determine the hourly owning and operating costs of a wheel loader.

Equipment = Wheel loader
Delivered Cost = $85,000
Tire Cost = $15,000 (Assume recapping one time)
Salvage Value = $10,000
Estimated Life = 5 years (2000 hr./yr.)
Depreciation Method = Straight-line method
Interest = 10% of AAV
Insurance = 3% of AAV
Taxes = 3% of AAV
Job Conditions = Average
Engine Power = 375 h.p.
Fuel Cost = $1.00/gallon
Cutting Edge Replacement Cost = $600
Operator's Wages = $14.28/hour (including labor burden)

**Solution:** The annual depreciation is:

$$D_{\text{Annual}} = \frac{\$85{,}000 - \$10{,}000 - \$15{,}000}{5\ \text{Yr.}} \qquad \text{(Eq. 11.4)}$$
$$= \$12{,}000/\text{year}$$

The hourly depreciation cost is:

$$\text{Hourly Depreciation Cost} = \frac{\$12{,}000/\text{Yr.}}{2000\ \text{Hr./Yr.}} \qquad \text{(Eq. 11.2)}$$
$$= \$6.00/\text{hour}$$

The average annual value is:

$$\text{AAV} = \frac{[\$85{,}000 \times (5 + 1)] + [\$10{,}000 \times (5 - 1)]}{2 \times 5\ \text{Yr.}} \qquad \text{(Eq. 11.13)}$$
$$= \$55{,}000$$

The total IIT percentage to be applied to the average annual value is:

$$\text{Total IIT \%} = 10\% + 3\% + 3\% \qquad \text{(Given)}$$
$$= 16\%$$

The yearly IIT cost is:

$$\text{Yearly IIT Cost} = 0.16 \times \$55{,}000 \qquad \text{(Eq. 11.8)}$$
$$= \$8800/\text{year}$$

The hourly IIT cost is:

$$\text{Hourly IIT Cost} = \frac{\$8800/\text{Yr.}}{2000\ \text{Hr./Yr.}} \qquad \text{(Eq. 11.2)}$$
$$= \$4.40/\text{hour}$$

The total hourly owning cost is:

$$\text{Total Hourly Owning Cost} = \$6.00/\text{Hr.} + \$4.40/\text{Hr.}$$
$$= \$10.40/\text{hour}$$

Fuel consumption rate is estimated to be:

$$\text{Fuel Consumption (Gal./Hr.)} = 0.057 \times 375\ \text{hp} \times 0.45 \qquad \text{(Eq. 11.15; Table 11-3)}$$
$$= 9.6\ \text{gallons/hour}$$

The hourly fuel cost is:

$$\text{Hourly Fuel Cost} = 9.6\ \text{Gal./Hr.} \times \$1.00/\text{Gal.} \qquad \text{(Eq. 11.16)}$$
$$= \$9.60/\text{hour}$$

The hourly cost for routine maintenance is:

$$\text{Routine Service Cost} = 1/3 \times \$9.60/\text{Hr.} = \$3.20/\text{hour} \qquad \text{(Eq. 11.18)}$$

The hourly cost for major repairs is:

$$\text{Hourly Repair Cost} = \frac{0.06 \times (\$85{,}000 - \$15{,}000)}{1000} \qquad \text{(Eq. 11.20; Table 11-7)}$$

$$= \$4.20/\text{hour}$$

The hourly cutting edge replacement cost is:

$$\text{Hourly Cutting Edge Cost} = \frac{\$600}{2000\ \text{Hr.}} = \$0.30/\text{hour} \qquad \text{(Eq. 11.24; Table 11-10)}$$

The hourly tire replacement cost is:

$$\text{Hourly Tire Replacement Cost} = \frac{0.83 \times \$15{,}000}{3250\ \text{Hr.}} = \$3.83/\text{hour} \qquad \text{(Eq. 11.28; Table 11-15)}$$

The hourly tire repair cost is:

$$\text{Hourly Tire Repair Cost} = 0.15 \times \$3.83/\text{Hr.} = \$0.57/\text{hour} \qquad \text{(Eq. 11.29; Table 11-16)}$$

The operator's wages are:

$$\text{Operator's Wages} = \$14.28/\text{hour} \qquad \text{(Given)}$$

The total hourly operating cost is:

$$\text{Total Hourly Operating Cost} = \$9.60 + \$3.20 + \$4.20 + \$0.30 + \$3.83 + \$0.57 + \$14.28 = \$35.98/\text{hour}$$

The total owning and operating cost is:

$$\text{Total “O \& O” Cost} = \$10.40/\text{Hr.} + \$35.98/\text{Hr.} = \$46.38/\text{hour}$$

**Example 11-25:** Assuming a total owning and operating cost of \$46.38 per hour, determine the unit production cost of a loader working under the conditions given in Example 3-2.

**Solution:** The hourly production rate is:

$$\text{Production} = 112 \text{ LCY/hour} \qquad \text{(Example 3-2)}$$

The unit production cost is:

$$\text{Unit Production Cost} = \frac{\$46.38/\text{Hr.}}{112 \text{ LCY/Hr.}} = \$0.41/\text{LCY} \qquad \text{(Eq. 11.1)}$$

**Example 11-26:** Assuming the machine is in operation only 1200 hours per year, determine the base hourly costs (excluding overhead and profit) that must be charged for the loader given in Example 11-24 in order to recapture owning and operating expenses.

**Solution:** Owning costs are the only costs that are affected, since operating costs are incurred only when the machine is actually working. Therefore, we must recalculate depreciation and IIT costs. The hourly depreciation cost that must be recovered is:

$$\text{Hourly Depreciation Cost} = \frac{\$12{,}000/\text{Yr.}}{1200 \text{ Hr./Yr.}} = \$10.00/\text{hour} \qquad \text{(Eq. 11.2; Example 11-24)}$$

The hourly IIT cost that must be recaptured is:

$$\text{Hourly IIT Cost} = \frac{\$8800/\text{Yr.}}{1200 \text{ Hr./Yr.}} = \$7.33/\text{hour} \qquad \text{(Eq. 11.2; Example 11-24)}$$

The total hourly owning cost that must be charged is:

$$\text{Total Hourly Owning Cost} = \$10.00/\text{Hr.} + \$7.33/\text{Hr.} = \$17.33/\text{hour}$$

Since the operating costs remain unchanged, the total owning and operating cost that must be charged is:

$$\text{Total "O \& O" Cost} = \$17.33/\text{Hr.} + \$35.98/\text{Hr.} = \$53.31/\text{hour} \qquad \text{(Example 11-24)}$$

The owning expense incurred during idle time is:

Idle-Time Expense = (2000 Hr. – 1200 Hr.) x $10.40/Hr. (Example 11-24)

= $8320/year

The contractor can recapture owning costs by charging them to the client at an hourly rate of $17.33 per hour (which includes idle time), based on 1200 hours of operation per year:

Total Owning Costs = $17.33/Hr. x 1200 Hr./Yr.
= $20,796/year*

Or, he can recapture the owning costs by charging them to the client at an hourly rate of $10.40 per hour, based on 1200 hours of operation per year, and account for the $8320 idle-time expense in the general overhead, which must also be charged to the client (Figure 11-17):

Total Owning Costs = ($10.40/Hr. x 1200 Hr./Yr.) + $8320
= $20,800/year*

The important thing to remember is that this expense *must* be recaptured if the contractor is to remain solvent.

**WHEN TO REPLACE EQUIPMENT**

Many contractors replace equipment when major repairs or an overhaul is required, when they have enough money to purchase new equipment, or when they are preparing to start a new project. However, we will review a method that can be used to determine when to replace equipment so that profit is maximized.

Of all the owning and operating costs previously discussed, only depreciation, investment cost and repair cost will be considered when making the replacement decision, since all other costs previously discussed (except for tire replacement) have little impact on the replacement decision. The effect of tire replacement costs on the equipment replacement decision is dependent upon the tire service life.

When making a *replacement analysis*, depreciation and investment costs should be calculated as realistically as possible. Thus, in the example that follows, we will use the sum-of-year digits method for determining depreciation, and we will calculate the IIT and repair costs on an individual yearly basis.

Other costs not previously mentioned must also be taken into consideration when making a replacement analysis, including *inflation cost*, *down-time cost* and *obsolescence cost.*

**The $4.00 difference is in a rounding-off error. The $17.33 is actually $17.333…*

**Inflation Cost**

The difference between the cost of a new machine and the trade-in or resale value of the present machine represents the capital outlay required at the time of replacement. Equipment prices are constantly undergoing inflation, and this must be taken into consideration when making a replacement analysis.

**Down-Time Costs**

Equipment breaks down from time to time, and the length of down time increases as the equipment ages. During down time, the contractor must rent replacement equipment, or production will be lost.

**Obsolescence costs**

As newer equipment is produced each year, some types produce more than older models, due to improved mechanical systems. Also, older equipment, as you and I, slow with age.

**The Replacement Analysis**

To make a replacement analysis, we will sum the applicable costs for each year of the equipment service life, and divide the yearly cost by the cumulative hours at the end of each year. The optimum time for replacement will be at the end of the period which yields the lowest cumulative cost per hour.

**Example 11-27:** Determine the optimum replacement time for a wheel loader under the following conditions:

Delivered Cost = $85,000
Tire Cost = $15,000
Salvage Value = $10,000
Depreciation Method = Sum-of-year digits method
Life = 5 years (2000 Hr./Yr.)
Investment cost = 15% of average yearly investment
Repair Factor = 60% of delivered cost, less tire cost
Down Time = 3% the first year, increasing by 3% for each subsequent year
Inflation = 5% per year (Also applies to rental equipment)
Obsolescence (Production loss) = 4% per year, after the first year
Rental Equipment Cost = $25.00 per hour

**Solution:** The depreciation for the first year is:

$$D_1 = 5/15 \times (\$85{,}000 - \$10{,}000 - \$15{,}000) = \$20{,}000 \quad \text{(Eq. 11.5)}$$

Continuing in this fashion, the depreciation during subsequent years is:

$$D_2 = \$16,000$$
$$D_3 = \$12,000$$
$$D_4 = \$8000$$
$$D_5 = \$4000$$

To determine the average investment on an individual yearly basis, we must determine the book values at the beginning and end of each individual year. To determine the book values, depreciation calculations must include the tire costs; therefore, depreciation during the first year is:

$$D_1 = 5/15 \times (\$85,000 - \$10,000)$$
$$= \$25,000$$

The book value at the end of the first year is:

$$\text{B.V. (End of Year 1)} = \$85,000 - \$25,000$$
$$= \$60,000$$

The average investment during the first year is:

$$\text{Ave. Investment (Year 1)} = \frac{\$85,000 + \$60,000}{2} \quad \text{(Eq. 11.14)}$$
$$= \$72,500$$

The total IIT cost during the first year is:

$$\text{IIT Cost (Year 1)} = 0.15 \times \$72,500 \quad \text{(Eq. 11.8)}$$
$$= \$10,875$$

Continuing in this fashion, the IIT cost during subsequent years is:

IIT Cost (Year 2) = \$7500
IIT Cost (Year 3) = \$4875
IIT Cost (Year 4) = \$3000
IIT Cost (Year 5) = \$1875

The lifetime repair cost is:

$$\text{Total Repair Cost} = 0.60 \times (\$85,000 - \$15,000) \quad \text{(Eq. 11.23)}$$
$$= \$42,000$$

The repair cost during the first year is:

Repair Cost (Year 1) = 1/15 x $42,000 (Eq. 11.22)
= $2800

Continuing in this fashion, the repair cost during subsequent years is:

Repair Cost (Year 2) = $5600
Repair Cost (Year 3) = $8400
Repair Cost (Year 4) = $11,200
Repair Cost (Year 5) = $14,000

The down-time cost during the first year is:

Down-Time Cost (Year 1) = 0.03 x 2000 Hr./Yr. x $25.00/Hr.
= $1500

The down-time cost during subsequent years is:

Down-Time Cost (Year 2) = 0.06 x 2000 x ($25.00 x 1.05*) = $3150
Down-Time Cost (Year 3) = 0.09 x 2000 x ($25.00 x 1.10) = $4950
Down-Time Cost (Year 4) = 0.12 x 2000 x ($25.00 x 1.15) = $6900
Down-Time Cost (Year 5) = 0.15 x 2000 x ($25.00 x 1.20) = $9000
** 5% inflation on rental equipment cost.*

The obsolescence cost during the first year is:

Obsolescence Cost (Year 1) = 0

The obsolescence cost during subsequent years is:

Obsolescence Cost (Year 2) = 0.04 x 2000 x $25.00 = $2000
Obsolescence Cost (Year 3) = 0.08 x 2000 x $25.00 = $4000
Obsolescence Cost (Year 4) = 0.12 x 2000 x $25.00 = $6000
Obsolescence Cost (Year 5) = 0.16 x 2000 x $25.00 = $8000

The increase in replacement cost due to inflation during any given year is:

Replacement cost (Any Year) = 0.05 x $85,000
= $4250

Yearly and cumulative costs are tabulated as follows:

| Cost Item | Year 1 | 2 | 3 | 4 | 5 |
|---|---|---|---|---|---|
| Depreciation | 20,000 | 16,000 | 12,000 | 8,000 | 4,000 |
| IIT | 10,875 | 7,500 | 4,875 | 3,000 | 1,875 |
| Repairs | 2,800 | 5,600 | 8,400 | 11,200 | 14,000 |
| Down Time | 1,500 | 3,150 | 4,950 | 6,900 | 9,000 |
| Obsolescence | 0 | 2,000 | 4,000 | 6,000 | 8,000 |
| Replacement | 4,250 | 4,250 | 4,250 | 4,250 | 4,250 |
| Yearly Total | 39,425 | 38,500 | 38,475 | 39,350 | 41,125 |
| Cumulative Total | 39,425 | 77,925 | 116,400 | 155,750 | 196,875 |
| Cumulative Hours | 2,000 | 4,000 | 6,000 | 8,000 | 10,000 |
| Cost/Cum. Hour | 19.71 | 19.48 | 19.40 | 19.47 | 19.69 |

The period that yields the lowest cumulative hourly cost is during the third year; therefore, we should sell or trade the machine at the end of the third year.

The decision to sell or trade at the end of the third year would also be influenced by the extent of tire wear at this point in time, because we want to avoid selling or trading a machine immediately after we have installed a new set of tires, unless we can be assured that the value of the new tires is equitably credited to the value of the machine.

**Example 11-28:** Determine the individual yearly tire cost under the following conditions:

Initial Tire Cost = \$15,000
(Assume 5% inflation for each new set of tires)
Tire Service Life = 3500 hours
Machine Service Life = 5 years (2000 hr./yr.)

**Solution:** During the 10,000-hour machine life, we will be required to replace the tires at 3500 and 7000 hours. The cost of the replacement sets of tires will be:

Tire Replacement Cost (2nd Set) = 1.05 x \$15,000
= \$15,750

Tire Replacement Cost (3rd Set) = 1.10 x \$15,000
= \$16,500

During the first year, we will be running on the original set of tires, with a yearly cost of:

$$\text{Tire Cost (Year 1)} = \frac{2000\ \text{Hr.}}{3500\ \text{Hr.}} \times \$15{,}000$$
$$= \$8571$$

During the second year, we will run for 1500 hours on the original set of tires, then 500 hours on the second set, with a yearly cost of:

$$\text{Tire Cost (Year 2)} = \left(\frac{1500}{3500} \times \$15{,}000\right) + \left(\frac{500}{3500} \times \$15{,}750\right)$$
$$= \$8679$$

During the third year, we will continue to run on the second of tires, with a yearly cost of:

$$\text{Tire Cost (Year 3)} = \frac{2000}{3500} \times \$15{,}750$$
$$= \$9000$$

During the fourth year, we will run for 1000 hours on the second set of tires, then 1000 hours on the third set, with a yearly cost of:

$$\text{Tire Cost (Year 4)} = \left(\frac{1000}{3500} \times \$15{,}750\right) + \left(\frac{1000}{3500} \times \$16{,}500\right)$$
$$= \$9214$$

During the fifth year, we will continue to run on the third set of tires, with a yearly cost of:

$$\text{Tire Cost (Year 5)} = \frac{2000}{3500} \times \$16{,}500$$
$$= \$9429$$

**Example 11-29:** Work Example 11-27 assuming a tire service life and cost as given in the previous example.

**Solution:** Referring to Example 11-28, we will be required to purchase a new set of tires during the second and fourth years of operation. Therefore, we can include the tire replacement costs in a tabulated form as follows:

| Cost Item | Year 1 | 2 | 3 | 4 | 5 |
|---|---|---|---|---|---|
| Totals (Example 11-27) | 39,425 | 38,500 | 38,475 | 39,350 | 41,125 |
| Tire Replacement Cost | 8,571 | 8,679 | 9,000 | 9,214 | 9,429 |
| Cumulative Total | 47,996 | 95,175 | 142,650 | 191,214 | 241,768 |
| Cumulative Hours | 2,000 | 4,000 | 6,000 | 8,000 | 10,000 |
| Cost/Cum. Hour | 24.00* | 23.79* | 23.78* | 23.90* | 24.18* |

Taking tire wear into consideration, we should sell or trade the machine at the end of the third year.

## RENT, OR PURCHASE?

A contractor might choose to rent, rather than purchase certain equipment for many reasons. Specialized equipment might be required for only a short period of time. The contractor might lack capital, or he might be pessimistic about obtaining future work. He could be so busy that his personal equipment is inadequate to do all the work required, or projects could be so scattered out that it is cheaper to rent equipment available near some of the project locations.

The terms of rental agreements vary, depending on local customs. As a general rule, the owner is responsible for overhauls and major repairs, while the renter is responsible for minor repairs and replacing fast-wearing items. Excluding normal wear, the machine is to be returned in as good condition as it was when it was taken.

Based on costs, only, the decision whether to rent or own a piece of equipment can be determined by tabulating owning costs versus rental costs on a cost-per-cumulative-hour basis.

**Example 11-30:** Assuming that the loader given in Examples 11-27 and 11-29 rents for $23.00 per hour for the first year, and increases at an inflation rate of 5% per year, determine the break-even point (point of indifference) when comparing renting versus owning costs. Assume that the rental service is responsible for major repairs and tire replacement.

**Solution:** The applicable cumulative hourly costs for owning the equipment is given in tabular form in Example 11-29. The cumulative hourly rental costs would be:

Rental Cost (Year 1) = $23.00/hour
Rental Cost (Year 2) = 1.05 x $23.00 = $24.15/hour
Rental Cost (Year 3) = 1.10 x $23.00 = $25.30/hour
Rental Cost (Year 4) = 1.15 x $23.00 = $26.45/hour
Rental Cost (Year 5) = 1.20 x $23.00 = $27.60/hour

**These cost-per-cumulative-hour values are substantially higher than those shown in Example 11-27 because tire replacement costs were not considered in Example 11-27, except for determining the average annual investment on a yearly basis.*

We can determine the point of indifference graphically as shown in Figure 11-16.

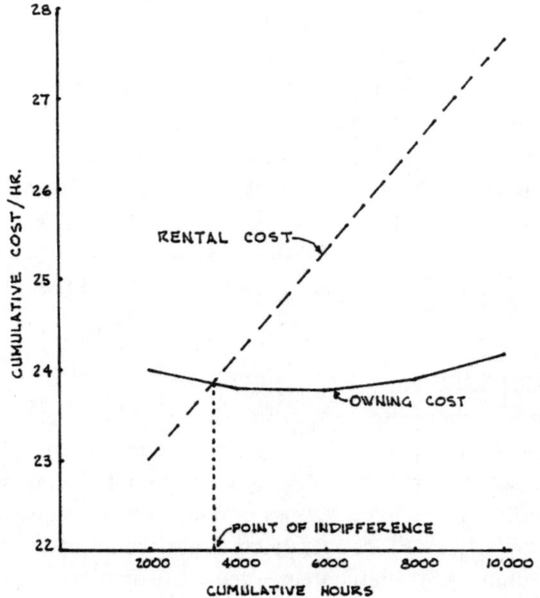


**Figure 11-16. Point of Indifference.**

In this particular situation, it is cheaper to rent the equipment up to 3450 hours, but after this point in time, it is cheaper to own the equipment.

## FIELD OVERHEAD VERSUS GENERAL (OFFICE) OVERHEAD

*General overhead* (also referred to as general operating overhead, office overhead, or indirect expenses) includes items whose costs can be distributed among all projects, whereas, *field overhead* (direct expenses) includes items whose costs are allocated to one particular project. Typical general overhead items are shown in Figure 11-17, and typical field overhead items are shown in Figure 11-18. Note in Figure 11-18 that the superintendent's salary and foreman's wages are included as field overhead. This is a matter of personal preference. Some contractors allocate superintendent's salary to general overhead and the foreman's wages to field overhead.

---

Principals' salaries
Employees' salaries
Office-related payroll taxes and social security
Consultants' fees
Rent
Utilities
Telephone
Duplication

Postage and shipping
Automobile expenses
Travel expenses
Office supplies
Subscriptions
Professional dues
Contributions
Insurance
Hospitalization
Advertising
Depreciation
Amortization
Miscellaneous

**Figure 11-17. List of general overhead expenses.**

Supervision
Foremen
Layout and survey
Building permit
Office and storage
Telephone
Utilities
Temporary electrical
Temporary heat
Portable toilets
Vehicle gas and oil
Special equipment
Cleanup and haul trash
Fence and barricades
Project sign
Hoisting
Contingencies
Safety (OSHA)
Testing
Bond on subcontractors
Job-related sales tax
Job-related payroll taxes and social security
Blueprints
Miscellaneous
Job-related insurance
Total project bond

**Figure 11-18. List of field overhead expenses.**

Through meticulous bookkeeping, the cost of any general overhead item (e.g., postage) can be accounted for on a job-by-job basis; however, for bidding purposes, all general overhead expenses are normally allocated to any given project as a percentage of gross annual income, and is calculated as follows:

$$\text{General Overhead (\%)} = \frac{\text{Total Annual General Overhead Expenses}}{\text{Total Annual Gross Income}}$$

**(Equation 11.30)**

It is obvious that the general overhead percentage can change from one year to the next, depending upon the relationship of gross income to general operating expense.

There are some "gray areas" encountered when determining precisely where a certain expense should be allocated. As an example, consider the gasoline expense for a superintendent's truck. For most projects built locally, the simplest way to account for this expense is to allocate it to general overhead. However, if the project is remote, but close enough that daily commuting is more practical than relocating the superintendent, the additional gasoline expense beyond those already included in general overhead should be accounted for as a field overhead expense.

**Example 11-31:** Determine the field overhead expense for gasoline used in a field superintendent's truck under the following conditions:

Project Duration = 260 working days
Round-Trip Distance = 150 miles
Gasoline Cost = $0.07 per mile

**Solution:** The field overhead gasoline expense is:

Gasoline Expense = 260 Days x 150 Mi./Day x $0.07/Mi.
= $2730

I cannot overemphasize the importance of the proper and realistic allocation of overhead expenses. As a general rule, there is an overlapping of field and general overhead expenses due to improper allocation, reducing the contractor's ability to be competitive. Unknowingly, some contractors make adjustments by reducing their labor unit prices, or their markup for profit in order to increase their competitiveness. This type of adjustment can be disastrous when the contractor bids a project requiring low material and high labor expenses. In Example 11-31, we have overlapping overheads, since the general overhead used for bidding the remote project already includes the cost of gasoline ordinarily used for traveling to and from local projects. The simplest way to correct this problem is to leave the general overhead as is, and make an adjustment in the field overhead expense.

**Example 11-32:** Work the previous example assuming the general overhead includes the expense of a superintendent's traveling an average round-trip distance of 10 miles to a locally-constructed project.
**Solution:** The distance traveled to and from the remote project, less the average distance ordinarily traveled to a local project (the expense of which is already included in the general overhead) is:

$$\text{Additional Distance Traveled} = 150 \text{ Mi.} - 10 \text{ Mi.} = 140 \text{ miles}$$

Therefore, the correct gasoline expense to be allocated to field overhead is:

$$\text{Gasoline Expense} = 260 \text{ Days} \times 140 \text{ Mi./Day} \times \$0.07/\text{Mi.} = \$2548$$

Note that no adjustments are made regarding the gasoline used for running job-related errands near the site, since similar errands would be necessary, whether the project were local, or remote. Therefore, this gasoline expense is accounted for in the general overhead.

In Examples 11-31 and 11-32, we separated (on paper) the gasoline expense allocated to general overhead from that allocated to field overhead for a remote project. If the bookkeeper is unaware of this method of allocation, all gasoline charged by the superintendent will be accounted for in general overhead during the current year, causing the general overhead to be unrealistically inflated in the coming year. To avoid such problems, the construction company must establish a clear line of communication between field personnel and the estimating and bookkeeping departments regarding the allocation of field and general overhead expenses.

We need to use caution when we apply general overhead to a project, and the best way to explain this is to give some examples of unwise applications. In the following examples, we will assume that general overhead is 10% of the annual gross income, based on a \$120,000 annual general overhead expense (approximately \$2308/week) while producing a \$1,200,000 annual work load.

**Example 11-33:** Assuming 10% general overhead, determine the weekly overhead charged by an electrical contractor installing a \$90,000 transformer while incurring a \$10,000 labor cost during a two-week period.
**Solution:** The overhead charged is:

$$\text{Weekly Overhead} = \frac{0.10 \times \$100{,}000}{2 \text{ Weeks}} = \$5000/\text{week}$$

We can see why there is a recent trend for the project owner to purchase expensive equipment and award a contract for the installation, only. The overhead expense charged is more than double the actual weekly general overhead expense, and the contractor would probably have other projects simultaneously contributing to the payment of the general overhead expenses. These additional projects would increase general overhead expenses; however, the increase would not be substantial. For instance, expenses for office supplies, copying, postage, etc. would increase, due to an expanded flow of paperwork; however, major overhead expenses such as salaries would not necessarily increase (but only the work load at the office).

**Example 11-34:** Assuming 10% general overhead, determine the weekly overhead charged for a project requiring \$10,000 for material and \$90,000 for labor if the project requires 26 weeks to complete.
**Solution:** The overhead charged is:

$$\text{Weekly Overhead} = \frac{0.10 \times \$100{,}000}{26 \text{ Weeks}}$$
$$= \$385/\text{week}$$

To recapture our general overhead costs, we would have to be simultaneously working 6 projects of this size and similarity.

A suggested solution to this dilemma is to base general overhead on field labor costs, only. We could express the general overhead cost as follows:

$$\text{General Overhead} = \frac{\text{General Overhead Costs/Year}}{\text{Total Manhours Worked/Year}} \qquad \textbf{(Equation 11.31)}$$

For example , if your field personnel accrue an annual total of 10,400 manhours per year (200 manhours per week), and your annual general overhead is \$120,000, your hourly general overhead cost will be:

$$\text{General Overhead} = \frac{\$120{,}000}{10{,}400 \text{ Hr.}} \qquad \text{(Eq. 11.31)}$$
$$= \$11.54/\text{hour}$$

**Example 11-35:** Determine the weekly general overhead cost for the project given in Example 11-33 assuming a total of 400 manhours is required with a general overhead cost of \$11.54 per manhour.
**Solution:** The weekly overhead cost is:

$$\text{Weekly Overhead} = \frac{\text{400 Manhours}}{\text{2 Weeks}} \times \$11.54/\text{Manhour}$$
$$= \$2308/\text{week}$$

**Example 11-36:** Determine the weekly general overhead cost for the project given in Example 11-34 assuming a total of 5200 manhours is required with a general overhead cost of $11.54 per manhour.
**Solution:** The weekly overhead cost is:

$$\text{Weekly Overhead} = \frac{\text{5200 Manhours}}{\text{26 Weeks}} \times \$11.54/\text{Manhour}$$
$$= \$2308/\text{week}$$

Just as recapturing general overhead, based on a *percentage* of total annual income as given in Examples 11-33 and 11-34 requires that you contract a minimum of $1,200,000 annually, recapturing general overhead based on total *manhours* worked as given in Examples 11-35 and 11-36 requires that your field personnel work a minimum of 10,400 manhours per year.

## OPTIMUM PROFIT

Before we get into detail regarding how much markup to apply to a bid, I would like to clear up a national controversy within the construction industry. The controversy involves whether or not to apply markup to general overhead expenses, as well as to the direct project costs. There is a difference, as the following example illustrates.

**Example 11-37:** Assuming a project base cost of $100,000, general overhead of 10% and profit of 5%, determine the total bid when: a) profit is not applied to the general overhead, and b) profit is applied to the general overhead.
**Solution:** a) When profit is not applied to the general overhead expense, the total bid is tabulated as follows:

$$\text{Total Bid} = \$100{,}000 \times 1.15$$
$$= \$115{,}000$$

b) When profit is applied to the general overhead expense, the total bid is tabulated as follows:

$$\text{Total Bid} = \$100{,}000 \times 1.10 \times 1.05$$
$$= \$115{,}500$$

The second tabulation is the correct way to tally the bid. The theory is that you are entitled to as much profit for a secretary's typing a letter and placing a stamp on an envelope as you are for a carpenter's nailing up a stud, or an operator's moving a cubic yard of earth.

Determining the markup for a project is almost an art form, based on intuition and knowledge regarding the local business climate (present and future), the desirability of the project, the size of the project, the amount of work subcontracted, and your company's ability to schedule and efficiently build the project.

A depressed market reduces the number of projects available for bid, and increases the number of competing bidders. Some "hungry" contractors will often bid a project with little or no markup to avoid laying off key personnel. Also, a large number of bidders increases the likelihood of a low bid resulting from an estimating mistake.

Some projects are undesirable due to the location, type of project, owner, designer, or the need for specialized equipment and personnel, etc. These types of projects are referred to as "*sleepers*," and you can oftentimes apply a greater markup to your bid, and still get the job. On the other hand, some contractors will bid a project with very little markup, knowing (from inside information) that the project will eventually be expanded, involving lucrative change orders.

There are instances when one project can be used to increase profits on another, especially in the earthmoving business. For instance, one project might require you to remove excess material (spoil) that can be used for fill (borrow) on another project in close proximity to the first. In such a scenario, a contractor will bid the "spoil project" with a low markup to ensure his being low bidder to reduce his costs on the "borrow project." Therefore, he will have a competitive advantage, and will make a profit on both projects.

As a general rule, the larger the project, the less markup that is applied. This is due mainly to the fact that a small project can be almost as time-consuming as a large project, with regard to the headaches and paperwork involved. This is especially true for general contractors who subcontract a majority of the work and have a "safety cushion" built into their estimate, due to markup applied to the cost of the subcontracted work. In fact, some general contractors obtain a competitive edge by applying markup that is proportional to the risk involved. For example, the markup might be calculated as follows:

| Cost Item | Markup |
|---|---|
| Materials by G.C. | 5% |
| Subcontract Work | 3% |
| Labor by G.C. | 12% |

Project *scheduling* (and the *implementation* of same) is one of the biggest money-saving and competitive bidding tools a contractor can have. The contractor who can predict the probable time required to finish a project will often be the low bidder, even though his bid includes a high markup. This is due to the fact that a contractor guessing at the project completion time will usually be conservative and will estimate a longer duration than is actually required, and time-related expenses such as supervision, field overhead, etc., will result in a high bid, even though a low markup is applied.

High markup will reduce your chances of being low bidder, but low markup will reduce your profit. The *optimum markup* is that which yields the greatest profit for a given yearly work load. To make a *profit analysis,* recall all bidding data in the past, and separate the projects into specific dollar ranges, since project size often determines the amount of markup. For example, assume that among the projects you bid last year, ten were around the one-million dollar amount ($10,000,000 yearly total) as tabulated below:

| % Markup | Percentage of time you would have been low bidder using the given markup |
|---|---|
| 5% | 40% |
| 10% | 30% |
| 15% | 20% |
| 20% | 10% |
| 25% | 5% |

From this data, you would construct the following table:

| % Markup | Yearly Work Load | Profit |
|---|---|---|
| 5% | 0.40 x $10,000,000 = $4,000,000 | x 0.05 = $200,000 |
| 10% | 0.30 x $10,000,000 = $3,000,000 | x 0.10 = $300,000 |
| 15% | 0.20 x $10,000,000 = $2,000,000 | x 0.15 = $300,000 |
| 20% | 0.10 x $10,000,000 = $1,000,000 | x 0.20 = $200,000 |
| 25% | 0.05 x $10,000,000 = $ 500,000 | x 0.25 = $125,000 |

Based on your historical bidding data, the optimum markup for a million-dollar project would be between 10% and 15%. At 12.5% markup, and assuming you would be low bidder 25% of the time, the profit would be:

Profit = $10,000,000 x 0.25 x 0.125
= $312,500

## OPERATOR PROFICIENCY

The old adage, "You get what you pay for" is true, especially when operating earthmoving equipment. That is, a skilled operator might cost you more than an unskilled operator, but he is often well worth the additional expense. The proficiency of an unskilled operator, as compared to that of a skilled operator, is referred to as *relative proficiency* and is defined as:

$$\text{Relative Proficiency} = \frac{\text{Production of Unskilled Operator}}{\text{Production of Skilled Operator}}$$

**(Equation 11.32)**

The *relative unit production cost* for any operator can be determined by:

$$\text{Relative Unit Production Cost} = \frac{\text{Total Hourly Cost}}{\text{Relative Proficiency}} \quad \textbf{(Equation 11.33)}$$

where: Total hourly cost includes equipment costs
and, Relative proficiency of the most skillful operator is 1.00.

**Example 11-38:** Determine the relative unit production cost under the following conditions:

Machine Cost = \$40.00/hour
Skilled Wages = \$15.00/hour
Unskilled Wages = \$7.50/hour
Skilled Production = 500 C.Y./hour
Unskilled Production = 300 C.Y./hour

**Solution:** The relative proficiency of the unskilled operator is:

$$\text{Relative Proficiency} = \frac{300 \text{ C.Y./Hour}}{500 \text{ C.Y./Hour}} \quad \text{(Eq. 11.32)}$$
$$= 0.60$$

The relative unit production costs are:

$$\text{Relative Unit Production Cost (Skilled Operator)} = \frac{\$55.00}{1.00} \quad \text{(Eq. 11.33)}$$
$$= \$55.00$$

$$\text{Relative Unit Production Cost (Unskilled Operator)} = \frac{\$47.50}{0.60} = \$79.17$$

Another way to look at this is to compare the unit production costs:

$$\text{Cost/C.Y. (Skilled)} = \frac{\$55.00}{500\ \text{C.Y.}} = \$0.11/\text{C.Y.} \qquad \text{(Eq. 11.1)}$$

$$\text{Cost/C.Y. (Unskilled)} = \frac{\$47.50}{300\ \text{C.Y.}} = \$0.16/\text{C.Y.}$$

The other side of the aforementioned adage is: "Your pay for what you get." That is, to be fair to an employee, and to avoid losing him to a higher-paying competitor, consider giving him raises on a performance-improvement basis. Based on performance, wages deserved can be determined by:

$$\text{Deserved Wages} = [\text{Relative Unit Cost (Skilled)} \times \text{Relative Proficiency}] - \text{Equipment Cost} \qquad \textbf{(Equation 11.34)}$$

**Example 11-39:** Assuming the unskilled operator given in the previous example increases his production rate consistently to 450 cubic yards per hour, determine the wages he should be paid.
**Solution:** The wages the operator deserves is:

$$\text{Deserved Wages} = (\$55.00 \times 450/500) - \$40.00 = \$9.50/\text{hour} \qquad \text{(Eq. 11.34; Example 11-38)}$$

**Example 11-40:** Determine the minimum production required for the operator given in Example 11-38 to be deserving of his $7.50/hour wages.
**Solution:** The minimum relative proficiency required is:

$$\text{Min. Relative Proficiency} = \frac{\text{Total Hourly Costs (Unskilled)}}{\text{Total Hourly Costs (Skilled)}} \qquad \textbf{(Equation 11.35)}$$

$$= \frac{\$47.50}{\$55.00} = 0.864 \qquad \text{(Example 11-38)}$$

The minimum production rate required is:

Min. Required Production (C.Y./Hr.) = 0.864 x 500 C.Y./Hr. (Example 11-38)

= 432 C.Y./hour

**OPTIMUM PROJECT DURATION**

Increased project duration increases the total project overhead costs; however, shortening project duration will require more crews (including equipment and personnel), and will increase the cost of crew mobilization. The optimum project duration can be determined by:

$$\text{Optimum Project Duration} = \sqrt{\frac{C_M}{C_{OH}}} \qquad \textbf{(Equation 11.36)}$$

where: $C_{OH}$ = Cost of time-dependent project overhead for one day
and, $C_M$ = Total cost to mobilize all crews necessary to complete the project in one day

$$= \frac{\text{Project Extent}}{\text{Crew Production}} \times C_1 \qquad \textbf{(Equation 11.37)}$$

where: Project extent is the total work required; e.g., total C.Y.
and, $C_1$ = Cost required to mobilize one crew.

The number of crews required to complete the project within the optimum project time is:

$$\text{Crews Required} = \frac{C_{OH} \times \text{Opt. Project Duration}}{C_1} \qquad \textbf{(Equation 11.38)}$$

where: $C_1$ = Cost required to mobilize one crew.

**Example 11-41:** Determine the optimum project time and crews required for a project under the following conditions:

Project Extent = 2,000,000 C.Y.
Production/Crew = 2000 C.Y./day
Mobilization Cost = $1000/crew
Time-Dependent Project Overhead = $300/day

**Solution:** The total cost to mobilize all crews necessary to complete the project in one day is:

$$C_M = \frac{2{,}000{,}000}{2000} \times \$1000 \qquad \text{(Eq. 11.37)}$$

$$= \$1{,}000{,}000$$

The optimum project duration is:

$$\text{Opt. Project Duration} = \sqrt{\frac{\$1{,}000{,}000}{\$300}} \qquad \text{(Eq. 11.36)}$$

$$= 58 \text{ days}$$

The number of crews required to complete the project in 58 days is:

$$\text{Crews Required} = \frac{\$300/\text{Day} \times 58 \text{ Days}}{\$1000/\text{Crew}} \qquad \text{(Eq. 11.38)}$$

$$= 18 \text{ crews}$$

# APPENDIX A

## DECIMAL EQUIVALENTS

**TABLE A-1: Converting inches to feet and meters**

| Inches | Approximate Decimal Equivalent (Feet) | Approximate Metric Equivalent (Meters) |
|---|---|---|
| ⅛ | 0.01 | 0.003 |
| ¼ | 0.02 | 0.006 |
| ½ | 0.04 | 0.013 |
| 1 | 0.08 | 0.025 |
| 1½ | 0.13 | 0.038 |
| 2 | 0.17 | 0.051 |
| 2½ | 0.21 | 0.064 |
| 3 | 0.25 | 0.076 |
| 3½ | 0.29 | 0.089 |
| 4 | 0.33 or 0.34* | 0.102 |
| 4½ | 0.38 | 0.114 |
| 5 | 0.42 | 0.127 |
| 5½ | 0.46 | 0.140 |
| 6 | 0.50 | 0.152 |
| 6½ | 0.54 | 0.165 |
| 7 | 0.58 | 0.178 |
| 7½ | 0.63 | 0.191 |
| 8 | 0.67 | 0.203 |
| 8½ | 0.71 | 0.216 |
| 9 | 0.75 | 0.229 |
| 9½ | 0.79 | 0.241 |
| 10 | 0.83 or 0.84* | 0.254 |
| 10½ | 0.88 | 0.267 |
| 11 | 0.92 | 0.279 |
| 11½ | 0.96 | 0.292 |
| 12 | 1.00 | 0.305 |

*For the sake of cautious estimating, I use the factors 0.34′ and 0.84′ for 4 inches and 10 inches, respectively. This results in quantities that are a bit long, but safe.

**TABLE A-2: Fractions of Inches Expressed as Decimals**

| Fraction (Inch) | Decimal Equivalent (Inch) | Approximate Metric Equivalent (Millimeters) |
|---|---|---|
| 1/16 | 0.0625 | 1.588 |
| 1/8 | 0.1250 | 3.175 |
| 3/16 | 0.1875 | 4.763 |
| 1/4 | 0.2500 | 6.350 |
| 5/16 | 0.3125 | 7.938 |
| 3/8 | 0.3750 | 9.525 |
| 7/16 | 0.4375 | 11.113 |
| 1/2 | 0.5000 | 12.700 |
| 9/16 | 0.5625 | 14.288 |
| 5/8 | 0.6250 | 15.875 |
| 11/16 | 0.6875 | 17.463 |
| 3/4 | 0.7500 | 19.050 |
| 13/16 | 0.8125 | 20.638 |
| 7/8 | 0.8750 | 22.225 |
| 15/16 | 0.9375 | 23.813 |
| 1 | 1.0000 | 25.400 |

EXAMPLE A-1:

$$7\text{-}3/8'' \div 2 = \frac{7.375''}{2} \quad \text{(Table A-2)}$$

$$= 3.6875''$$

$$= 3\text{-}11/16 \text{ inches} \quad \text{(Table A-2)}$$

EXAMPLE A-2:

$$3\text{-}7/16'' + 4\text{-}3/8'' + 5\text{-}3/4'' + 2\text{-}1/2''$$

$$= 3.4375'' + 4.375'' + 5.750'' + 2.500'' \quad \text{(Table A-2)}$$

$$= 16.0625''$$

$$= 16\text{-}1/16'' \quad \text{(Table A-2)}$$

$$= 1' - 4\text{-}1/16''$$

EXAMPLE A-3: Convert 1½ inches into terms of a decimal-foot equivalent.

Solution:

$$1.5 \text{ inches} \times \frac{\text{L.F.}}{12 \text{ inches}} = 0.125 \text{ feet}$$

EXAMPLE A-4: Convert 0.12 lineal feet into terms of inches.

Solution:

$$0.12 \text{ L.F.} \times 12 \text{ inches/L.F.} = 1.44 \text{ inches}$$

EXAMPLE A-5: Convert 0.44 inches into an approximate fractional equivalent of 16th's of an inch.

Solution:

$$0.44'' = \frac{44''}{100}$$

$$\frac{44}{100} = \frac{x}{16}$$

$$x = \frac{44 \times 16}{100}$$

$$= 7.04$$

Therefore,

$$0.44'' = \frac{7.04''}{16}$$

$$\simeq \frac{7}{16} \text{ inches}$$

EXAMPLE A-6: Express 47.625 inches in terms of feet and inches.

Solution: We can simplify the problem by deducting the nearest whole foot from the given value; i.e.,

$$47.625'' - 36'' = 3' \text{ - } 11.625''$$

Table A-2 can now be used to simplify the "inch" portion of the result:

$$3' \text{ - } 11.625'' = 3' \text{ - } 11\text{-}5/8'' \qquad \text{(Table A-2)}$$

EXAMPLE A-7: An American standard drill bit has a diameter of 0.203125 inches. Express the diameter in terms of a fraction of an inch.

Solution: This type of bit is manufactured only in diameter increments of half, quarter, 8th, 16th, 32nd and 64th inches. Since the given decimal value cannot be found in Table A-2, the bit will have a diameter in terms of 32nd's, or 64th's of an inch. Through research, I have found the following to always be true:

| Diameter Increments of an Inch | Number of Digits to the Right of the Decimal Point |
|---|---|
| Halves | 1 |
| Quarters | 2 |
| Eighths | 3 |
| Sixteenths | 4 |
| Thirty-Second's | 5 |
| Sixty-Fourth's | 6 |

Since the given decimal has 6 digits to the right of the decimal, the fraction will be expressed in terms of 64th's of an inch; i.e.,

$$x/64 = 0.203125$$

Solving for the numerator of the fraction, we have:

$$\begin{aligned} x &= 64 \times 0.203125 \\ &= 13 \end{aligned}$$

Thus, the diameter of the drill bit is 13/64 inches.

EXAMPLE A-8: Architectural brick has a height of 2¼ inches and is normally laid up vertically so that three brick and three mortar joints equal eight inches. Determine the thickness of each mortar joint.

Solution: The total mortar joint thickness is:

$$\begin{aligned} \text{Total Joint Thickness} &= 8'' - (3 \times 2.25'') \\ &= 1.25 \text{ inches} \end{aligned}$$

The thickness of each joint is:

$$\begin{aligned} \text{Joint Thickness} &= \frac{1.25''}{3} \\ &= 0.416666\ldots\ldots\text{etc. inches} \end{aligned}$$

For simplification, we will round this answer off to 0.416 inches; therefore, we can express the decimal as a reducible fraction:

$$0.416'' = \frac{416''}{1000} = \frac{208''}{500} = \frac{104''}{250} = \frac{52''}{125}$$

This is as far as the fraction can be reduced. If we wish to approximate the value of the fraction in terms of standard fractional inch subdivisions, we can change the fraction thusly:

$$\begin{aligned} \frac{52''}{125} &\simeq \frac{26''}{64} \\ &= 13/32 \text{ inches} \end{aligned}$$

## APPENDIX B
## Metric Conversion Factors

**Table B-1: Conversion Factors—U.S. to Metric**

| U.S. Unit | Metric Equivalent |
|---|---|
| **Length** | |
| mile (mi.) = 5,280 ft. = 1,760 yd. | 1.609 km. |
| yard (yd.) = 3 ft. = 36 in. | 0.9144 m. |
| foot (ft.) = 12 in. | 0.3048 m. = 30.48 cm. |
| inch (in.) | 2.54 cm. |
| **Area** | |
| square mile (sq. mi.) = 640 acres | 2.59 sq. km. |
| acre (a) = 4840 sq. yd. | 4047 sq. m. |
| square yard (sq. yd.) = 9 sq. ft. | 0.836 sq. m. |
| square foot (sq. ft.) = 144 sq. in. | 0.093 sq. m |
| square inch (sq. in.) | 6.451 sq. cm. |
| square (SQ) = 100 sq. ft. | 9.29 sq. m. |
| **Volume (cubic)** | |
| cubic yard (cu. yd. or C.Y.) = 27 cu. ft. | 0.765 cu. m. |
| cubic foot (cu. ft.) = 1728 cu. in. | 0.028 cu. m. |
| cubic inch (cu. in.) | 16.387 cu. cm. |
| board foot (B.F.) = 0.083 cu. ft. = 144 cu. in. | 2360 cu. cm. |
| **Capacity (liquid)** | |
| gallon (gal.) = 4 qt. = 231 cu. in. = 0.134 cu. ft. | 3.785 l. |
| quart (qt.) = 2 pt. = 57.75 cu. in. = 32 fl. oz. | 0.946 l. |
| pint (pt.) | 0.473 l. |
| **Weight** | |
| short ton (ton) = 2000 lb. = 20 cwt. | 0.907 metric tons |
| short hundredweight (cwt.) = 100 lb. | 45.359 kg. |
| pound (lb.) = 16 oz. | 0.454 kg. = 453.59 g. |
| ounce (oz.) | 28.349 g. |
| **Density** | |
| pounds per cubic yard | 0.5929 kg/cu. m. |
| density ($H_20$) = 8.345 lb./gal. = 62.428 lb./cu. ft. | 1.0 kg/1. |
| **Pressure** | |
| pounds per sq. ft. (psf) | 0.04783 kilopascals |
| pounds per sq. in. (psi) | 6.8951 kilopascals |
| atmosphere (atm) | 1.014 bar |
| **Velocity** | |
| miles per hour (mph) = 88 ft./min. | 1.61 km./hr |

**Table B-2: Conversion Factors—Metric to U.S.**

| Metric Unit | Approximate U.S. Equivalent |
|---|---|
| **Length** | |
| kilometer (km.) = 1000 m. | 0.62 mi. |
| meter (m.) = 0.001 km. = 100 cm. = 1000 mm. | 3.28 ft. = 39.37 in. |
| centimeter (cm.) = 0.01 m. = 10 mm. | 0.39 in. |
| millimeter (mm.) = 0.001 m. | 0.039 in. |
| **Area** | |
| square kilometer (sq. km.) | 0.3861 sq. mi. |
| hectare (ha.) | 2.47 acres |
| square meter (sq. m.) | 1.196 sq. yd. = 10.76 sq. ft. |
| square centimeter (sq. cm.) | 0.155 sq. in. |
| **Volume (cubic)** | |
| cubic meter (cu. m.) | 1.31 cu. yd. = 35.37 cu. ft. |
| cubic centimeter ($cm^3$) | 0.034 fl. oz. |
| **Capacity (liquid)** | |
| kiloliter (kl.) = 1000 1. | 35.37 cu. ft. |
| liter (1.) = 0.001 kl. = 1000 ml. | 0.264 U.S. gal. = 1.057 qt. = 61.02 cu. in. |
| milliliter (ml.) = 0.001 1. = 1 cu. cm. | 0.06 cu. in. |
| **Weight (mass)** | |
| metric ton (MT) = 1000 kg. | 1.1 tons = 2204.62 lb. |
| kilogram (kg.) = 1000 g. | 2.2046 lb. = 35.37 oz. |
| gram (g.) = 0.001 kg. | 0.035 oz. |
| **Density** | |
| kilograms per cubic meter | 1.687 lb./cu. yd. = 0.062 lb./cu. ft. |
| **Pressure** | |
| kilopascal | 20.91 psf = 0.145 psi |
| bar | 14.5 psi = 0.987 atm. |
| **Velocity** | |
| kilometers per hour (km/h) | 0.621 mph |

# APPENDIX C

## GRADE RESISTANCE (ASSISTANCE) EQUATION

For each one-percent increase in grade, the tractive effort of a vehicle is increased (while ascending), or decreased (while descending) by approximately 20 pounds per ton of gross vehicle weight. Figure C-1 illustrates why this is so.

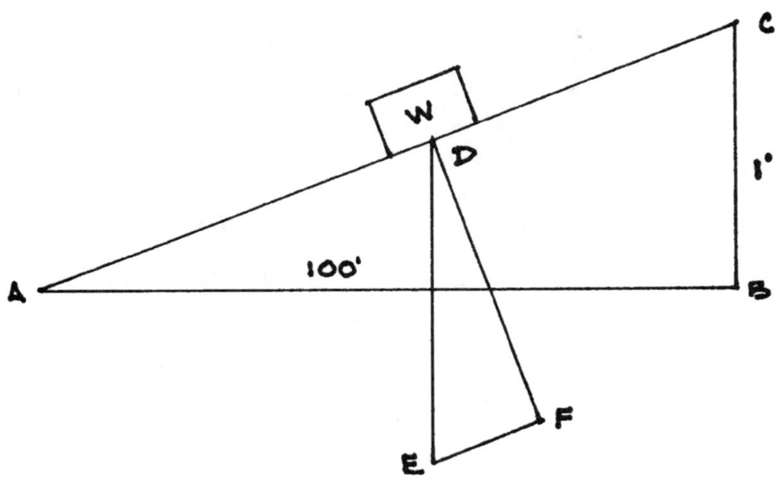


**Figure C-1. Grade resistance force diagram.**

Triangle ABC represents a one-percent slope and Triangle DEF is similar to Triangle ABC, where DE is perpendicular to AB, DF is perpendicular to AC and EF is parallel to AC. Since the triangles are similar, we have:

$$\frac{EF}{DE} = \frac{BC}{AC}$$

Since grade resistance is the component of vehicle weight that acts parallel to an inclined surface (EF), we have:

$$\text{Grade Resistance} = EF = \frac{DE \times BC}{AC}$$

Line DE represents the gross vehicle weight. For slopes up to 15 degrees (33% slope), AC can be considered to be 100 feet long without significant error. Line BC is increased by 1 foot for each increase in % slope; therefore, we can express grade resistance as follows:

$$\text{Grade Resistance (lb.)} = \frac{\text{Gross Vehicle Weight (lb.) x 1 Ft./\% Slope}}{\text{100 Ft.}}$$

**(Equation C.1)**

Assuming a gross vehicle weight of 1 ton (2000 pounds), the grade resistance imposed by a 1% grade is:

$$\text{Grade Resistance (lb.)} = \frac{\text{2000 lb. x 1'}}{\text{100'}} \qquad \text{(Eq. C.1)}$$

$$= \text{20 pounds}$$

If BC is increased to 2 feet (2% slope), the grade resistance is increased to:

$$\text{Grade Resistance (lb.)} = \frac{\text{2000 lb. x 2'}}{\text{100'}}$$

$$= \text{40 pounds}$$

The exact value of grade resistance can be obtained by:

$$\text{Grade Resistance (lb.)} = \text{Gross Vehicle Weight (lb.) x Sin } \varnothing$$

**(Equation C.2)**

where: $\varnothing$ = Slope expressed in degrees.

**Example C-1:** Determine the precise grade resistance imposed on a 1-ton vehicle ascending a 6% grade.

**Solution:** We must convert percent grade into degrees as follows:

$$\varnothing = \text{Tan}^{-1} \text{(\% Slope)}$$

**(Equation C.3)**

Thus, a 6% slope equates to:

$$\varnothing = \text{Tan}^{-1} (0.06) \qquad \text{(Eq. C.3)}$$

$$= 3.434°$$

Therefore, the grade resistance is:

$$\text{Grade Resistance (lb.)} = \text{2000 lb. x Sin } 3.434° \qquad \text{(Eq. C.2)}$$

$$= \text{119.8 pounds}$$

I have provided Table C-1 for grade resistance values given in pounds per ton of gross vehicle weight for various slopes.

**Table C-1: Grade Resistance (lb./Ton) for Various Slopes**

| Slope (%) | G.R. (lb./Ton) | Slope (%) | G.R. (lb./Ton) |
|---|---|---|---|
| 1 | 20.0 | 12 | 238.4 |
| 2 | 40.0 | 13 | 257.8 |
| 3 | 60.0 | 14 | 277.4 |
| 4 | 80.0 | 15 | 296.6 |
| 5 | 100.0 | 20 | 392.3 |
| 6 | 119.8 | 25 | 485.2 |
| 7 | 139.8 | 30 | 574.7 |
| 8 | 159.2 | 35 | 660.6 |
| 9 | 179.2 | 40 | 742.8 |
| 10 | 199.0 | 45 | 820.8 |
| 11 | 218.0 | 50 | 894.4 |

# APPENDIX D*

## AREAS AND VOLUMES

### AREAS

**Square**

$A = a^2$

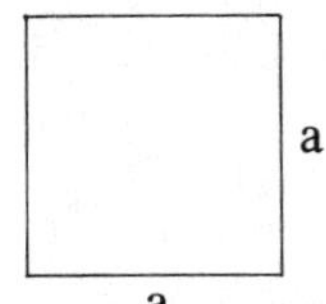


**Rectangle**

$A = bh$



**Triangle**

$A = \frac{1}{2}bh$

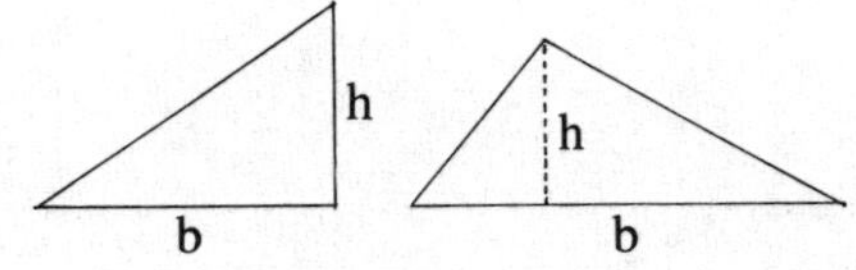


**Parallelogram**

$A = bh = ab\ \mathrm{Sin}\phi$

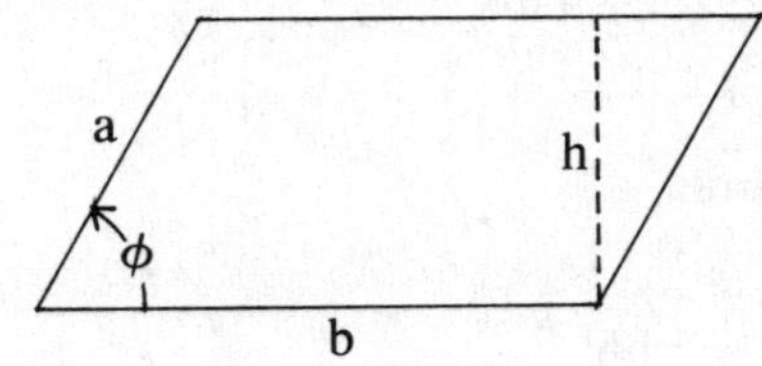


**Trapezoid**

$A = \left(\frac{a + b}{2}\right)h$

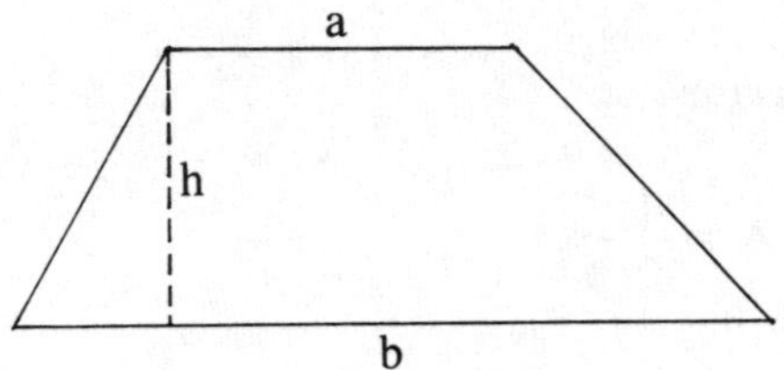


**Trapezium**
(No sides parallel)

$A = ½(bh + b'h')$

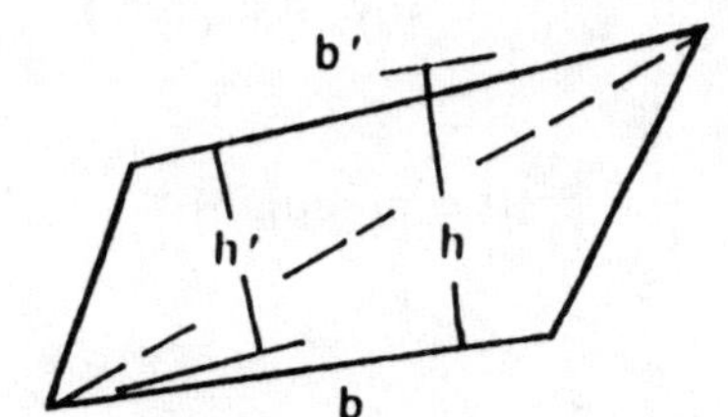


*Appendix D illustrated by Tami Danielle Atcheson

**Polygon**
(hexagon shown)

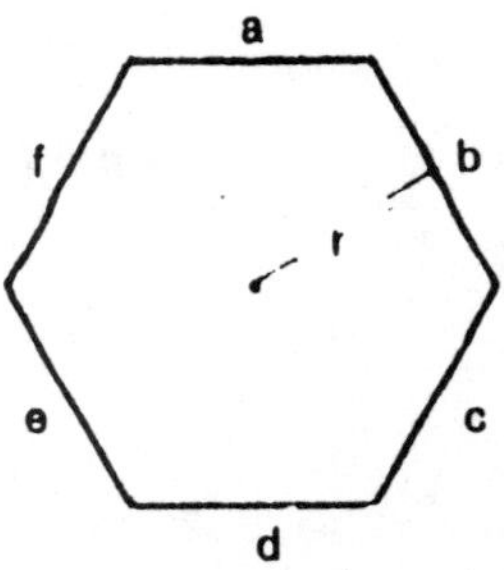


A = r × ½ sum of sides

A = r/2 (a + b + c + d + e + f)

**Circle**

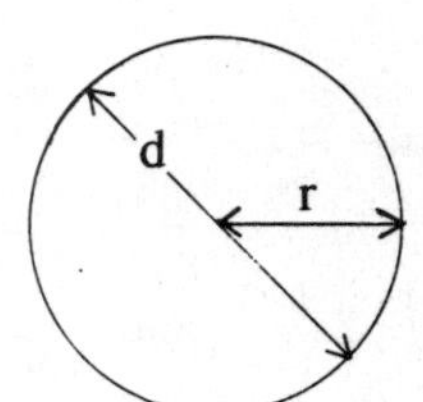


$$A = \pi r^2 = \frac{\pi d^2}{4}$$

Circumference = C = $2\pi r = \pi d$

where $\pi \simeq 3.14 \ldots$

**Ellipse**

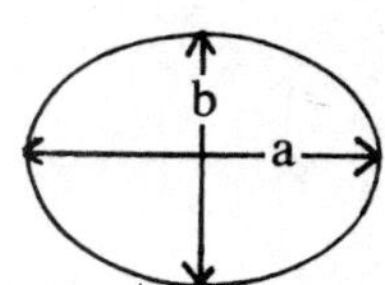


A = 0.7854 ab

**Parabola**

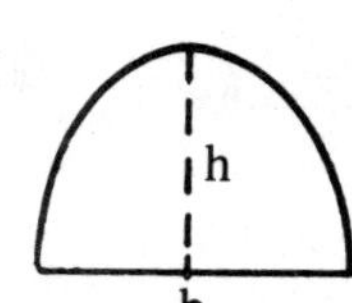


$$A = \frac{2}{3}bh$$

**Sector**

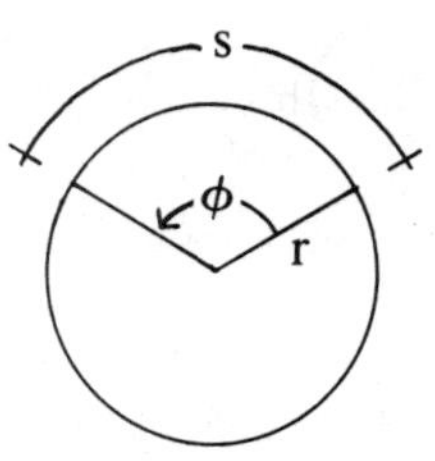


$$A = \frac{1}{2}rs$$

$$= \frac{\phi}{360}\pi r^2 \text{ (If } \phi \text{ in degrees)}$$

$$= \frac{\phi r^2}{2} \text{ (If } \phi \text{ in radians)*}$$

**Segment of Circle**

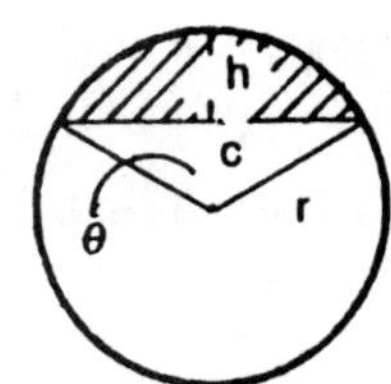


$$A = \frac{\pi r^2 \theta}{360} - \frac{c\,(r - h)}{2}$$

*A radian is a central angle of a circle that subtends an arc equal in length to the radius of the circle.

$\phi = 1$ radian $\simeq 57.3°$

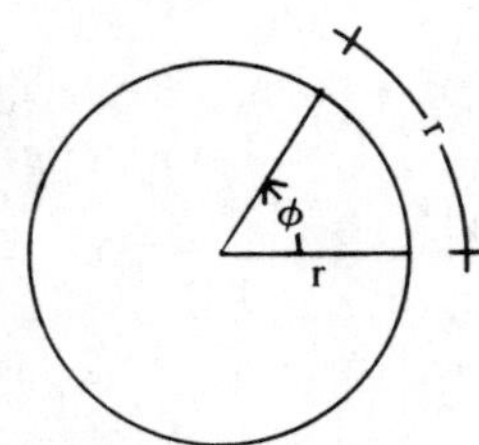


$360° = 2\pi$ radians $\simeq 6.283$ radians

$180° = 1\pi$ radians $\simeq 3.142$ radians

$90° = \frac{\pi}{2}$ radians $\simeq 1.571$ radians

## VOLUMES

**Cube**

$V = a^3$

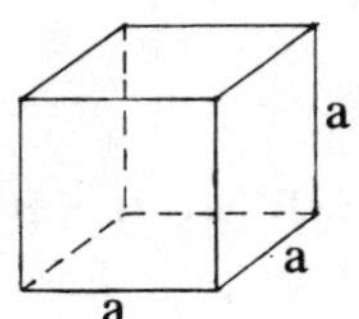


**Rectangular Parallelopiped**

$V = abc$

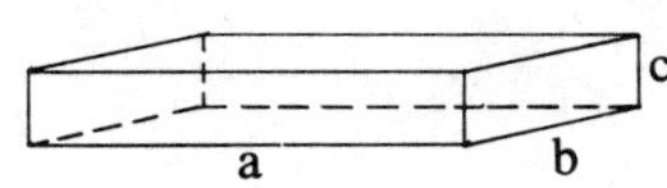


**Wedge**

$V = \frac{1}{2}abc$

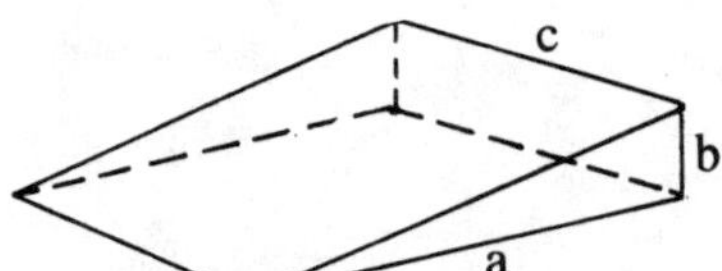


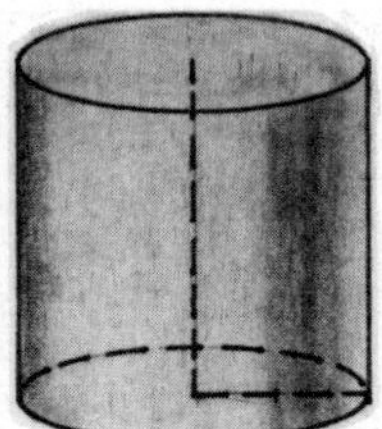

**Cylinder**

$V = \pi r^2 h$

**Pyramid**

$V = \frac{1}{3}(\text{Base})\ h$

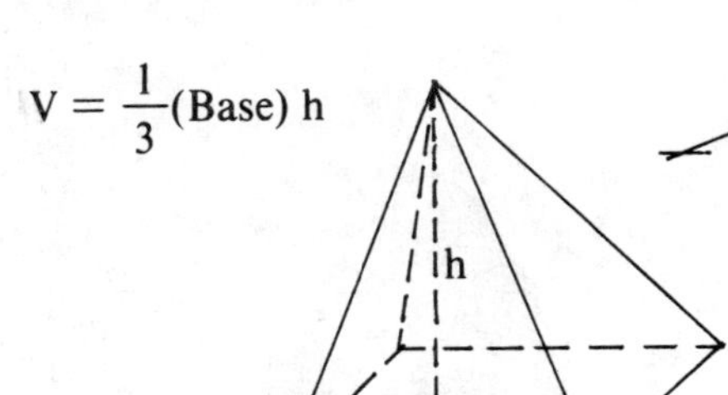


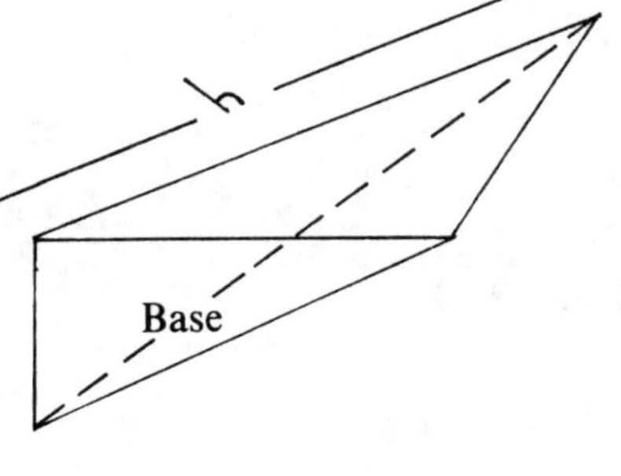

**Cone**

$$V = \frac{1}{3}\pi r^2 h$$

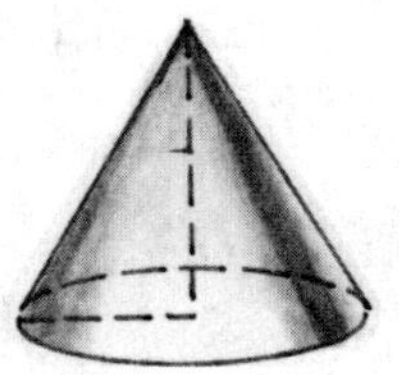

**Frustums**

$$V = \frac{1}{3}h\,(B + B' + \sqrt{B \times B'})$$

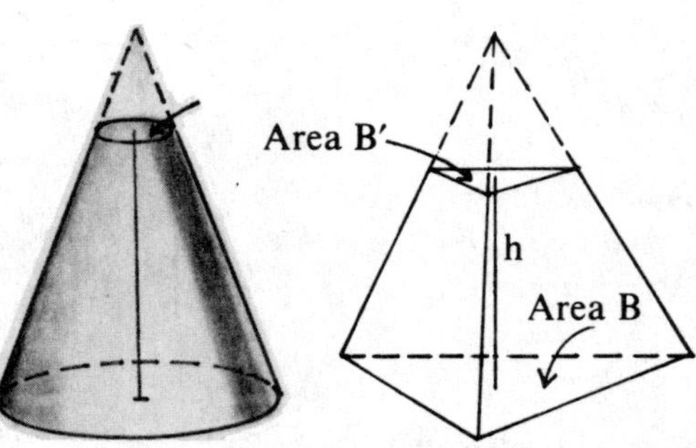


**Sphere**

$$V = \frac{4}{3}\pi r^3 = \frac{1}{6}\pi d^3$$

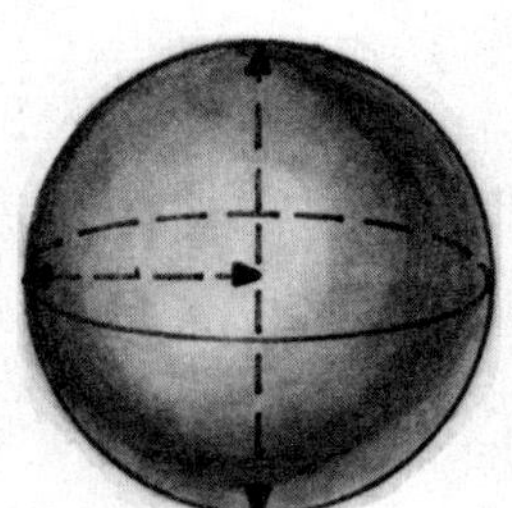

**Ellipsoid**

$$V = \frac{\pi}{6}abc$$

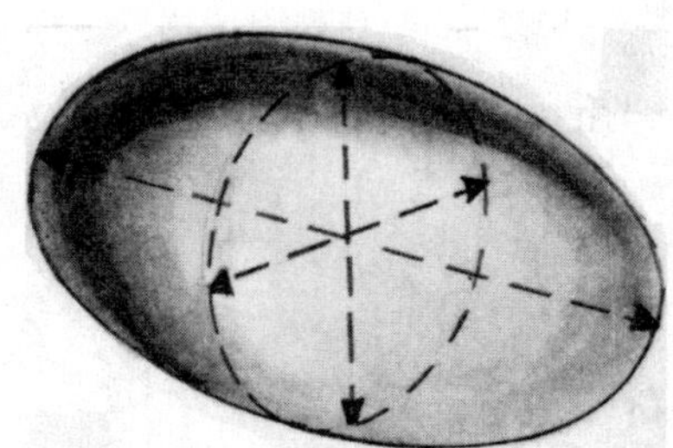

**Paraboloid**

$$V = \frac{1}{2}\pi r^2 h$$

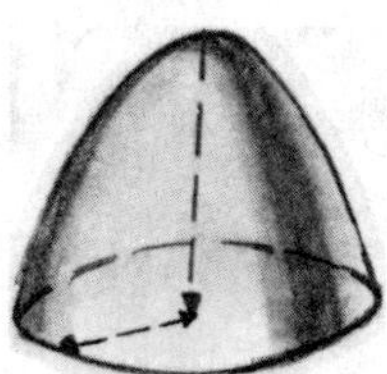

# IRREGULAR SOLIDS

**Cross-Section Method**

$$V = \frac{a + b + c + d}{4} \times \text{Base}$$

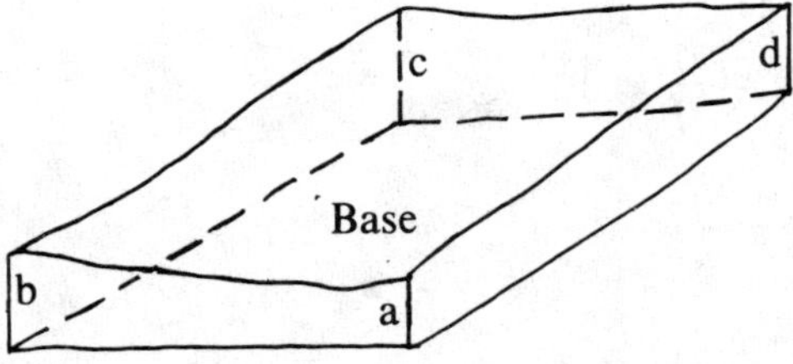


**Average End Area Method**

$$V = \frac{A + B}{2} \times L$$

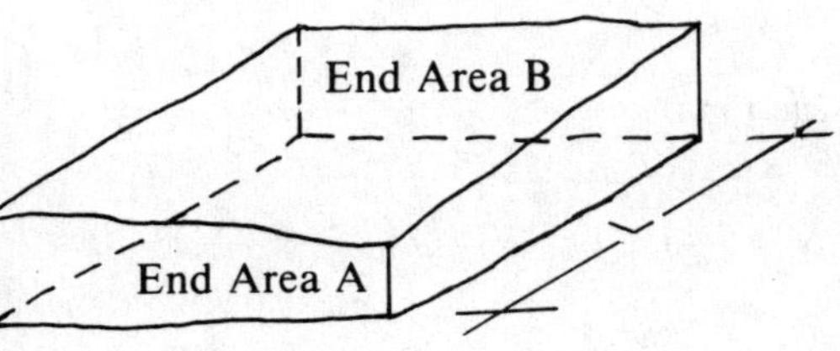


**Prismoidal Formula**

$$V = \frac{A + 4B + C}{6} \times L$$

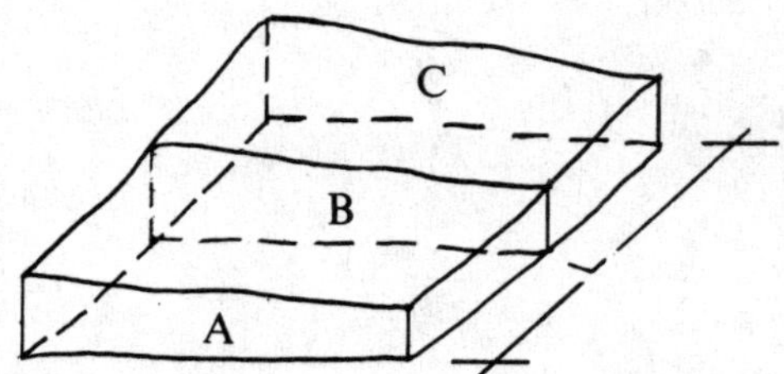

# INDEX

# NOTES